Advanced Technology for Solar Thermal Cooling, Heating, and Energy Storage

Advanced Technology for Solar Thermal Cooling, Heating, and Energy Storage

Guest Editors

Jay Wang
Dingli Duan
Oguzhan Kazaz

MDPI

Basel • Beijing • Wuhan • Barcelona • Belgrade • Novi Sad • Cluj • Manchester

Guest Editors

Jay Wang
Engineering, Computer & Mathematical Sciences
Auckland University of Technology
Auckland
New Zealand

Dingli Duan
Department of Thermal Energy and Power Engineering
Yantai University
Yantai
China

Oguzhan Kazaz
Mechanical & Nuclear Engineering
Khalifa University
Dhabi
United Arab Emirates

Editorial Office
MDPI AG
Grossepeteranlage 5
4052 Basel, Switzerland

This is a reprint of the Special Issue, published open access by the journal *Energies* (ISSN 1996-1073), freely accessible at: https://www.mdpi.com/journal/energies/special_issues/WU8532P4I2.

For citation purposes, cite each article independently as indicated on the article page online and as indicated below:

Lastname, A.A.; Lastname, B.B. Article Title. *Journal Name* **Year**, *Volume Number*, Page Range.

ISBN 978-3-7258-5459-2 (Hbk)
ISBN 978-3-7258-5460-8 (PDF)
https://doi.org/10.3390/books978-3-7258-5460-8

Contents

About the Editors

Jay Wang

Jay Wang is a Lecturer at Auckland University of Technology (AUT), specializing in eco-friendly refrigeration, heat pump systems, solar thermal cooling, thermal energy storage, and other sorts of energy conversion systems. He earned his Ph.D. from the University of South Australia and was a Postdoctoral researcher at the University of Southern Denmark (SDU). He is also a member of esteemed academic and industrial organizations, including Engineering New Zealand (ENZ), Engineers Australia (EA), and the International Institute of Refrigeration (IIR). He is committed to advancing energy efficiency and environmental sustainability in the industrial and commercial sectors.

Dingli Duan

Dingli Duan is a Lecturer at Yantai University, and he earned his Ph.D. from the Dalian University of Technology. His research focuses on the intersections between ejector refrigeration system, ejector performance enhancement, and thermal energy storage.

Oguzhan Kazaz

Oguzhan Kazaz is a postdoctoral researcher at Khalifa University. He earned his BSc and MSc in Mechanical Engineering from Cukurova University (Turkey) and completed his PhD at the University of Glasgow (United Kingdom) in 2024, focusing on nanofluids and phase change materials for solar thermal energy using experimental and computational methods. He has delivered invited talks in London and received a prestigious PhD Fellowship from the Turkish Ministry of Education. His research is the intersection of thermofluids, thermal energy storage, and nanotechnology in clean energy, desalination, and hydrogen storage. He also serves as a reviewer and editorial board member.

energies

Article

Two-Phase Lattice Boltzmann Study on Heat Transfer and Flow Characteristics of Nanofluids in Solar Cell Cooling

Hui Liu , Minle Bao , Luyuan Gong, Shengqiang Shen and Yali Guo *

Key Laboratory of Ocean Energy Utilization and Energy Conservation of Ministry of Education, Dalian University of Technology, Dalian 116024, China; liuhui1994@mail.dlut.edu.cn (H.L.); mlbao@dlut.edu.cn (M.B.); lygong@dlut.edu.cn (L.G.); zzbshen@dlut.edu.cn (S.S.)
* Correspondence: ylguo@dlut.edu.cn

Abstract: During solar cell operation, most light energy converts to heat, raising the battery temperature and reducing photoelectric conversion efficiency. Thus, lowering the temperature of solar cells is essential. Nanofluids, with their superior heat transfer capabilities, present a potential solution to this issue. This study investigates the mechanism of enhanced heat transfer by nanofluids in two-dimensional rectangular microchannels using the two-phase lattice Boltzmann method. The results indicate a 3.53% to 22.40% increase in nanofluid heat transfer, with 0.67% to 6.24% attributed to nanoparticle–fluid interactions. As volume fraction (φ) increases and particle radius (R) decreases, the heat transfer capability of the nanofluid improves, while the frictional resistance is almost unaffected. Therefore, the performance evaluation criterion (PEC) of the nanofluid increases, reaching a maximum value of 1.225 at φ = 3% and R = 10 nm. This paper quantitatively analyzes the interaction forces and thermal physical parameters of nanofluids, providing insights into their heat transfer mechanisms. Additionally, the economic feasibility of nanofluids is examined, facilitating their practical application, particularly in solar cell cooling.

Keywords: nanofluids; heat transfer; friction factor; solar cell cooling; two-phase lattice Boltzmann method

1. Introduction

Solar cells play a crucial role in the shift toward renewable energy by directly turning sunlight into electricity. However, a large portion of the absorbed light energy is converted into heat during operation, resulting in an increase in the temperature of the solar cell. As the temperature of solar cells increases, the higher intrinsic carrier concentration causes more electron-hole recombination, reducing the number of charge carriers and decreasing efficiency. Additionally, prolonged exposure to elevated temperatures can degrade materials like the sealant, anti-reflective coating, and semiconductor, leading to a decline in both mechanical and electrical properties, further reducing the cell's efficiency and lifespan [1–3]. Consequently, effective cooling mechanisms are essential to maintain optimal performance and prolong the operational life of solar cells. Various cooling methods are used to cool solar cells, such as deploying visibly clear metamaterials to reduce the heat absorbed by the solar cells [4], employing thermal rectification [5] to direct heat from the sensitive area (the active layer) of the cell to the radiator, phase change material cooling [6], and traditional air-cooled or water-cooled radiators.

Improving the cooling effect of traditional radiators has attracted the interest of researchers. Microchannels, which provide a significant surface-area-to-volume ratio that enables more efficient transfer of heat from fluids to the surrounding environment, have been used in cooling solar cells [7–9]. Nanofluids are a type of stable colloid engineered by dispersing nanometer-sized particles into a base fluid and generally exhibit higher thermal conductivity than the base fluids [10,11]. Many studies [12–14] have shown that the use of nanofluids within microchannels can further enhance heat transfer rates. Wang et al. [15] designed a new microchannel heat sink with V-ribs using a water-based nanofluid as a coolant for micro solar cells to improve the heat transfer efficiency. Yang et al. [16] used hybrid nanofluids containing carbon nanotubes and microchannels to cool batteries connected to solar systems. They were successful in maintaining the batteries' temperature within the allowed range. Karami and Rahimi [17] experimentally assessed the cooling performance of water-based Boehmite nanofluid in a hybrid photovoltaic (PV) cell with microchannels. They found that 0.01 wt.% nanofluid reduced average PV surface temperature from 62.29 °C to 32.5 °C and improved electrical efficiency by 27% compared to water at a 300 mL/min flow rate. Radwan et al. [18] proposed a new cooling technique for low-concentrated photovoltaic–thermal systems using a microchannel heat sink with Al_2O_3–water and SiC–water nanofluids. The study showed significant solar cell temperature reduction, particularly at high nanoparticle volume fractions.

The research studies mentioned above focused on reducing the temperature of solar cells. In-depth research into the mechanisms by which nanofluids enhance heat transfer will help design more efficient cooling schemes. Research studies [19–21] showed that the interaction forces (such as Brownian force, drag force, buoyancy, gravity forces, etc.) between nanoparticles and base fluids were important factors in enhancing heat transfer. Farahani and Nezhad [22] highlighted that continuous-scale computational fluid dynamics (CFD) methods, such as finite difference (FDM), finite element (FEM), and finite volume (FVM), were computationally costly and difficult to capture the fluid–solid interactions. Fortunately, the lattice Boltzmann method (LBM) exhibits exceptional flexibility in dealing with complex force terms [23]. By considering nanofluids as two distinct phases, Xuan et al. [24] first utilized the two-phase lattice Boltzmann method (TPLBM) coupled with forces to simulate the complicated fluid dynamics inside nanofluids. Subsequently, Guo et al. [25], as well as Ahmed and Eslamian [26,27], employed the TPLBM to simulate the natural convection of nanofluids within cavities. Meanwhile, Zarghami et al. [28], Saberi and Kalteh [29], and Ahmed and Eslamian [30] utilized TPLBM to investigate the forced convection of nanofluids in channels. Their studies primarily focused on enhancing heat transfer rates but ignored the impact of friction factors. However, the friction factor is crucial for accurately assessing energy consumption in fluid flow processes. Ignoring this factor leads to an incomplete understanding of the system's performance, as it directly influences the pressure drop and overall energy efficiency. Our study advanced the current research by simultaneously analyzing both fluid friction and heat transfer rates. This comprehensive approach allowed for a more accurate evaluation of nanofluid performance, ensuring that both thermal and hydraulic considerations were taken into account.

This study aims to explore the heat transfer and flow characteristics of nanofluids in cooling microchannels using the TPLBM. In particular, the effect of nanoparticle–base fluid interactions on enhanced heat transfer is included. The performance evaluation criteria (PEC) will be calculated to evaluate the overall heat transfer performance and flow resistance under different volume fractions (φ = 0~3%) and nanoparticle radii (R = 10~50 nm). The study is expected to provide valuable insights into the heat transfer and flow characteristics of nanofluids in solar cell cooling systems. This will contribute to the development of optimized cooling strategies that maximize the performance and sustainability of solar cell systems.

2. Materials and Methods

2.1. Problems Statement

The geometry of the microchannel is illustrated in Figure 1. The height of the microchannel is H = 30 μm, and the length is L = 600 μm. Constant temperature boundary conditions (θ_w = 323 K) are imposed on the top and down walls. The nanofluid flows through the microchannel with a uniform temperature of θ_{in} = 283 K and a uniform velocity of u_{in}. Fully developed conditions for the velocity and extrapolation scheme for temperature fields are considered at the microchannel outlet. The simulations are conducted at a

Reynolds number of *Re* = 50 to ensure relevance to microchannel flows, which typically exhibit low *Re* due to small dimensions and low velocities [31,32]. Under these conditions, the effects of particle volume fraction, particle size, and interaction forces between particles and base liquid on flow and heat transfer are investigated.

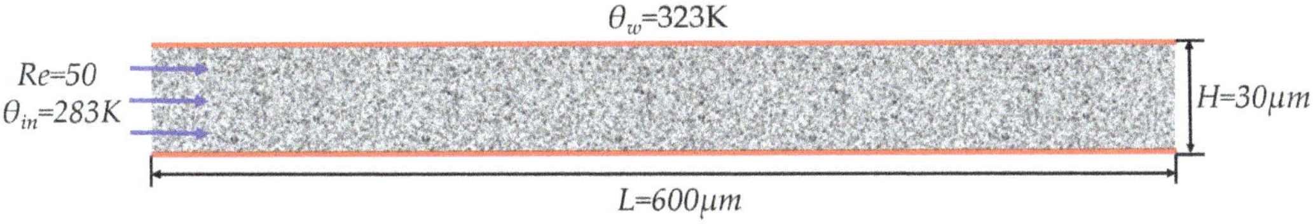

Figure 1. Schematic of the of the microchannel.

2.2. Physical Properties of Nanofluids

Table 1 lists the physical properties of water, the Al_2O_3 nanoparticle, and nanofluids. The density (ρ) and specific heat capacity (c_p) of the nanofluid can be obtained by the relationship between the two-phase mixture:

$$\rho_{nf} = (1-\varphi)\rho_{bf} + \varphi\rho_{np} \tag{1}$$

$$c_{p,nf} = (1-\varphi)c_{p,bf} + \varphi c_{p,np} \tag{2}$$

where φ is the volume fraction of nanoparticles. The subscripts *bf*, *np*, and *nf* represent the base fluid, nanoparticles, and nanofluids, respectively.

Corcione model [33] proposed an empirical formula to predict the viscosity (μ) of nanofluids based on extensive experimental data encompassing various nanoparticle volume fractions, ranging from 0.01% to 7.1%:

$$\mu_{nf} = \frac{\mu_{bf}}{1 - 34.87\left(d_{np}/d_{bf}\right)^{-0.3}\varphi^{1.03}} \tag{3}$$

Maxwell [34] proposed an equation of thermal conductivity applicable to particle suspensions in 1873. It has been widely used to calculate the thermal conductivity (k) of nanofluid:

$$k_{nf} = k_{bf}\left[\frac{(k_{np}+2k_{bf}) - 2\varphi(k_{bf}-k_{pp})}{(k_{np}+2k_{bf}) + \varphi(k_{bf}-k_{np})}\right] \tag{4}$$

Using the obtained values for c_p, μ, and k, the Prandtl number for the nanofluid is calculated as:

$$Pr = \frac{c_{p,nf}\cdot\mu_{nf}}{k_{nf}} \tag{5}$$

Table 1. Physical properties of water and Al_2O_3 nanofluid [35].

	k/(W·m^{-1}·K^{-1})	c_p/(J·kg^{-1}·K^{-1})	ρ/(kg·m^{-3})	μ/(Pa·s)	Pr
water	0.618	4174.0	995.7	0.8015×10^{-3}	5.41
Al_2O_3	46.0	769.0	3970.0	——	——
nanofluids (φ = 1%, R = 10 nm)	0.636	4042.2	1025.4	0.8835×10^{-3}	5.62
nanofluids (φ = 2%, R = 10 nm)	0.654	3917.8	1055.2	0.9888×10^{-3}	5.92
nanofluids (φ = 3%, R = 10 nm)	0.673	3800.2	1084.9	1.1252×10^{-3}	6.35
nanofluids (φ = 3%, R = 30 nm)	0.673	3800.2	1084.9	1.0106×10^{-3}	5.71
nanofluids (φ = 3%, R = 50 nm)	0.673	3800.2	1084.9	0.9745×10^{-3}	5.50

The following non-dimensional parameters X, Y, T, and U are used:

$$X = \frac{x}{H}, \ Y = \frac{y}{H}, \ T = \frac{\theta - \theta_{in}}{\theta_w - \theta_{in}}, \ U = \frac{u}{u_{in}} \tag{6}$$

where T and U are dimensionless temperature and velocity, respectively. $u_{in} = Re \cdot v_{lattice}/2m$, $v_{lattice}$ is the lattice dimensional kinematic viscosity of nanofluids, and m is the number of grids divided along channel height.

2.3. Base Fluid–Nanoparticles Interaction Forces

Nanoparticles suspended in base fluid suffer various interactions: Brownian force $\boldsymbol{F}_B$ [36], drag force $\boldsymbol{F}_D$, buoyancy and gravity forces $\boldsymbol{F}_H$, and interaction potential force $\boldsymbol{F}_A$ [37]. The corresponding calculation equations are as follows:

$$\frac{\boldsymbol{F}_B}{m_{np}} = G_i \sqrt{\frac{\pi S_0}{\Delta t}} \tag{7}$$

$$\boldsymbol{F}_{\mathrm{D}} = -6\pi \mu_{bf} R \left(\boldsymbol{u}_{np} - \boldsymbol{u}_{bf} \right) \tag{8}$$

$$\boldsymbol{F}_H = -\frac{4\pi R^3}{3} \boldsymbol{g} \left(\rho_{np} - \rho_{bf} \right) \tag{9}$$

$$\boldsymbol{F}_A = \sum_{i=1}^{8} n_i \frac{\partial V_A}{\partial \boldsymbol{r}_i} \tag{10}$$

where m_{np} is the mass of the nanoparticle. G_i is a Gaussian random number satisfying normal distribution. $\boldsymbol{g}$ denotes the gravitational acceleration. $n_i = \rho V / m_{np}$ is the number of particles in an adjacent lattice. V represents the lattice volume. S_0 and V_A are the spectral intensity function and the interaction potential, respectively. They can be derived as:

$$S_0 = (216 v_{bf} k_B T) / [\pi^2 \rho_{bf} d_{np}^5 (\rho_{np}/\rho_{bf})^2 C_C] \tag{11}$$

$$V_A = -\frac{1}{6} A \left(\frac{2R^2}{L_{cc}^2 - 4R^2} + \frac{2R^2}{L_{cc}^2} + \frac{L_{cc}^2 - 4R^2}{L_{cc}^2} \right) \tag{12}$$

where A is the Hamaker constant. R and L_{cc} denote the radius of nanoparticles and the average distance between two adjacent nanoparticles, respectively. k_B is the Boltzmann constant ($k_B = 1.38 \times 10^{-23}$ J/K). The Cunningham coefficient Cc can be calculated using the equations:

$$Cc = 1 + \frac{2\lambda}{d_{np}} \left[1.257 + 0.4 e^{\left(\frac{-1.1 d_{np}}{\lambda} \right)} \right] \tag{13}$$

Hence, the total force acting on the nanoparticle and base fluid can be written, respectively, as:

$$\boldsymbol{F}_{np} = n(\boldsymbol{F}_H + \boldsymbol{F}_D + \boldsymbol{F}_A + \boldsymbol{F}_B)/V \tag{14}$$

$$\boldsymbol{F}_{bf} = -n(\boldsymbol{F}_D + \boldsymbol{F}_B) \tag{15}$$

2.4. Governing Equations

Continuous equation:

$$\nabla \cdot \boldsymbol{u} = 0 \tag{16}$$

Momentum equation:

$$\frac{\partial u}{\partial t} + \nabla \cdot (\boldsymbol{uu}) = -\frac{\nabla p}{\rho_0} + v \nabla^2 \boldsymbol{u} + \boldsymbol{F} \tag{17}$$

where p is the pressure and F represents the forces between the fluid and nanoparticles.

Energy equation:

$$\frac{\partial T}{\partial t} + \nabla \cdot (\boldsymbol{u}T) = \nabla \cdot (\chi \nabla T) \tag{18}$$

where χ is the thermal diffusion coefficient.

2.5. Lattice Boltzmann Model

The two-phase lattice Boltzmann method (TPLBM) conjugated with interaction forces proposed by Xuan et al. [24] is used. The lattice Boltzmann equations for nanofluid are written as follows:

$$f_i^{\sigma}(\boldsymbol{x} + \boldsymbol{e}_i \Delta t, t + \Delta t) - f_i^{\sigma}(\boldsymbol{x}, t) = -\frac{1}{\tau_f^{\sigma}}\left(f_i^{\sigma}(\boldsymbol{x}, t) - f_i^{\sigma,eq}(\boldsymbol{x}, t)\right) + \left(\frac{2\tau_f^{\sigma} - 1}{2\tau_f^{\sigma}}\right)\left(\frac{\boldsymbol{F}_i^{\sigma} \cdot \boldsymbol{e}_i \Delta t}{B_i c^2}\right) \tag{19}$$

$$g_i^{\sigma}(\boldsymbol{x} + \boldsymbol{e}_i \Delta t, t + \Delta t) - g_i^{\sigma}(\boldsymbol{x}, t) = -\frac{1}{\tau_g^{\sigma}}\left(g_i^{\sigma}(\boldsymbol{x}, t) - g_i^{\sigma,eq}(\boldsymbol{x}, t)\right) \tag{20}$$

where f and g are density and temperature distribution functions, respectively. The superscript eq denotes the equilibrium state and σ = bf, np denotes base fluid and nanoparticles. The subscript i denotes the discrete lattice direction. Δx and Δt represent the lattice length step and time step, respectively. $c = \Delta x/\Delta t$ standing for the lattice speed. τ_f and τ_g are dimensionless collision relaxation times for flow and temperature, respectively. B is the mass coefficient ($B_{1\text{–}4} = 3$ and $B_{5\text{–}8} = 12$) to ensure the mass and momentum conservation in the calculation. $\boldsymbol{F}^{\sigma}$ is the resultant force acting on the σ phase. $\boldsymbol{e}_i$ is the discrete lattice velocity vector. In the D2Q9 model, $\boldsymbol{e}_i$ takes the following value:

$$\boldsymbol{e}_i = \begin{cases} (0,0) & i = 0 \\ c\left(\cos\frac{(i-1)\pi}{2}, \sin\frac{(i-1)\pi}{2}\right) & i = 1,\ldots,4 \\ \sqrt{2}c\left(\cos\frac{(2i-9)\pi}{4}, \sin\frac{(2i-9)\pi}{4}\right) & i = 5,\ldots,8 \end{cases} \tag{21}$$

The flow and temperature equilibrium distribution functions are:

$$f_i^{\sigma,eq} = \omega_i \rho^{\sigma}\left[1 + 3\frac{(\boldsymbol{e}_i \cdot \boldsymbol{u}^{\sigma,eq})}{c^2} + \frac{9}{2}\frac{(\boldsymbol{e}_i \cdot \boldsymbol{u}^{\sigma,eq})^2}{c^4} - \frac{3}{2}\frac{(\boldsymbol{u}^{\sigma,eq})^2}{c^2}\right] \tag{22}$$

$$g_i^{\sigma,eq} = \omega_i T^{\sigma}\left[1 + 3\frac{(\boldsymbol{e}_i \cdot \boldsymbol{u}^{\sigma,eq})}{c^2} + \frac{9}{2}\frac{(\boldsymbol{e}_i \cdot \boldsymbol{u}^{\sigma,eq})^2}{c^4} - \frac{3}{2}\frac{(\boldsymbol{u}^{\sigma,eq})^2}{c^2}\right] \tag{23}$$

where ω_i is the weight coefficients ($\omega_0 = 4/9$, $\omega_{1\text{–}4} = 1/9$; $\omega_{5\text{–}8} = 1/36$).

The Navier–Stokes equation can be recovered by the multi-scale technique (Chapman–Enskog expansion) [38], and the kinematic viscosity coefficient and thermal diffusion coefficient for the σ phase can be determined, respectively:

$$\nu^{\sigma} = \frac{\tau_f^{\sigma} - 0.5}{3} c^2 \Delta t \tag{24}$$

$$\chi^{\sigma} = \frac{\tau_g^{\sigma} - 0.5}{3} c^2 \Delta t \tag{25}$$

While the average kinematic viscosity coefficient and thermal diffusion coefficient for the nanofluid are:

$$\nu = \frac{\sum_{\sigma} C^{\sigma} \tau_f^{\sigma} - 0.5}{3} c^2 \Delta t \tag{26}$$

$$\chi = \frac{\sum_{\sigma} C^{\sigma}\tau_g^{\sigma} - 0.5}{3} c^2 \Delta t \tag{27}$$

where $C^{\sigma} = \rho^{\sigma}/(\sum_{\sigma}\rho)$.

The macro density ρ, the velocity $\boldsymbol{u}$, and the temperature T for the σ phase can be written, respectively, as:

$$\rho^{\sigma} = \sum_{i=0}^{8} f_i^{\sigma} \tag{28}$$

$$\rho^{\sigma}\boldsymbol{u}^{\sigma} = \sum_{i=0}^{8} f_i^{\sigma}\boldsymbol{e}_i Po(x) = f(x)\cdot Re \tag{29}$$

$$T^{\sigma} = \sum_{i=0}^{8} g_i^{\sigma} \tag{30}$$

In the absence of force exerted on the σ phase, particles and base liquid move at the same speed and temperature:

$$\boldsymbol{u} = \frac{\sum_{\sigma} \rho^{\sigma}\boldsymbol{u}^{\sigma}/\tau_f^{\sigma}}{\sum_{\sigma} \rho^{\sigma}/\tau_f^{\sigma}} \tag{31}$$

$$T = \frac{\sum_{\sigma} \rho^{\sigma} c_p^{\sigma} T^{\sigma}/\tau_g^{\sigma}}{\sum_{\sigma} \rho^{\sigma} c_p^{\sigma}/\tau_g^{\sigma}} \tag{32}$$

In fact, particles and base fluid are subjected to different forces. The action of the $\boldsymbol{F}^{\sigma}$ is considered as the correction of the local equilibrium velocity according to Shan and Chen [39]:

$$\boldsymbol{u}_{new}^{\sigma,eq} = \boldsymbol{u}^{\sigma} + \frac{\boldsymbol{F}^{\sigma}\tau_f^{\sigma}\Delta t}{\rho^{\sigma}} \tag{33}$$

Energy between nanoparticles and base fluid is exchanged due to the internal forces, and the macroscopic temperature for nanoparticles and base fluid is then given as:

$$T_{new}^{\sigma} = T^{\sigma} + \Delta t \tau_g^{\sigma} \frac{dT^{\sigma}}{dt} \tag{34}$$

The local Nusselt number $Nu(x)$, local fanning friction factor $f(x)$, and the local Poiseuille $Po(x)$ number along the lower wall are defined as:

$$Nu(x) = \frac{k_{nf}}{k_{bf}} \frac{(\partial T/\partial y)|_{y=0}}{T_w - T_m} \tag{35}$$

$$f(x) = \frac{2\mu_{nf}(\partial u/\partial y)|_{y=0}}{\rho_{nf} u_m^2} \tag{36}$$

$$Po(x) = f(x)\cdot Re \tag{37}$$

where $T_m = \left(\int_A \rho c_p u T dA\right)/\left(\int_A \rho c_p u dA\right)$, $u_m = \left(\int_A \rho u dA\right)/\left(\int_A \rho dA\right)$.

The average Nusselt number (Nu_{avg}) and the average fanning friction factor (f_{avg}) are:

$$Nu_{avg} = \frac{1}{H}\int_0^H Nu(x)dx \tag{38}$$

$$f_{avg} = \frac{1}{H}\int_0^H f(x)dx \tag{39}$$

To assess the comprehensive performance of nanofluids, the term performance evaluation criterion (PEC) is introduced:

$$PEC = \frac{Nu_{avg,nf}/Nu_{avg,water}}{\left(f_{avg,nf}/f_{avg,water}\right)^{1/3}} \quad (40)$$

3. Results

3.1. LBM Code Validation

A series of sized grids, from 80 × 1600 to 120 × 2400, are used with constant Re = 50 and Pr = 5.41 to confirm grid independence firstly. The Nusselt number and the fanning friction factor at the exit with various grid numbers are compared and summarized in Table 2. Considering the calculating speed and the numerical accuracy compared, a grid size of 100 × 2000 is employed for the simulation.

Table 2. Grid independency study for water at Re = 50.

	60 × 1400	80 × 1600	100 × 2000	120 × 2400
Nu_{avg}	8.478	8.396	8.345	8.301
f_{avg}	0.471	0.474	0.475	0.476

In order to verify the reliability of the TPLBM model, a simulation was conducted to analyze the heat transfer results of Cu–water nanofluids through two parallel plates, with one side constant heat flux and the other insulated. The simulation conditions were consistent with those of Saberi and Kalteh [29], and the comparison of the average Nusselt number is illustrated in Figure 2. The maximum difference between the present result and the Saberi and Kalteh [29] date was 3.1%, which indicates that the method of this paper was credible.

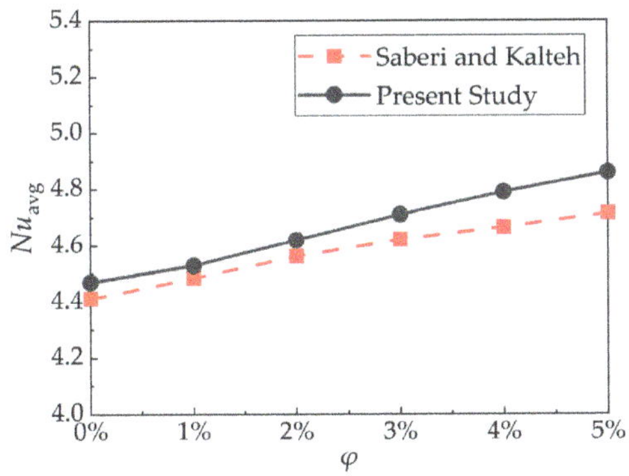

Figure 2. The average Nusselt number (Nu_{avg}) compared with the results of Saberi and Kalteh [29].

3.2. Effects of Interaction Forces

A set of interaction forces, including Brownian force F_B, buoyancy and gravity force F_H, drag force F_D, and van der Waals force F_A for different nanoparticle volume fractions and radii are calculated, and their importance in the forced convective flow and heat transfer is evaluated in this study. Figure 3a shows the range of forces acting on nanoparticles at various volume fractions and particle sizes. According to this figure, the magnitude of Brownian force is the largest, followed by drag force, while buoyancy and gravity and van der Waals forces are negligible: $F_B > F_D > F_H > F_A$. Moreover, it is clear that the force increases with the volume fraction and decreases with the particle size.

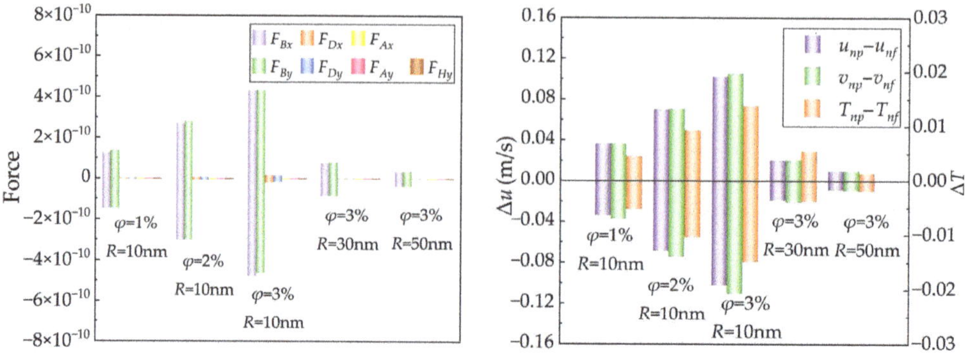

(a) **(b)**

Figure 3. The range of (**a**) forces acting on the nanoparticle and (**b**) the velocity difference and temperature difference between particle and base solution with different volume fractions and particle sizes.

Particles are displaced by interaction forces, resulting in a difference in velocity and temperature between particles and the base fluid. This means that the micro-convection occurs between particles and base fluid as a result of the interaction force. Figure 3b illustrates the range of the velocity difference and dimensionless temperature difference between suspended particles and base fluid under various volume fractions and particle sizes. For Re = 50, φ = 3% and R = 10 nm, the range of velocity difference between particle and base fluid is $\Delta u_x = -0.1019\sim0.1021$ m/s and $\Delta u_y = -0.1104\sim0.1053$ m/s, the range of dimensionless temperature difference is $\Delta T = -0.0148\sim0.0138$. It is important to note that these differences are non-negligible when compared to inlet velocity and inlet dimensionless temperature for nanofluids. It is also evident from the figure that the difference in velocity and dimensionless temperature between particles and the base liquid is positively correlated with the magnitude of interaction forces, i.e., the micro-convection between particles and base liquid is more intense at high volume fractions and small particle sizes.

3.3. Flow Characteristics of Nanofluids

Figure 4 illustrates the velocity (u) of pure water and Al_2O_3–water nanofluids with various volume fractions at the center horizontal and vertical lines of the channel. It is evident that the velocity of nanofluids increases with volume fraction. This is due to the fact that the increase in particle volume fraction leads to an increase in density and viscosity at the same time. Based on Table 1, it is apparent that the increment of density is smaller than that of the viscosity. In accordance with the definition of Re number ($Re = u_{in}D_h/\nu$, where $D_h = 2H$ is the hydraulic diameter of the microchannel), a higher inlet velocity of the nanofluid is required to maintain a constant Reynolds number.

Figure 5 illustrates the velocity (u) of pure water and Al_2O_3–water nanofluids with various particle radii at the center horizontal and vertical lines of the channel. It is obvious that the velocity of nanofluids slows down with the increase in particle size. This is because the viscosity of the nanofluid decreases with the particle radius, and the inlet velocity of the nanofluid should be reduced to keep the Re number constant. Therefore, in the range of working conditions studied in this paper, the velocity of nanofluid reaches the maximum when φ = 3% and R = 10 nm. In addition, there is a slight fluctuation in the velocity of the nanofluid with φ = 3%, R = 10 nm at Y = 0.5, as shown in Figures 4a and 5a. This is because the Brownian force leads to the irregular movement of the particles, and the higher the concentration and the smaller the particle size, the larger the Brownian force. However,

Figures 4b and 5b indicate that the velocity fluctuation is not obvious at X = 10, which is owing to the large velocity gradient.

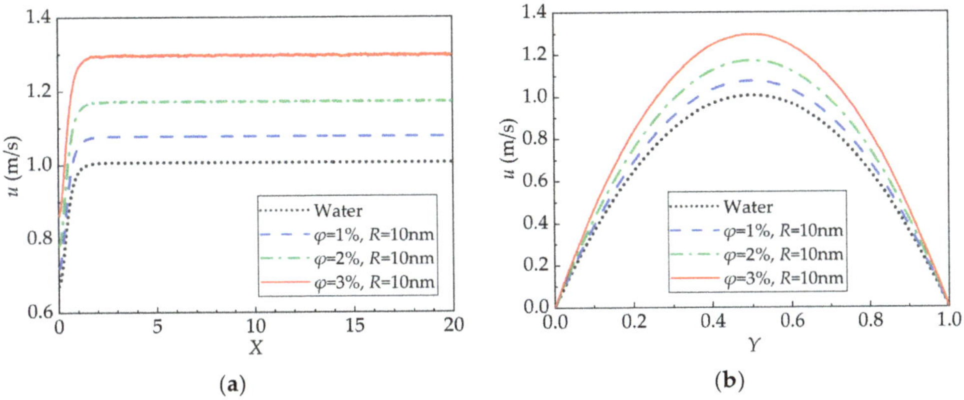

Figure 4. Velocity profiles for various nanoparticle volume fractions at (**a**) $Y = 0.5$ and (**b**) $X = 10$.

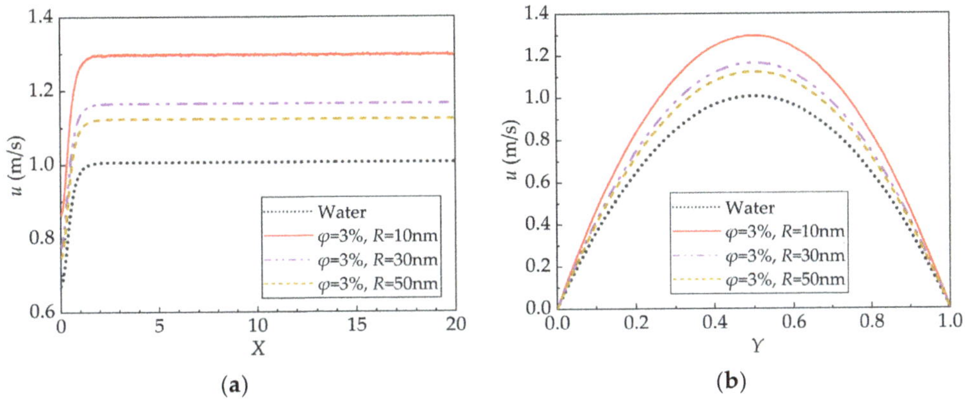

Figure 5. Velocity profiles for various nanoparticle radii at (**a**) $Y = 0.5$ and (**b**) $X = 10$.

An analysis of the dimensionless velocity is necessary since its gradient reflects the magnitude of flow resistance. Figures 6 and 7 exhibit the dimensionless velocity (U) for various nanoparticle volume fractions and nanoparticle radii, respectively. Figures 6a and 7a illustrate that the maximum dimensionless velocity of water along the center horizontal of the channel $U_{max,w} \approx 1.5$, while that of nanofluids fluctuates around $U_{max,w}$. The amplitude of fluctuation augments with the increase in volume fraction and reduces with the increase in particle size. The reason is that the irregular motion of particles caused by Brown force intensified with the increase in concentration and the decrease in particle size. Figures 6b and 7b show that in the fully developed region, the dimensionless velocity distribution of nanofluids with arbitrary volume fraction and particle size along the vertical line of the channel almost coincides and satisfies the formula $U = 1.5(4Y - 4Y^2)$. Furthermore, the dimensionless velocity fluctuations of the nanofluids are invisible at X = 10. This is due to the fact that the velocity fluctuations caused by the Brownian motion of particles are too poor compared to the drastic changes in velocity along the longitudinal section of the channel.

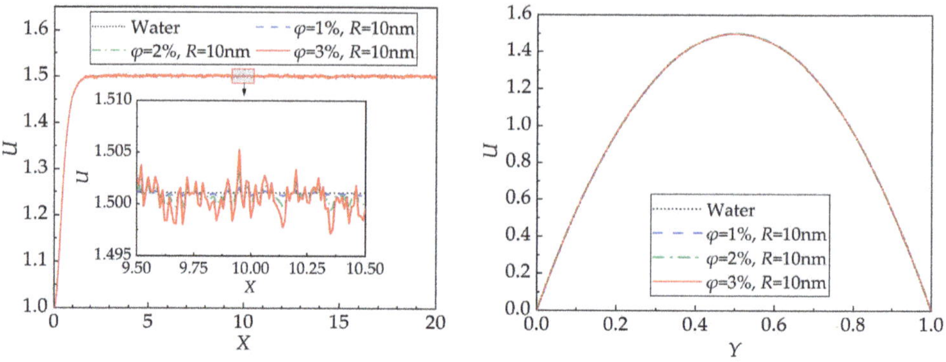

(a) (b)

Figure 6. Dimensionless velocity profiles for various nanoparticle volume fractions at (**a**) $Y = 0.5$ and (**b**) $X = 10$.

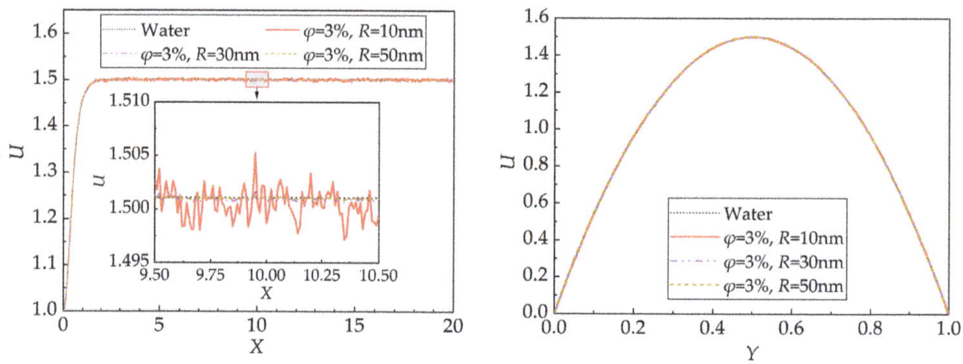

(a) (b)

Figure 7. Dimensionless velocity profiles for various nanoparticle radii at (**a**) $Y = 0.5$ and (**b**) $X = 10$.

The Poiseuille number ($Po = f_{app} \cdot Re$) can directly reflect frictional resistance as fluid flows. In the case of a constant Re number, the higher the Po number, the greater the friction coefficient. The local Po numbers of nanofluids along the lower wall with different nanoparticle volume fractions and nanoparticle radii are presented in Figure 8a,b, respectively. It is obvious that in the fully developed region, the Po number of water is $Po_{full,w} \approx 24$. The Po number of nanofluid fluctuates, and the fluctuation amplitude enhances with the increase in volume fraction and weakens with the growth in particle size, which is caused by the change in Brownian force. Moreover, the Po number of nanofluids fluctuates around $Po_{full,w}$, just as the nanofluid velocity fluctuates around $U_{max,w}$. In order to explain this, we derive the relationship between the Po number and the dimensionless velocity U. Combine Equations (36) and (37):

$$Po(x) = \frac{4(\partial U/\partial Y)|_{Y=0} \cdot U_{nf,in}}{U_m^2} \tag{41}$$

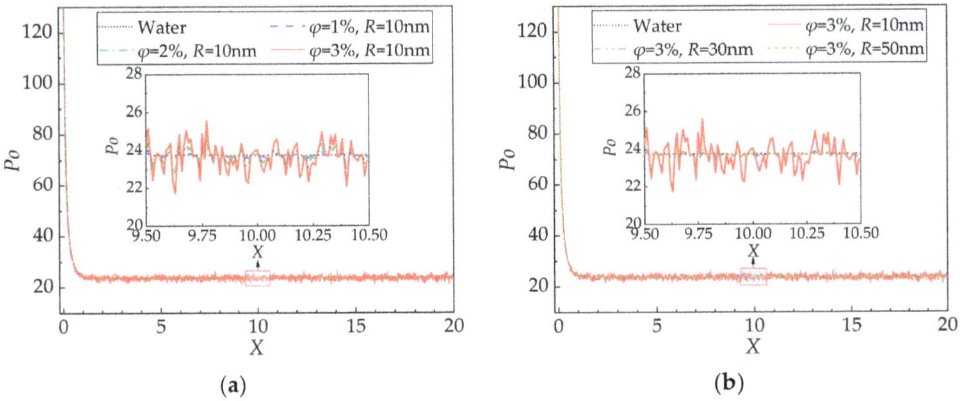

Figure 8. The local Poiseuille number profiles along the lower wall for different (**a**) nanoparticle volume fractions and (**b**) nanoparticle radii.

Equation (41) shows that the *Po* number is uniquely related to dimensionless velocity distributions; therefore, their distributions are similar. According to Figures 6 and 7, the dimensionless velocity distributions of nanofluids with varying volumes and particle sizes almost overlap. Consequently, the difference in the average *Po* number is slight between them. These also indicate that the particle size and volume fraction have little influence on the average friction coefficient of nanofluids.

In order to explore the effect of interaction forces on the flow of nanofluids, the average friction coefficient (f_{avg}) with and without interaction forces at different volume fractions and sizes are compared, as shown in Figure 9. It can be seen that the f_{avg} of the nanofluid ranges from 0.504 to 0.506. In addition, the f_{avg} of nanofluids without considering the interaction force increases slightly with the increasing volume fraction but decreases slightly with increasing particle size. For nanofluids, considering the interaction force, f_{avg}, decreases slightly with increasing volume fraction and particle size. However, the f_{avg} varies within 0.3% under different parameters. Therefore, the effects of volume fraction, particle size, and interaction forces on f_{avg} can be ignored.

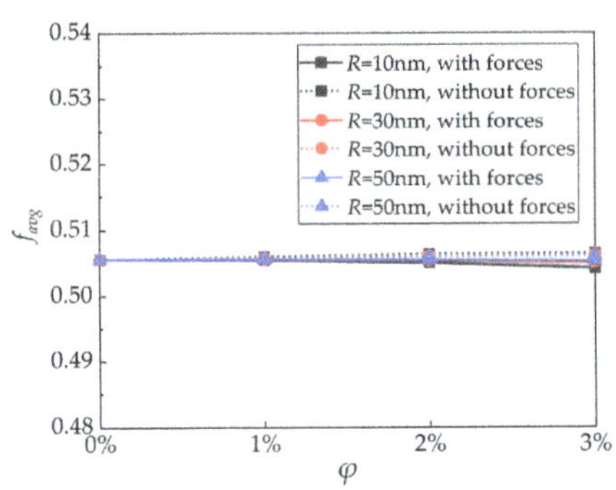

Figure 9. The average friction coefficient at the bottom wall for different volume fractions and nanoparticle radii.

3.4. Heat Transfer Characteristics of Nanofluids

Figure 10 shows the dimensionless temperature distribution of pure water and Al_2O_3–water nanofluid with various volume fractions at $X = 10$ and $Y = 0.5$. It indicates that the temperature of nanofluids decreases with the increase in volume fraction. The combined effect of nanofluid physical properties and interaction forces results in this phenomenon.

With the increment of the volume fraction, the micro-convection induced by interaction forces, the thermal conductivity and velocity of nanofluids are improved. The first two can enhance the heat transfer and increases the temperature of the nanofluid, while the last reduces the heat transfer time between the fluid and the hot wall and decreases the temperature of the nanofluid. The flow rate of nanofluid increases sharply with volume fraction and plays a dominant role in affecting the temperature.

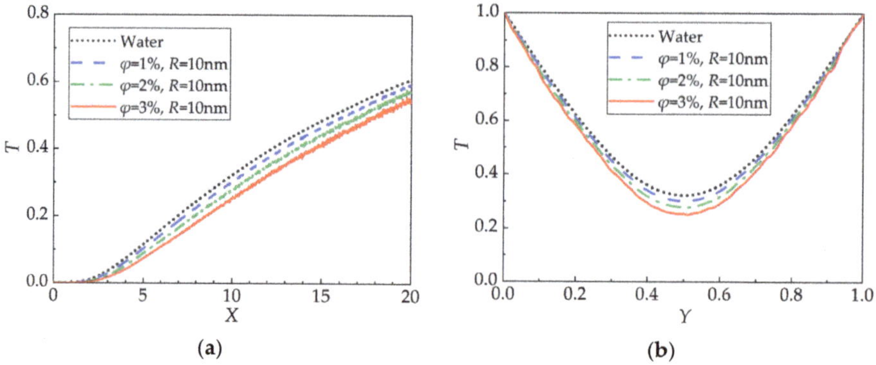

Figure 10. Non-dimensional temperature profiles for various nanoparticle volume fractions at (**a**) $Y = 0.5$ and (**b**) $X = 10$.

Figure 11 shows the dimensionless temperature distribution of pure water and Al_2O_3–water nanofluid with various particle radii at $X = 10$ and $Y = 0.5$. It demonstrates that the temperature of nanofluids increases with the increment of particle size. This is due to the drops in velocity of the nanofluids as the particle size increases, resulting in a longer heat transfer time between the nanofluid and the wall, which, in turn, raises the nanofluids' temperature. To sum up, the temperature of nanofluids in the microchannel is lowest at $\varphi = 3\%$ and $R = 10$ nm. In addition, the temperature field of the nanofluid appears perturbed, which is more prominent near the outlet, as shown in Figures 10a and 11a. This can be explained by the fact that the higher nanofluid temperature near the exit leads to a higher Brownian force, causing the particles to move more irregularly. Furthermore, the temperature of the nanofluid is disturbed, and the degree of disturbance strengthens with the increase in volume fraction and the decrease in particle radius. The reason is that the Brownian force is multiplied by particle number; thereby, the velocity difference and temperature difference between particles and the base fluid increase, resulting in enhanced micro-convection. However, a larger particle size causes the Brownian force to weaken, leading to the opposite outcome.

Figure 12 shows the local Nusselt number along the lower wall for different nanoparticle volume fractions and nanoparticle radii. The local Nusselt number decreases sharply at the entrance and then trends to flat. The reason is that the velocity changes drastically at the entrance, and convective heat transfer is extremely efficient. The Nusselt number of nanofluid increases with the increasing nanoparticle volume fraction, as illustrated in Figure 12a. This is because the thermal conductivity and flow velocity of nanofluids are enhanced as the volume fraction increases, as well as the random motion of the particles. The improved thermal conductivity allows nanofluids to absorb more heat, and the increased flow velocity makes nanofluids disperse heat efficiently. Moreover, micro-convection induced by the irregular motion of particles benefits the heat transfer. In addition, the Nusselt number perturbances irregularly under the action of the Brownian force. With the increase in volume fraction, the Brownian force increases, and the perturbation intensifies. Figure 12b indicates that the increasing particle size leads to the lessening of the Nusselt number. One of the reasons is that the velocity of nanofluid slows down, and the heat transfer is weakened at larger particle sizes. The other is that the increasing particle size plays a role in attenuating the Brownian motion of particles, which diminishes the micro-convection between particles and base fluid. Moreover, the particle size also plays a role in the amplitude of Nusselt number fluctuations. A flat Nusselt number perturbation results from the weak Brownian motion of large particles.

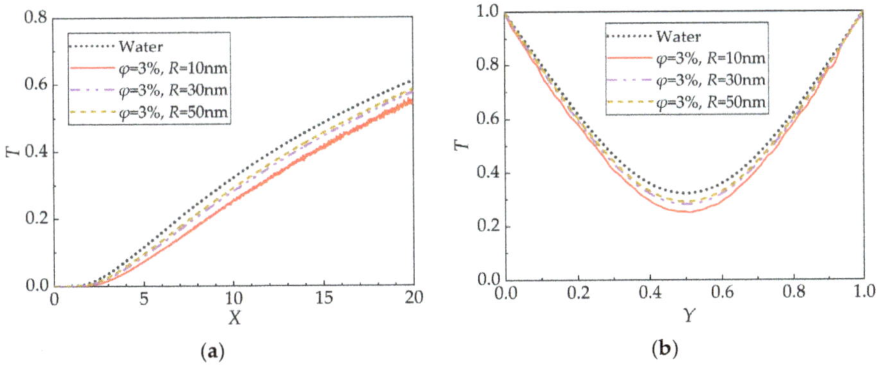

Figure 11. Non-dimensional temperature profiles for various nanoparticle radii at (**a**) $Y = 0.5$ and (**b**) $X = 10$.

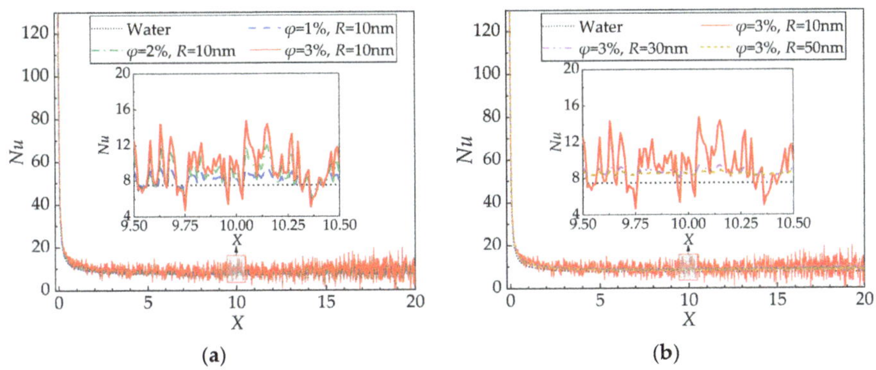

Figure 12. Local Nusselt number along the lower wall for different (**a**) nanoparticle volume fractions and (**b**) nanoparticle radii.

To visually demonstrate the contribution of thermal physical properties and particle–base fluid interaction forces on heat transfer, Figure 13 compares the average Nusselt number (Nu_{avg}) of nanofluid with and without forces under different volume fractions and particle sizes. This figure suggests that the interaction force and volume fraction improve the Nu_{avg}, while the particle radius decreases it. The Nu_{avg} of water is 8.348. For R = 10 nm and φ = 3%, the Nu_{avg} of nanofluids considering the interaction forces reaches 10.214, while the Nu_{avg} is 9.693 if the interaction forces inside the nanofluid are ignored. It indicates that the micro-convection caused by the interaction forces between particles and base fluid increases the Nu_{avg} by 6.24%, and the improvement of thermal physical properties increases Nu_{avg} by 16.16%. For the particle radius R = 10 nm, an increase in the volume fraction of nanoparticles from 0 to 3% increases the Nu_{avg} by approximately 22.40%. This increase can be attributed to the increased volume fraction improving the thermal conductivity, velocity, and micro-convection in a nanofluid. However, the rise in Nu_{avg} drops to 14.15% when the particle size increases to R = 50 nm. This phenomenon can be explained by the fact that an increase in particle size results in a low flow rate and weak micro-convection.

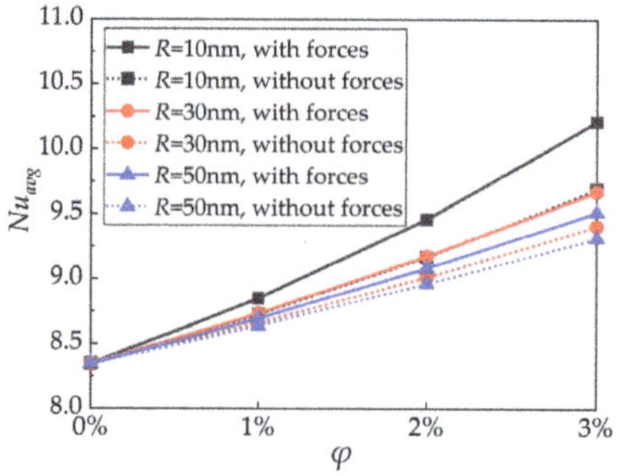

Figure 13. The average Nusselt number along the bottom wall for different volume fractions and nanoparticle radii.

In order to compare the comprehensive thermodynamic and hydrodynamic properties of nanofluids with various volume fractions and particle sizes, the performance evaluation criterion (PEC) is studied, and the result is illustrated in Figure 14. The PEC represents the combined influence of heat transfer efficiency and flow resistance of nanofluids. Its variation is almost identical to that in Nu_{avg} since the particle size, volume fraction, and the interaction force between particles and base fluid have little impact on f_{avg}. The nanofluid has a PEC greater than 1.0, which shows that the heat transfer performance is better under the same flow resistance. Furthermore, the comprehensive performance of nanofluids reaches the highest value (PEC = 1.225) when the volume fraction is the largest, and the particle size is the smallest (φ = 3%, R = 10 nm).

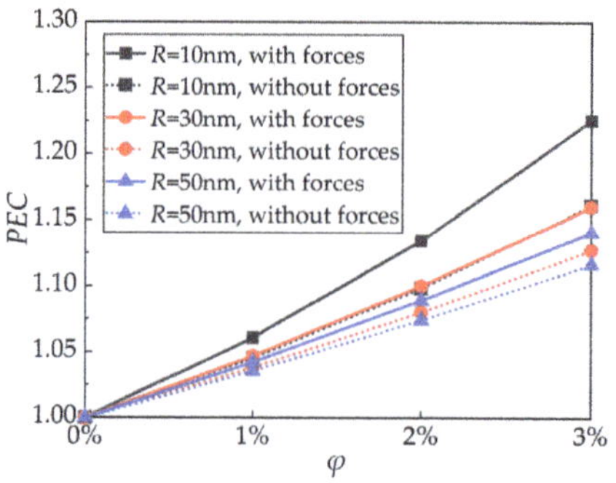

Figure 14. The PEC of nanofluids for different volume fractions and nanoparticle radii.

4. Conclusions

Al_2O_3–water nanofluid flow inside a microchannel has been studied with various nanoparticle volume fractions and radii by the two-phase lattice Boltzmann method. The physical properties and internal interaction forces of nanofluids are calculated, and their effects on the velocity, temperature, Nusselt number, Poiseuille number, and performance evaluation criterion are investigated. The obtained results can be summarized as:

(1) As the volume fraction of nanoparticles increases and the radius decreases, the velocity of nanofluids increases while the temperature decreases.

(2) It appears that the viscosity plays a greater role in determining temperature than the thermal conductivity in nanofluids at a Re = 50.

(3) An increase in nanoparticle concentration has a positive effect, and an increase in the size of nanoparticles has a negative effect on the rate of heat transfer.

(4) The interaction forces between particles and base liquid increases the Nu_{avg} number of nanofluids by 0.67~6.95%, and the physical property parameters increase the Nu_{avg} number by 3.53~14.15%.
(5) The interaction forces, volume fractions, and nanoparticle sizes have little effect on the f_{avg}.
(6) The comprehensive performance of the nanofluid is PEC = 1.0~1.225, which indicates that it is an excellent heat transfer medium.

This study demonstrates that nanofluids can significantly enhance heat transfer in solar cell cooling, providing a feasible method for maintaining lower operating temperatures and preventing efficiency losses in photovoltaic systems. The economic analysis suggests that this technology could be both cost-effective and efficient, which is crucial for large-scale adoption in the solar energy industry, where cost considerations are paramount.

Future research should focus on exploring the long-term stability of nanofluids, addressing potential issues such as particle agglomeration, sedimentation, or degradation over time. Understanding these factors will be essential for evaluating the practical longevity and maintenance needs of nanofluid-based cooling systems. Additionally, 3D simulations may provide a more accurate representation of real physical systems, potentially leading to more refined results and a better understanding of nanofluid behavior in practical applications. We expect that future research can improve the design and optimization of thermal management systems by providing a more comprehensive understanding of the nanofluids' performance in practical applications.

Author Contributions: Conceptualization, H.L. and M.B.; methodology, Y.G. and H.L.; software, H.L.; validation, H.L.; formal analysis, Y.G., H.L., M.B. and L.G.; investigation, H.L.; resources, Y.G., L.G. and S.S.; data curation, Y.G. and H.L.; writing—original draft preparation, H.L.; writing—review and editing, M.B. and Y.G.; visualization, H.L.; supervision, Y.G.; project administration, L.G.; funding acquisition, Y.G. and S.S. All authors have read and agreed to the published version of the manuscript.

Funding: This research was funded by the National Natural Science Foundation of China, grant number 51936002.

Data Availability Statement: The data presented in this study are available on request from the corresponding author, since the data are not publicly available due to privacy or ethical restrictions.

Conflicts of Interest: The authors declare no conflicts of interest.

Nomenclature

Abbreviations	
BGK	Bhatnagar–Gross–Krook
DDF	Double distribution functions
PEC	Performance evaluation criterion
PV	Photovoltaic
TPLBM	Two-phase lattice Boltzmann method
English symbols	
A	Area [m^2]
B	Mass coefficient
c	Lattice velocity
C	Mass fraction
c_p	Specific heat [J kg^{-1} K^{-1}]
d	Diameter [m]
e_i	Direction of lattice velocity
f_i	Density distribution function
f_{app}	Fanning friction factor
$\boldsymbol{F}$	Force vector [N]
F_A	Interaction potential force [N]
F_B	Brownian force [N]
F_{bf}	The total force acting on the base fluid [N]

F_D	Drag force [N]
F_B	Buoyancy force [N]
F_{np}	The total force acting on the nanoparticle [N]
g_i	Temperature distribution function
H	Microchannel height [m]
h	Heat transfer coefficient [W m^{-2} K^{-1}]
k	Thermal conductivity [W m^{-1} K^{-1}]
L	Microchannel length [m]
m_{np}	Nanoparticle mass [kg]
n	The nanoparticle number in a lattice
Nu	Nusselt number
PEC	Performance evaluation criterion
Po	Poiseuille number
Pr	Prandtl number
R	Nanoparticles radius [m]
Re	Reynolds number
t	Time [s]
T	Dimensionless Temperature
u	Velocity [m/s]
U	Dimensionless velocity
x, y	Axial and vertical Cartesian coordinate
X,Y	Dimensionless axial and vertical Cartesian coordinate
Δt	Lattice length
Δx	Time step
Greek Symbols	
γ	Surface tension [N]
θ	Temperature (K)
μ	Dynamics viscosity [kg m^{-1} s^{-1}]
v	Kinematic viscosity coefficient [m^{-2} s^{-1}]
ρ	Density [kg m^{-3}]
τ_f	Dimensionless collision-relaxation time for flow
τ_g	Dimensionless collision-relaxation time for temperature
φ	Nanoparticle volume fraction
χ	Thermal diffusion coefficient [m^{-2} s^{-1}]
ω	Weight coefficients
Subscripts	
avg	Average value
bf	Base fluid
in	Inlet
m	Mean value
nf	Nanofluids
np	Nanoparticle
w	Wall
X, Y	X and Y directions
Superscripts	
eq	Equilibrium state
i	Discrete lattice directions
σ	Base fluid or nanoparticle phase

References

1. Ebaid, M.S.Y.; Ghrair, A.M.; Al-Busoul, M. Experimental investigation of cooling photovoltaic (PV) panels using (TiO_2) nanofluid in water -polyethylene glycol mixture and (Al_2O_3) nanofluid in water-cetyltrimethylammonium bromide mixture. *Energy Convers. Manag.* **2018**, *155*, 324–343. [CrossRef]
2. Faheem, M.; Abu Bakr, M.; Ali, M.; Majeed, M.A.; Haider, Z.M.; Khan, M.O. Evaluation of Efficiency Enhancement in Photovoltaic Panels via Integrated Thermoelectric Cooling and Power Generation. *Energies* **2024**, *17*, 2590. [CrossRef]
3. Tuncer, A.D.; Gürbüz, E.Y.; Şahinkesen, İ.; Georgiev, A.; Keçebaş, A. Passive cooling of photovoltaic panels with latent heat storage unit: Analyzing the effects of using fins and iron nanoparticles on the performance, economy and environmental impact. *Energy* **2024**, *288*, 129860. [CrossRef]

4. Lee, K.W.; Lim, W.; Jeon, M.S.; Jang, H.; Hwang, J.; Lee, C.H.; Kim, D.R. Visibly Clear Radiative Cooling Metamaterials for Enhanced Thermal Management in Solar Cells and Windows. *Adv. Funct. Mater.* **2021**, *32*, 2105882. [CrossRef]
5. Malik, F.K.; Fobelets, K. A review of thermal rectification in solid-state devices. *J. Semicond.* **2022**, *43*, 103101. [CrossRef]
6. Zhang, C.; Wang, N.; Xu, H.; Fang, Y.; Yang, Q.; Talkhoncheh, F.K. Thermal management optimization of the photovoltaic cell by the phase change material combined with metal fins. *Energy* **2023**, *263*, 125669. [CrossRef]
7. Abo-Zahhad, E.M.; Ookawara, S.; Esmail, M.F.C.; El-Shazly, A.H.; Elkady, M.F.; Radwan, A. Thermal management of high concentrator solar cell using new designs of stepwise varying width microchannel cooling scheme. *Appl. Therm. Eng.* **2020**, *172*, 115124. [CrossRef]
8. Hong, S.; Zhang, B.; Dang, C.; Hihara, E. Development of two-phase flow microchannel heat sink applied to solar-tracking high-concentration photovoltaic thermal hybrid system. *Energy* **2020**, *212*, 118739. [CrossRef]
9. Chen, L.; Deng, D.; Ma, Q.; Yao, Y.; Xu, X. Performance evaluation of high concentration photovoltaic cells cooled by microchannels heat sink with serpentine reentrant microchannels. *Appl. Energy* **2022**, *309*, 118478. [CrossRef]
10. Mehta, B.; Subhedar, D.; Panchal, H.; Said, Z. Synthesis, stability, thermophysical properties and heat transfer applications of nanofluid—A review. *J. Mol. Liq.* **2022**, *364*, 120034. [CrossRef]
11. Gupta, M.; Singh, V.; Kumar, R.; Said, Z. A review on thermophysical properties of nanofluids and heat transfer applications. *Renew. Sustain. Energy Rev.* **2017**, *74*, 638–670. [CrossRef]
12. Ho, C.J.; Peng, J.; Yang, T.; Rashidi, S.; Yan, W. On the assessment of the thermal performance of microchannel heat sink with nanofluid. *Int. J. Heat Mass Trans.* **2023**, *201*, 123572. [CrossRef]
13. Heidarshenas, A.; Azizi, Z.; Peyghambarzadeh, S.M.; Sayyahi, S. Experimental investigation of heat transfer enhancement using ionic liquid-Al_2O_3 hybrid nanofluid in a cylindrical microchannel heat sink. *Appl. Therm. Eng.* **2021**, *191*, 116879. [CrossRef]
14. Jahanbakhshi, A.; Nadooshan, A.A.; Bayareh, M. Cooling of a lithium-ion battery using microchannel heatsink with wavy microtubes in the presence of nanofluid. *J. Energy Storage* **2022**, *49*, 104128. [CrossRef]
15. Wang, R.; Wang, J.; Yuan, W. Analysis and Optimization of a Microchannel Heat Sink with V-Ribs Using Nanofluids for Micro Solar Cells. *Micromachines* **2019**, *10*, 620. [CrossRef]
16. Yang, X.; Zhao, Z.; Liu, Y.; Xing, R.; Sun, Y. Simulation of nanofluid-cooled lithium-ion battery during charging: A battery connected to a solar cell. *Int. J. Mech. Sci.* **2021**, *212*, 106836. [CrossRef]
17. Karami, N.; Rahimi, M. Heat transfer enhancement in a hybrid microchannel-photovoltaic cell using Boehmite nanofluid. *Int. Commun. Heat Mass* **2014**, *55*, 45–52. [CrossRef]
18. Radwan, A.; Ahmed, M.; Ookawara, S. Performance enhancement of concentrated photovoltaic systems using a microchannel heat sink with nanofluids. *Energy Convers. Manag.* **2016**, *119*, 289–303. [CrossRef]
19. Pinto, R.V.; Fiorelli, F.A.S. Review of the mechanisms responsible for heat transfer enhancement using nanofluids. *Appl. Therm. Eng.* **2016**, *108*, 720–739. [CrossRef]
20. Khoshtarash, H.; Siavashi, M.; Ramezanpour, M.; Blunt, M.J. Pore-scale analysis of two-phase nanofluid flow and heat transfer in open-cell metal foams considering Brownian motion. *Appl. Therm. Eng.* **2023**, *221*, 119847. [CrossRef]
21. Mahian, O.; Kolsi, L.; Amani, M.; Estellé, P.; Ahmadi, G.; Kleinstreuer, C.; Marshall, J.S.; Siavashi, M.; Taylor, R.A.; Niazmand, H.; et al. Recent advances in modeling and simulation of nanofluid flows-Part I: Fundamentals and theory. *Phys. Rep.* **2019**, *790*, 1–48. [CrossRef]
22. Farahani, M.V.; Nezhad, M.M. On the effect of flow regime and pore structure on the flow signatures in porous media. *Phys. Fluids* **2022**, *34*, 115139. [CrossRef]
23. Vasheghani Farahani, M.; Foroughi, S.; Norouzi, S.; Jamshidi, S. Mechanistic study of fines migration in porous media using lattice Boltzmann method coupled with rigid body physics engine. *J. Energy Resour. Technol.* **2019**, *141*, 123001. [CrossRef]
24. Xuan, Y.; Qiang, L.; Yao, Z. Application of lattice Boltzmann scheme to nanofluids. *Sci. China* **2004**, *47*, 129–140. [CrossRef]
25. Guo, Y.; Qin, D.; Shen, S.; Bennacer, R. Nanofluid multi-phase convective heat transfer in closed domain: Simulation with lattice Boltzmann method. *Int. Commun. Heat Mass* **2012**, *39*, 350–354. [CrossRef]
26. Ahmed, M.; Eslamian, M. Natural convection in a differentially-heated square enclosure filled with a nanofluid: Significance of the thermophoresis force and slip/drift velocity. *Int. Commun. Heat Mass* **2014**, *58*, 1–11. [CrossRef]
27. Ahmed, M.; Eslamian, M. Numerical Simulation of Natural Convection of a Nanofluid in an Inclined Heated Enclosure Using Two-Phase Lattice Boltzmann Method: Accurate Effects of Thermophoresis and Brownian Forces. *Nanoscale Res. Lett* . **2015**, *10*, 296. [CrossRef] [PubMed]
28. Zarghami, A.; Ubertini, S.; Succi, S. Finite-volume lattice Boltzmann modeling of thermal transport in nanofluids. *Comput. Fluids* **2013**, *77*, 56–65. [CrossRef]
29. Saberi, A.H.; Kalteh, M. Numerical investigation of nanofluid convection heat transfer in a microchannel using two-phase lattice Boltzmann method. *J. Therm. Anal. Calorim.* **2019**, *138*, 1761–1777. [CrossRef]
30. Ahmed, M.; Eslamian, M. Laminar forced convection of a nanofluid in a microchannel: Effect of flow inertia and external forces on heat transfer and fluid flow characteristics. *Appl. Therm. Eng.* **2015**, *78*, 326–338. [CrossRef]
31. Li, L.; Liu, J.; Pan, Z. Enhancement of heat transfer and mixing with two side-by-side freely rotatable cylinders in microchannel. *Int. J. Heat Mass Trans.* **2022**, *189*, 122717. [CrossRef]
32. Gupta, S.; Subbarao, P.M.V. Empirical investigation of nanofluid performance in a microchannel heat sink for various attributes in low Reynolds number range. *Int. J. Heat Fluid Flow* **2024**, *105*, 109258. [CrossRef]

33. Corcione, M. Heat transfer features of buoyancy-driven nanofluids inside rectangular enclosures differentially heated at the sidewalls. *Int. J. Therm. Sci.* **2010**, *49*, 1536–1546. [CrossRef]
34. Maxwell, J.C. *A Treatise on Electricity and Magnetism;* Clarendon Press: Oxford, UK, 1873.
35. Yang, S.; Tao, W. *Heat Transfer;* Higher Education Press: Beijing, China, 2006.
36. Li, A.; Ahmadi, G. Dispersion and Deposition of Spherical Particles from Point Sources in a Turbulent Channel Flow. *Aerosol Sci. Technol.* **1992**, *16*, 209–226. [CrossRef]
37. Russel, W.B.; Saville, D.A.; Schowalter, W.R. *Colloidal Dispersion;* Cambridge University Press: Cambridge, UK; New York, NY, USA, 1989.
38. He, X.; Luo, L.S. Lattice Boltzmann Model for the Incompressible Navier–Stokes Equation. *J. Stat. Phys.* **1997**, *88*, 927–944. [CrossRef]
39. Shan, X.; Chen, H. Lattice Boltzmann Model for Simulating Flows with Multiple Phases and Components. *Phys. Rev. E* **1993**, *47*, 1815–1819. [CrossRef]

energies

MDPI

Article

Scaling Law of Flow and Heat Transfer Characteristics in Turbulent Radiative Rayleigh-Bénard Convection of Optically Thick Media †

Jiajun Song [1], Panxin Li [1,2], Lu Chen [1,3], Yuhang Zhao [1], Fengshi Tian [1] and Benwen Li [1,*]

1 Key Laboratory of Ocean Energy Utilization and Energy Conservation of Ministry of Education, School of Energy & Power Engineering, Dalian University of Technology, Dalian 116024, China; songjjdut@163.com (J.S.); wslpx2012@163.com (P.L.); chenluneu@163.com (L.C.); zhaoyuhang1203@163.com (Y.Z.); 13134342044@163.com (F.T.)
2 Institute of Thermodynamics and Fluid Mechanics, Technische Universität Ilmenau, 98684 Ilmenau, Germany
3 MCC Capital Engineering & Research Inc., Ltd., Beijing 100176, China
* Correspondence: heatli@dlut.edu.cn

† Correlations for the Transport Characteristics of Radiative Rayleigh-Bénard Convection in Optically Thick Media, Shanghai, China, 12–14 August.

Abstract: Radiative natural convection is of vital importance in the process of energy storage, power generation, and thermal storage technology. As the attenuation coefficients of many heat transfer media in these fields are high enough to be considered as optically thick media, like nanofluids or molten salts in concentrated solar power or phase change thermal storage, Rosseland approximation is commonly used. In this paper, we delve into the impact of thermal radiation on the Rayleigh-Bénard (RB) convection. Theoretical analysis has been conducted by modifying the Grossmann-Lohse (GL) model. Based on turbulent dissipation theory, the corresponding scaling laws in four main regimes are proposed. Direct numerical simulation (DNS) was also performed, revealing that radiation exerts a notable influence on both flow and heat transfer, particularly on the formation of large-scale circulation. By comparing with DNS results, it is found that due to the presence of radiation, the modified *Nu* scaling law in small *Pr* range of the GL model is more suitable for predicting the transport characteristics of optical thick media with large *Pr*. The maximum deviation between the results of DNS and prediction model is about 10%, suggesting the summarized scaling law can effectively predict the *Nu* of radiative RB convection.

Keywords: solar energy; energy storage power generation; Radiative Rayleigh-Bénard convection; optically thick media; scaling law

1. Introduction

The increased uptake of the renewable energy has led to an increased need for updating solar thermal energy utilization and thermal storage technology. For example, by combining concentrated solar power with photovoltaics, both the heat and electricity can be provided at the same system, making the energy maximally utilized [1]. What is more, the combined heat and power system, when paired with a phase change thermal storage system [2,3], can achieve a continuous supply of energy. Obviously, understanding and optimizing these fluid transport characteristics is essential for the development and improvement of these energy conversion and storage systems.

It is worth noting that, in applications of the above fields, the operating temperature is usually high and the temperature difference in the system may also be considerable. This makes the effect of thermal radiation non-negligible, which is a typical radiative convection problem. Different from thermal conduction and convection, the radiative energy can penetrate into the interior of the fluid, and transfer between all the elements, as it propagates as an electromagnetic wave. Besides, the directional and spectral natures of radiation increase the dimensionality of the problem, making the determination of radiative heat flux complex. Therefore, many simplified models have been proposed, and the Rosseland diffusion approximation is the most widely used in the radiative convection of optical thick media, in which the radiative heat flux is supposed to be a function of temperature gradient, i.e., $q^{\mathrm{R}} \sim -\frac{16\sigma T_{\mathrm{ref}}^3}{3\beta_r}\frac{\partial T}{\partial x}$ (linear approximation) or $q^{\mathrm{R}} \sim -\frac{16\sigma T^3}{3\beta_r}\frac{\partial T}{\partial x}$ (nonlinear approximation). For example, by using linear Rosseland approximation, Sheikholeslami et al. [4] studied the nanofluid flow and heat transfer under the effect of electric field in a porous lid driven cavity. They found thermal radiation can enhance the temperature gradient near the positive electrode, as well as the convective mode. Similar conclusions were obtained by many other researchers [5–7]. Besides, with the help of nonlinear Rosseland approximation, Shah et al. [8] proposed an artificial neural network model for heat transfer problem in carbon nanotubes-based nanofluid, which shows excellent performances by comparing with numerical simulation. Yousuf Ali et al. [9] investigated the effect of radiation on the heat transfer of nanofluids flowing through a stretching sheet by using both linear and nonlinear Rosseland approximations, and found that the thermal performance is enhanced.

Indeed, the influence of radiation on natural convection has been a subject of extensive research for many years [10–12]. Kolsi et al. [10] studied the radiative natural convection for $Ra = 10^5$ and $Pr = 13.6$, and proposed that, compared to peripheral spiraling motion, the principal flow structure is more sensitive to radiation effect. What is more, in absence of radiation there exists a zone where the flow structure exhibits quasi-2D characteristics, while the effect of radiation on the 3D behavior of the flow is significant, especially in the core of cavity. Wang et al. [13] discovered that thermal radiation can reduce the spatial spreading of the plumes but has little impact on the kinetic field. However, some other scholars suggested that the thermal radiation shows a stabilizing effect on natural convection. Both the convection and convective heat transfer are dampened by the surface radiation [14–17]. In fact, it has been observed that radiation emanating from the medium exhibits a similar dampening influence on the convective heat transfer [18,19].

The aforementioned studies hold crucial significance, but most of them have derived qualitative conclusions. In order to quantitatively determine the influence of control parameters on the transport characteristics of natural convection, some prediction models for radiative natural convection have been developed. Gad et al. [20] proposed the scaling law of heat transfer in laminar natural convection with radiation for the range of $5 \times 10^3 \leq Ra \leq 10^5$, $1 \leq \Gamma \leq 5$, $0.1 \leq \varepsilon_e \leq 0.85$. By applying k-ε turbulence model, Miroshnichenko et al. [16] investigated the heat transport characteristics for turbulent flow in a cavity, which is heated below and cooled by the environment on the other walls. They also summarized the correlations between Nu and (Ra, ε_e) for $10^9 \leq Ra \leq 10^{10}$ and $0 \leq \varepsilon_e \leq 0.6$. One of the most prominent works is the POD-based low-order model derived by Soucasse et al. [21,22] in the range of $10^6 \leq Ra \leq 10^8$, $Pr = 0.7$. Both the observed and predicted models they derived can capture the dynamics and energy transport features of the natural convection coupled with radiation. More relevant researches can be found in the review provided by Song et al. [23]. Nevertheless, one can easily see that most of the existing scaling laws are applicable within a narrow range of parameters and are insufficient to meet the needs of industrial design.

In fact, in the specific case of Rayleigh-Bénard (RB) convection, several scaling laws have been proposed, including the marginal stability theory [24], Chicago mixing zone model [25–27], Shraiman-Siggia model (SS model) [28,29], Grossmann-Lohse theory (GL model) [30–33], and so on. The GL model is widely recognized due to its wide applicability and high accuracy. Besides, it is used as a basis for many other correction models to describe the transport characteristics of RB convection under additional effects. Chakraborty [34] suggested that, by introducing the concept of magnetic energy dissipation, the arguments and assumptions in GL model can be extended to the case of turbulent magnetohydrodynamic (MHD) RB convection with a vertical magnetic field. Based on this idea, Zürner et al. [35,36] have given a set of implicit scaling equations between (Nu, Re) and (Ra, Pr,

Ha) in a limited parameter range ($Pr \leq 3 \times 10^{-2}$, $Pm = \mu\sigma_b v \leq 10^{-5}$). Moreover, the scaling behaviors of heat and momentum transport in MHD-RB convection were analyzed by Yan et al. [37] on the basis of the ratio of magnetic energy to kinetic energy and the contribution of viscous and ohmic dissipation. Therefore, analogous methodologies were adopted in this study for modeling the transport characteristics of radiative RB convection for the first time.

In this paper, a comprehensive study of the flow and heat transfer characteristics in radiative Rayleigh-Bénard convection within optically thick media has been conducted by using both theoretical analysis and direct numerical simulation. Based on the GL model, the relevant scaling laws have been derived according to the turbulent dissipation theory and the relative contribution of bulk and boundary layers to dissipation. In order to verify the prediction models, direct numerical simulation for the flow in the range of $Ra = 1 \times 10^8 \sim 5 \times 10^8$, $Pr = 7, 30$, $\mathbb{R} = 1.0 \sim 6.0$ (The definition of dimensionless parameters is depicted in Equation (6)) has also been conducted. The results show the proposed scaling law can effectively predict the Nu of radiative Rayleigh-Bénard convection, especially for situations with large optical thicknesses. The conclusions drawn from this paper can furnish a theoretical underpinning for industrial design pertaining to the realms of energy storage, power generation, and thermal storage technology.

Section 2 introduces the model and assumptions used in this theoretical analysis. Section 3 shows the scaling law in different regimes and the boundaries between them. Section 4 displays the direct numerical simulation results used to validate theoretical analysis. Finally, the conclusions of this study are presented in Section 5.

2. The Model

Before the theoretical analysis and numerical simulation, some assumptions need to be introduced first:

- An incompressible and optically thick fluid is considered;
- The Boussinesq approximation, as well as Rosseland diffusion approximation are invoked in this work;
- The flow is in a turbulent state and there exists a large-scale circulation (LSC);
- The kinetic BL is assumed to be of Prandtl-Blasius type.

Therefore, considering a fluid layer differentially heated by two rigid plates, the following governing equations can be obtained, shown as Equations (1)–(3):

$$\nabla \cdot \boldsymbol{u} = 0 \tag{1}$$

$$\partial_\tau \boldsymbol{u} + (\boldsymbol{u} \cdot \nabla)\boldsymbol{u} = -\frac{1}{\rho}\nabla p + \nu\nabla^2\boldsymbol{u} + g\beta\Delta e_z \tag{2}$$

$$\partial_\tau T + \boldsymbol{u} \cdot \nabla T = \kappa\nabla^2 T - \frac{1}{\rho c_p}\nabla q^R \tag{3}$$

The radiative heat flux by employing Rosseland approximation is defined as Equation (4):

$$q^R = -\frac{4}{3}\frac{\sigma}{\beta_r}\nabla T^4 \tag{4}$$

When neglecting the high-order terms from the Taylor series of T^4 about T_{ref} (here the linear Rosseland approximation is considered), the radiative heat source in Equation (3) can be obtained as Equation (5):

$$\nabla q^R = -\frac{16}{3}\frac{\sigma T_{\text{ref}}^3}{\beta_r}\nabla^2 T \tag{5}$$

Here three important criterion numbers, say, the Rayleigh number *Ra*, Prandtl number *Pr* and radiation parameter $\mathbb{R}$, as well as two response criterion numbers, Nusselt number *Nu* and Reynold number *Re*, to characterize the RB convection, are defined as Equation (6):

$$\begin{aligned} Ra &= \tfrac{g\beta\Delta h^3}{\nu\kappa}, Pr = \tfrac{\nu}{\kappa}, \mathbb{R} = 1 + \tfrac{4}{3}\tfrac{1}{\tau_r Pl} \\ Nu &= \tfrac{\langle wT\rangle_A - \kappa\partial_z\langle T\rangle_A}{\kappa\Delta h^{-1}}, Re = \tfrac{\overline{u}h}{\nu} \end{aligned} \quad (6)$$

In above equations, $\boldsymbol{e}_z$, $\boldsymbol{u} = \{u, v, w\}$, $t, p, h, T, \Delta = T_h - T_c, g, \beta, \kappa, \lambda, \overline{u}, \beta_r$, and σ are the unit vector in the **z** direction, velocity, time, pressure, distance between two differential heated plates, temperature, temperature difference between the two plates, gravitational acceleration, thermal expansion coefficient, thermal diffusivity, thermal conductivity, the mean large-scale velocity, spectral extinction coefficient, and Stefan-Boltzmann constant ($\sigma = 5.67 \times 10^{-8}\text{W}\cdot\text{m}^{-2}\cdot\text{K}^{-4}$), respectively. The determination of characteristic quantities is illustrated in the Figure 1. Subscript h and c stand for the hot bottom and cold top plates, individually. $\tau_r = \beta_r h$ is the optical thickness. $Pl = \frac{\lambda/h}{4\sigma T_{\text{ref}}^3}$ is the Planck number. T_{ref} is the reference temperature, which will be discussed later. $\langle\cdot\rangle_A$ denotes the average over any horizonal plane (z-plane).

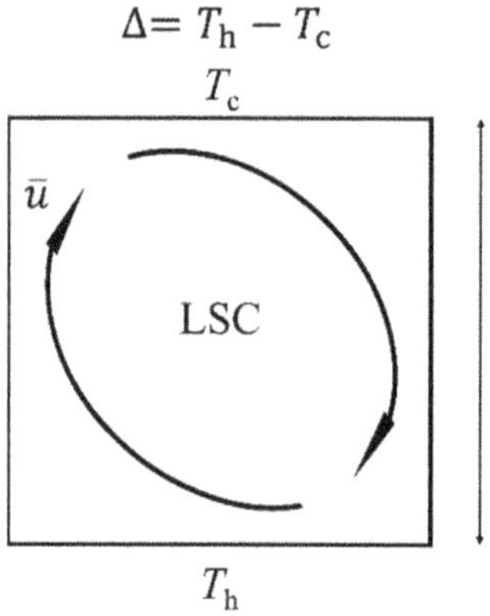

Figure 1. The dependence of characteristic quantities.

Generally, the mean temperature, $T_{\text{m-ref}} = (T_h + T_c)/2$ is usually considered as the reference temperature. However, due to the dependence of radiative heat transfer on $|T_h^4 - T_c^4|$, another strategy has been proposed [38] as Equation (7):

$$T_{\text{rad-ref}} = \left(\frac{T_h^4 + T_c^4}{2}\right)^{\frac{1}{4}} \quad (7)$$

By comparing these two reference temperatures, the relative error between them can be gained, shown as Equation (8):

$$O = \frac{T_{\text{rad-ref}} - T_{\text{m-ref}}}{T_{\text{m-ref}}} = \left[8 \cdot \frac{1 + (T_h/T_c)^4}{(1 + T_h/T_c)^4}\right]^{\frac{1}{4}} - 1 \quad (8)$$

Figure 2 depicts the dependence of the temperature difference (Δ and Δ/T_c) and O upon T_c. It can be easily seen that even if the relative temperature difference Δ/T_c reaches 50%, O is only 5.56%. It should be noticed that the basic premise for applying the Boussinesq approximation is that the physical properties of fluid do not change significantly except for the density. As shown in Figure 1, Δ has exceeded 100 K for $T_c > 200$ K and $\Delta/T_c = 50\%$. Indeed, such a large temperature difference usually leads to significant changes in the physical properties of fluid. Under such conditions, the Boussinesq approximation is no longer valid. Therefore, either of these two reference temperatures can usually be considered for most cases with small temperature differences.

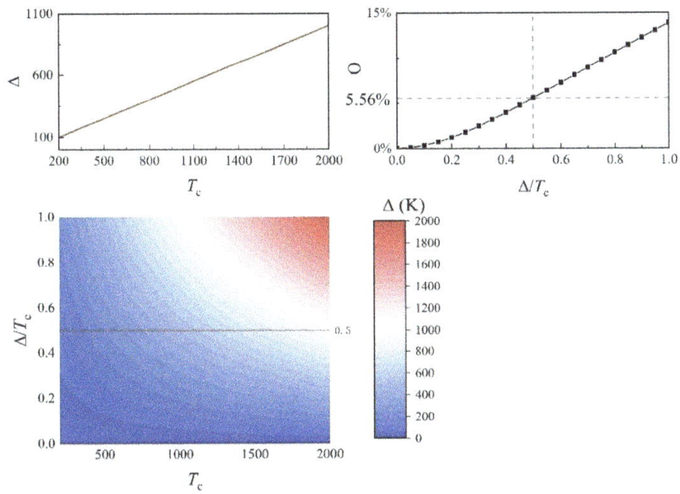

Figure 2. The dependence of the temperature difference (Δ and Δ/T_c) and O upon T_c.

3. Scaling Laws

In correspondence with classical RB convection, one can derive the following rigorous relations for the global averaged kinetic and thermal dissipation rates, ε_u and ε_T, according to the theory of Shraiman and Siggia [28,29], as shown in Equations (9) and (10):

$$\varepsilon_u = \langle \varepsilon_u(x,t) \rangle_V = \frac{\nu^3}{h^4}(Nu-1)RaPr^{-2} \tag{9}$$

$$\varepsilon_T = \langle \varepsilon_T(x,t) \rangle_V = \kappa \frac{\Delta^2}{h^2} Nu \tag{10}$$

where $\langle \cdot \rangle_V$ denotes the volume average. $\varepsilon_u(x,t) = \nu \left[\partial_i u_j(x,t)\right]^2$ and $\varepsilon_T(x,t) = \kappa \left[\partial_i T_j(x,t)\right]^2$ are the local kinetic and thermal dissipation rates.

The main idea of Grossmann-Lohse theory [30–33] is that the global averaged energy dissipations can be decomposed into the bulk and boundary layer (BL) contributions, as shown in Equation (11):

$$\varepsilon = \varepsilon_{\text{bulk}} + \varepsilon_{\text{BL}} \tag{11}$$

By means of dimensional analysis, the different contributions can be estimated as Equations (12)–(17):

$$\varepsilon_{u-\text{bulk}} \sim \frac{\overline{u}^3}{h} = \frac{\nu^3}{h^4} Re^3 \tag{12}$$

$$\varepsilon_{T-\text{bulk}} \sim \frac{\overline{u}\Delta^2}{h} = \kappa \frac{\Delta^2}{h^2} PrRe(\text{for } \frac{\delta_u}{\delta_T} < 1) \tag{13}$$

$$\varepsilon_{T-\text{bulk}} \sim \overline{u}\frac{\delta_T}{\delta_u}\frac{\Delta^2}{h} = \kappa \frac{\Delta^2}{h^2} \mathbb{R} PrRe^{\frac{3}{2}} Nu^{-1}(\text{for } \frac{\delta_u}{\delta_T} > 1) \tag{14}$$

$$\varepsilon_{u-\text{BL}} \sim \nu \frac{\overline{u}^2}{\delta_u^2}\frac{\delta_u}{h} \sim \frac{\nu^3}{h^4} Re^{\frac{5}{2}} \tag{15}$$

$$\varepsilon_{T-\mathrm{BL}} \sim \kappa \frac{\Delta^2}{\delta_T^2}\frac{\delta_T}{h} \sim \kappa \frac{\Delta^2}{h^2}\mathbb{R}^{\frac{1}{2}} Pr^{\frac{1}{2}} Re^{\frac{1}{2}} (\text{for} \frac{\delta_u}{\delta_T} < 1) \tag{16}$$

$$\varepsilon_{T-\mathrm{BL}} \sim \kappa \frac{\Delta^2}{\delta_T^2}\frac{\delta_T}{h} \sim \kappa \frac{\Delta^2}{h^2}\mathbb{R}^{\frac{2}{3}} Pr^{\frac{1}{3}} Re^{\frac{1}{2}} (\text{for} \frac{\delta_u}{\delta_T} > 1) \tag{17}$$

where $\delta_u \sim hRe^{-\frac{1}{2}}$ and $\delta_T \sim \mathbb{R}hNu^{-1}$ are the thickness of the kinetic and thermal BL, respectively. Their ratio is related to Pr, i.e., $Pr \sim \delta_u/\delta_T$. It should be emphasized that the kinetic BL is assumed to be of Prandtl-Blasius type in Grossmann-Lohse theory.

Utilizing the aforementioned relations, four main regimes can be easily identified as shown in Table 1, contingent upon whether the bulk or BL is the primary contributor to energy dissipation:

Table 1. The four main regimes and the primary contributors.

	ε_u	ε_T
Regime I	$\varepsilon_{u-\mathrm{BL}}$	$\varepsilon_{T-\mathrm{BL}}$
Regime II	$\varepsilon_{u-\mathrm{bulk}}$	$\varepsilon_{T-\mathrm{BL}}$
Regime III	$\varepsilon_{u-\mathrm{BL}}$	$\varepsilon_{T-\mathrm{bulk}}$
Regime IV	$\varepsilon_{u-\mathrm{bulk}}$	$\varepsilon_{T-\mathrm{bulk}}$

Additionally, based on the relative thickness of the kinetic and thermal BL, each main regime can be divided into two sub-regions, i.e., $\delta_u/\delta_T < 1$ or $\delta_u/\delta_T > 1$. The scaling laws are given in each sub-region as following:

- *Regime I-a:*$\varepsilon_u \sim \varepsilon_{u-\mathrm{BL}}, \varepsilon_T \sim \varepsilon_{T-\mathrm{BL}}, \delta_u/\delta_T < 1$

Substituting Equations (8) and (9) into Equations (15) and (16), respectively, one can get the following equations as shown in Equations (18) and (19):

$$Re = \mathbb{R}^{\frac{1}{4}} Ra^{\frac{1}{2}} Pr^{-\frac{3}{4}} \tag{18}$$

$$Nu = \mathbb{R}^{\frac{5}{8}} Ra^{\frac{1}{4}} Pr^{\frac{1}{8}} \tag{19}$$

- *Regime I-b:*$\varepsilon_u \sim \varepsilon_{u-\mathrm{BL}}, \varepsilon_T \sim \varepsilon_{T-\mathrm{BL}}, \delta_u/\delta_T > 1$

By combining Equations (8) with (15) and (9) with (17), the following relations can be obtained as shown in Equations (20) and (21):

$$Re = \mathbb{R}^{\frac{1}{3}} Ra^{\frac{1}{2}} Pr^{-\frac{5}{6}} \tag{20}$$

$$Nu = \mathbb{R}^{\frac{5}{6}} Ra^{\frac{1}{4}} Pr^{-\frac{1}{12}} \tag{21}$$

- *Regime II-a:*$\varepsilon_u \sim \varepsilon_{u-\mathrm{bulk}}, \varepsilon_T \sim \varepsilon_{T-\mathrm{BL}}, \delta_u/\delta_T < 1$

Using Equations (8), (9), (12) and (16), one can obtain the following equations as shown in Equations (22) and (23):

$$Re = \mathbb{R}^{\frac{1}{5}} Ra^{\frac{2}{5}} Pr^{-\frac{3}{5}} \tag{22}$$

$$Nu = \mathbb{R}^{\frac{3}{5}} Ra^{\frac{1}{5}} Pr^{\frac{1}{5}} \tag{23}$$

- *Regime II-b:*$\varepsilon_u \sim \varepsilon_{u-\mathrm{bulk}}, \varepsilon_T \sim \varepsilon_{T-\mathrm{BL}}, \delta_u/\delta_T > 1$

Using Equation (17) instead of (16) for (9), we can get the following equations as shown in Equations (24) and (25):

$$Re = \mathbb{R}^{\frac{4}{15}} Ra^{\frac{2}{5}} Pr^{-\frac{2}{3}} \tag{24}$$

$$Nu = \mathbb{R}^{\frac{4}{5}} Ra^{\frac{1}{5}} \tag{25}$$

- *Regime III-a:* $\varepsilon_u \sim \varepsilon_{u-\mathrm{BL}}, \varepsilon_T \sim \varepsilon_{T-\mathrm{bulk}}, \delta_u/\delta_T < 1$

Combining Equations (8), (9), (15) and (13), one can derive the following relations as shown in Equations (26) and (27):

$$Re = Ra^{\frac{2}{3}} Pr^{-\frac{2}{3}} \tag{26}$$

$$Nu = Ra^{\frac{2}{3}} Pr^{\frac{1}{3}} \tag{27}$$

- *Regime III-b:* $\varepsilon_u \sim \varepsilon_{u-\mathrm{BL}}, \varepsilon_T \sim \varepsilon_{T-\mathrm{bulk}}, \delta_u/\delta_T > 1$

Similarly, using Equation (14) instead of (13) for (9), the following relations can be obtained as shown in Equations (28) and (29):

$$Re = \mathbb{R}^{\frac{2}{7}} Ra^{\frac{4}{7}} Pr^{-\frac{6}{7}} \tag{28}$$

$$Nu = \mathbb{R}^{\frac{5}{7}} Ra^{\frac{3}{7}} Pr^{-\frac{1}{7}} \tag{29}$$

- *Regime IV-a:* $\varepsilon_u \sim \varepsilon_{u-\mathrm{bulk}}, \varepsilon_T \sim \varepsilon_{T-\mathrm{bulk}}, \delta_u/\delta_T < 1$

Substituting Equations (12) and (13) into Equations (8) and (9), respectively, one can get the following relations as shown in Equations (30) and (31):

$$Re = Ra^{\frac{1}{2}} Pr^{-\frac{1}{2}} \tag{30}$$

$$Nu = Ra^{\frac{1}{2}} Pr^{\frac{1}{2}} \tag{31}$$

- *Regime IV-b:* $\varepsilon_u \sim \varepsilon_{u-\mathrm{bulk}}, \varepsilon_T \sim \varepsilon_{T-\mathrm{bulk}}, \delta_u/\delta_T > 1$

Combining Equations (8), (9), (12) and (14), the following relations can be determined as shown in Equations (32) and (33):

$$Re = \mathbb{R}^{\frac{2}{9}} Ra^{\frac{4}{9}} Pr^{-\frac{2}{3}} \tag{32}$$

$$Nu = \mathbb{R}^{\frac{2}{3}} Ra^{\frac{1}{3}} \tag{33}$$

The boundaries between the two different regimes or sub-regions can be derived by considering $\varepsilon_{\mathrm{bulk}} \sim \varepsilon_{\mathrm{BL}}$ or $\delta_u \sim \delta_T$, respectively. Thereby, the following correlations can be obtained as shown in Equations (34)–(43):

$$Ra_{1a-2a} \sim \mathbb{R}^{-\frac{1}{2}} Pr_{1a-2a}^{\frac{3}{2}} \tag{34}$$

$$Ra_{1a-3a} \sim \mathbb{R}^{\frac{3}{2}} Pr_{1a-3a}^{-\frac{1}{2}} \tag{35}$$

$$Ra_{3a-4a} \sim \mathbb{R}^{0} Pr_{3a-4a}^{1} \tag{36}$$

$$Ra_{2a-4a} \sim \mathbb{R}^{2} Pr_{2a-4a}^{-1} \tag{37}$$

$$Ra_{1b-3b} \sim \mathbb{R}^{\frac{2}{3}} Pr_{1b-3b}^{\frac{1}{3}} \tag{38}$$

$$Ra_{3b-4b} \sim \mathbb{R}^{-\frac{1}{2}} Pr_{3b-4b}^{\frac{3}{2}} \tag{39}$$

$$Pr_{1a-b} \sim \mathbb{R}^{1} Ra^{0} \tag{40}$$

$$Pr_{2a-b} \sim \mathbb{R}^{1} Ra^{0} \tag{41}$$

$$Pr_{3a-b} \sim \mathbb{R}^{\frac{3}{2}} Ra^{-\frac{1}{2}} \tag{42}$$

$$Pr_{4a-b} \sim \mathbb{R}^{\frac{4}{3}} Ra^{-\frac{1}{3}} \tag{43}$$

It should be noticed that the specific correlations for the boundaries require a massive amount of experimental or simulated results, which is currently lacking. As a result, it is recommended to refer to the original model at present. What is more, one can easily see that the effect of aspect ratio, Γ, is not shown in the above scaling laws. This is due to the fact that no significant dependence of the scaling exponent on the aspect ratio is anticipated, in contrast to the pre-factors [30,39].

4. Direct Numerical Simulations

In order to verify the above scaling laws derived from theorical analysis, direct numerical simulation (DNS) has been conducted at the range of $Pr = (7, 30)$, $Ra = (1, 3, 5) \times 10^8$, $\mathbb{R} = 1.0 \sim 6.0$, which are common ranges of fluid parameters in practical applications [40]. The effect of thermal radiation on flow structure and heat transfer has been analyzed.

It should be noticed that $\mathbb{R} = 1.0$ stands for the case without considering radiation effects, while $\mathbb{R} = 6.0$ does not represent the limit case. Actually, according to Equation (7) and the research of Zhang et al. [41,42], the radiation parameter $\mathbb{R}$ depends on the reference temperature, characteristic length of cavity, and attenuation coefficient of salt.

4.1. Numerical Model

The convection cell adopted in this paper is a 2D square cavity with length h and aspect ratio $\Gamma = 1$, as shown in Figure 3. The temperatures of the top and bottom walls are fixed at T_c and T_h. The sidewalls are taken to be perfectly adiabatic. No-slip velocity boundary conditions are considered for all the walls. By setting the length h, free-fall velocity $u_f = \sqrt{g\beta\Delta h}$, and temperature difference Δ to the corresponding reference scales, the following dimensionless variables can be obtained, shown as Equation (44).

$$\begin{gathered} \{X, Y, Z\} = \frac{\{x,y,z\}}{h}, \boldsymbol{U} = \frac{u}{u_f}, \tau = \frac{t}{h/u_f}, P = \frac{p}{\rho u_f^2} \\ \theta = \frac{T - T_{\mathrm{ref}}}{\Delta}, \widetilde{q^R} = -\frac{16}{3}\frac{\sigma T_{\mathrm{ref}}^3}{\beta_r \lambda}\sqrt{\frac{1}{RaPr}}\nabla\theta \end{gathered} \tag{44}$$

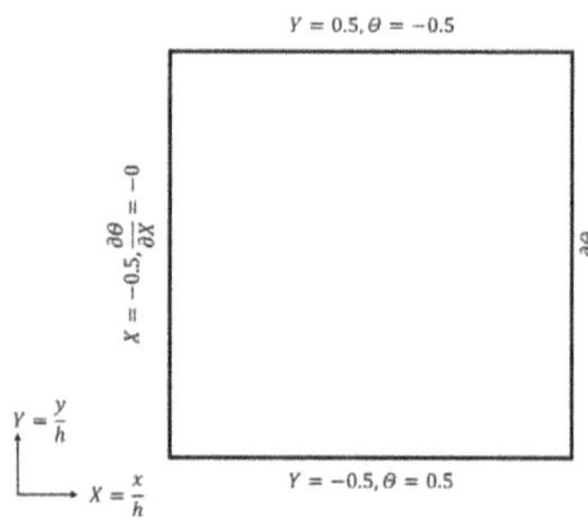

Figure 3. The convection cell and dimensionless boundary conditions.

Therefore, the dimensionless governing equations can be conventionally obtained as shown in Equations (45)–(47) (refer to Equations (1)–(3)).

$$\nabla \cdot \boldsymbol{U} = 0 \tag{45}$$

$$\partial_\tau \boldsymbol{U} + (\boldsymbol{U} \cdot \nabla)\boldsymbol{U} = -\nabla P + \sqrt{\frac{Pr}{Ra}} \nabla^2 \boldsymbol{U} + \theta e_z \tag{46}$$

$$\partial_\tau \theta + \boldsymbol{U} \cdot \nabla \theta = \sqrt{\frac{1}{RaPr}} \nabla^2 \theta - \nabla \widetilde{q^R} \tag{47}$$

where $\boldsymbol{U}$, P, θ and $\widetilde{q^R}$ are the dimensionless velocity, pressure, temperature, and radiative heat flux, respectively.

The dimensionless boundary conditions are as shown in Equation (48):

$$\begin{array}{l} X = \pm 0.5 \quad -\frac{\partial \theta}{\partial X} = 0 \\ Y = -0.5 \quad \theta = 0.5 \\ Y = 0.5 \quad \theta = -0.5 \end{array} \tag{48}$$

The initial conditions are set to $\tau = 0$, $U = 0$, $\theta = 0$ (for fluid and sidewall), $P = 1$. In order to strike the balance between capturing system dynamics accurately and maintaining computational efficiency, the size of the time step is set to be adjustable, and the corresponding maximum Courant number is limited to 0.9. The temporal average is performed over a duration of more than 5000 dimensionless time units after the convective flow has been fully developed, to allow for the averaging out of transient fluctuations.

4.2. Numerical Method and Validation

The governing equations are discretized by adopting the finite volume method. The pressure-velocity coupling is handled by the PISO algorithm. The Crank-Nicolson scheme is used for the transient terms, the second order upwind scheme is applied to treat the convective terms, and the second order central difference scheme is adopted for diffusion terms.

As shown in Figure 4, non-equidistant meshes are applied and the computational meshes are refined close to all the four walls so that the transport characteristics in BLs can be captured. The grid independence test is carried out at $Ra = 5 \times 10^8$, $Pr = 1$. The average Nusselt numbers for different grids have been displayed in Table 2. It turns out that the solution becomes independent of grid size for 256 × 256, since the deviation is less than 0.5%.

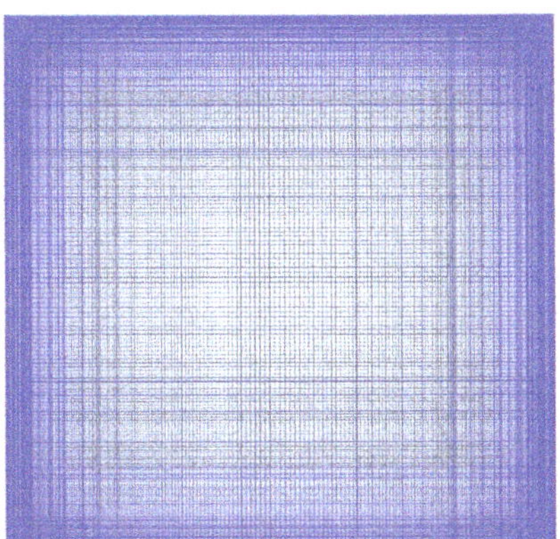

Figure 4. The sketch of grids with refined meshes near walls.

Table 2. The averaged Nusselt numbers for different grids.

Grid Size	*Nu*	Error (%)
128 × 128	39.17710	-
256 × 256	39.84992	1.72%
384 × 384	40.03957	0.48%

In order to validate the present code, the results for $Ra = 1.0 \times 10^8$, $Pr = 4.3$ were compared with those in References [43,44], as shown in Table 3. It can be seen that the maximum relative error for Nu is less than 3.5%. Therefore, the correctness of present numerical method code is verified.

Table 3. Comparisons with public results from References.

	Nu	Error (%)
Present	24.9	-
Huang et al. [43]	25.38	1.89%
Yang et al. [44]	258.0	3.49%

4.3. Effect of Radiation on Flow Structure

Figure 5 shows the time-evolution of isotherm contours (background) and streamlines (blue lines) at different $\mathbb{R}$ when $Pr = 7, 30$ and $Ra = (1, 5) \times 10^8$. It can be seen that the large-scale circulation (LSC) has been formed when the effect of radiation is not taken into account ($\mathbb{R} = 1.0$) or weak ($\mathbb{R} = 1.8$). The shape of LSC is approximately elliptical, which follows the diagonal of the cavity as the major axis and is accompanied by two small vortices in the short axis direction. The two small vortices may expand, causing the large-scale circulation to gradually collapse and change its tilt direction. Interestingly, with the increasing $\mathbb{R}$, the collapse and redirection frequencies of LSC decrease. However, for the case of $\mathbb{R} = 6.0$, there are two small vortices arranged vertically. This makes it difficult for heat to exchange between the cold fluid at the top and the hot fluid at the bottom. The flow structure is similar to that of small Ra, indicating that thermal radiation suppresses flow.

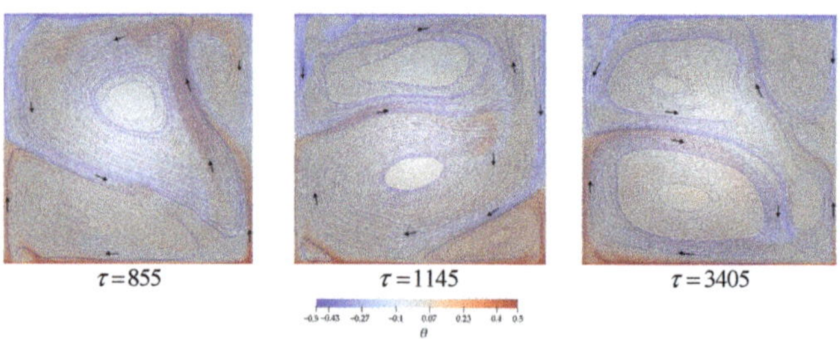

(a) $Pr = 7$, $Ra = 1 \times 10^8$, $\mathbb{R} = 1.0$

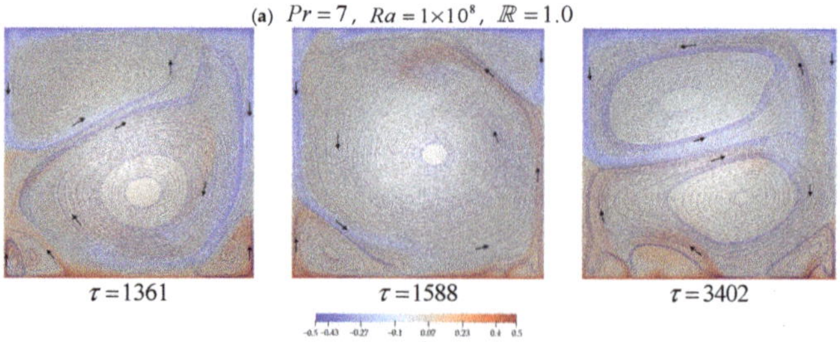

(b) $Pr = 7$, $Ra = 1 \times 10^8$, $\mathbb{R} = 1.8$

Figure 5. *Cont.*

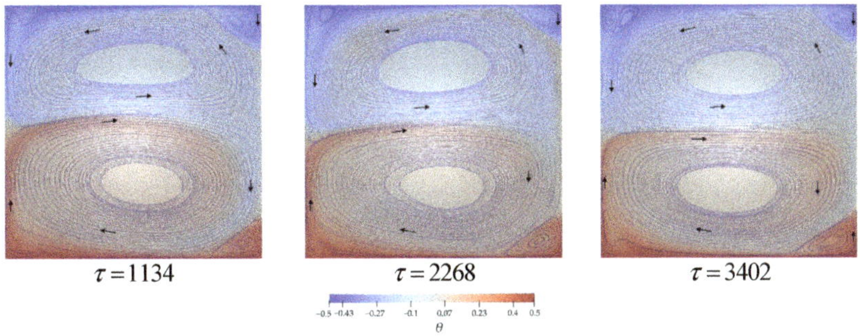

(**c**) $Pr = 7$, $Ra = 1\times10^8$, $\mathbb{R} = 6.0$

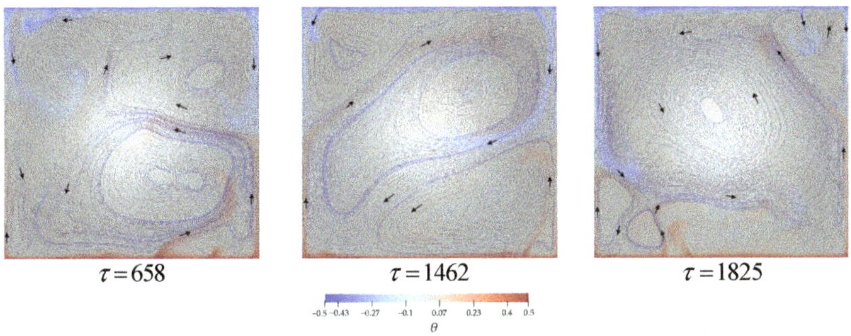

(**d**) $Pr = 30$, $Ra = 1\times10^8$, $\mathbb{R} = 1.0$

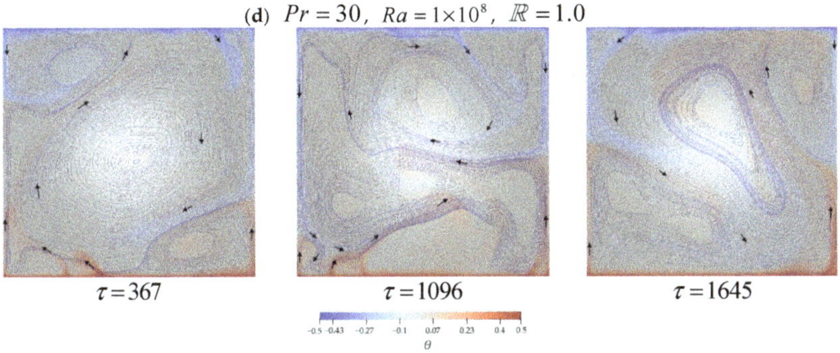

(**e**) $Pr = 30$, $Ra = 1\times10^8$, $\mathbb{R} = 1.8$

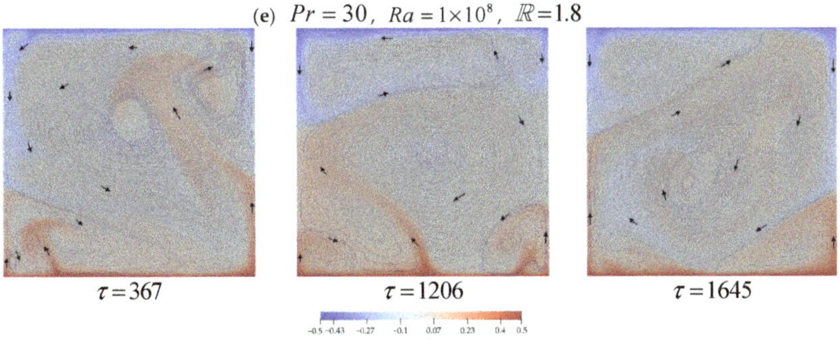

(**f**) $Pr = 30$, $Ra = 1\times10^8$, $\mathbb{R} = 6.0$

Figure 5. *Cont.*

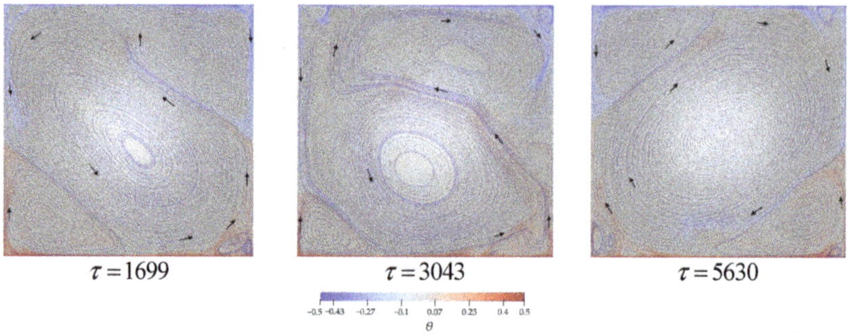

(g) $Pr=7,\ Ra=5\times10^8,\ \mathbb{R}=1.0$

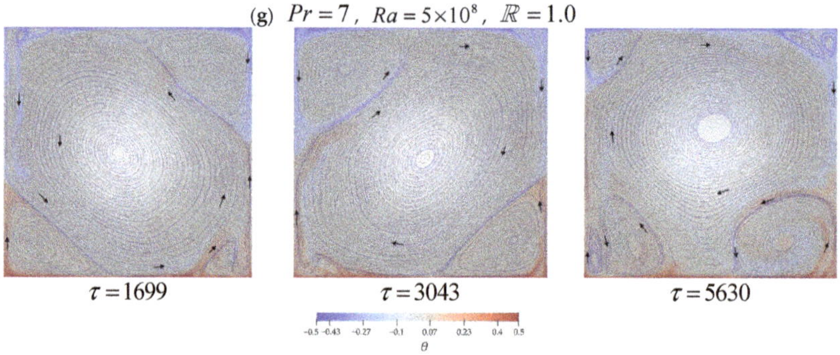

(h) $Pr=7,\ Ra=5\times10^8,\ \mathbb{R}=1.8$

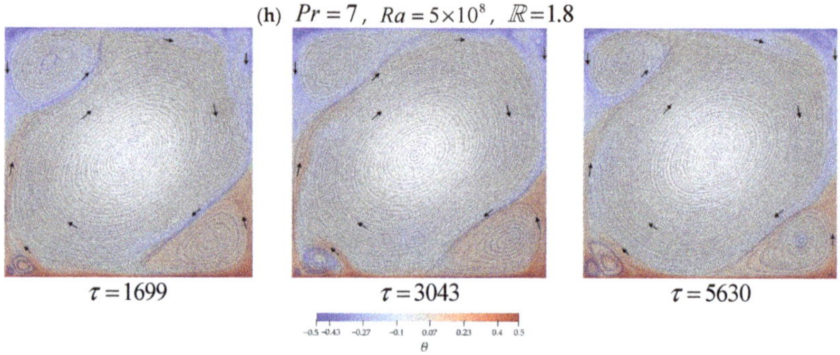

(i) $Pr=7,\ Ra=5\times10^8,\ \mathbb{R}=6.0$

Figure 5. The time-evolution of isotherm contours (background) and streamlines (blue lines) for $Ra = (1,5) \times 10^8$, $Pr = 7,30$ at different $\mathbb{R}$. The direction of flow is indicated by the black arrows.

When Pr increases to 30, LSC still exists even at $\mathbb{R} = 6.0$, suggesting that the effect of thermal radiation weakens for larger Pr. Moreover, as Ra increases, the effect of thermal radiation is not sufficient to suppress the formation of LSC, but its stabilizing effect will cause a decrease in the redirection frequency of LSC.

4.4. Comparisons between Scaling Law and DNS Results

In order to quantitatively verify the correctness of Equations (19) and (21) for examples, the comparison of time average Nu between DNS results and the theoretical analysis has been performed, as shown in Figure 6. Consistent with the results of many other literatures, the heat transfer is significantly enhanced with an increase in $\mathbb{R}$.

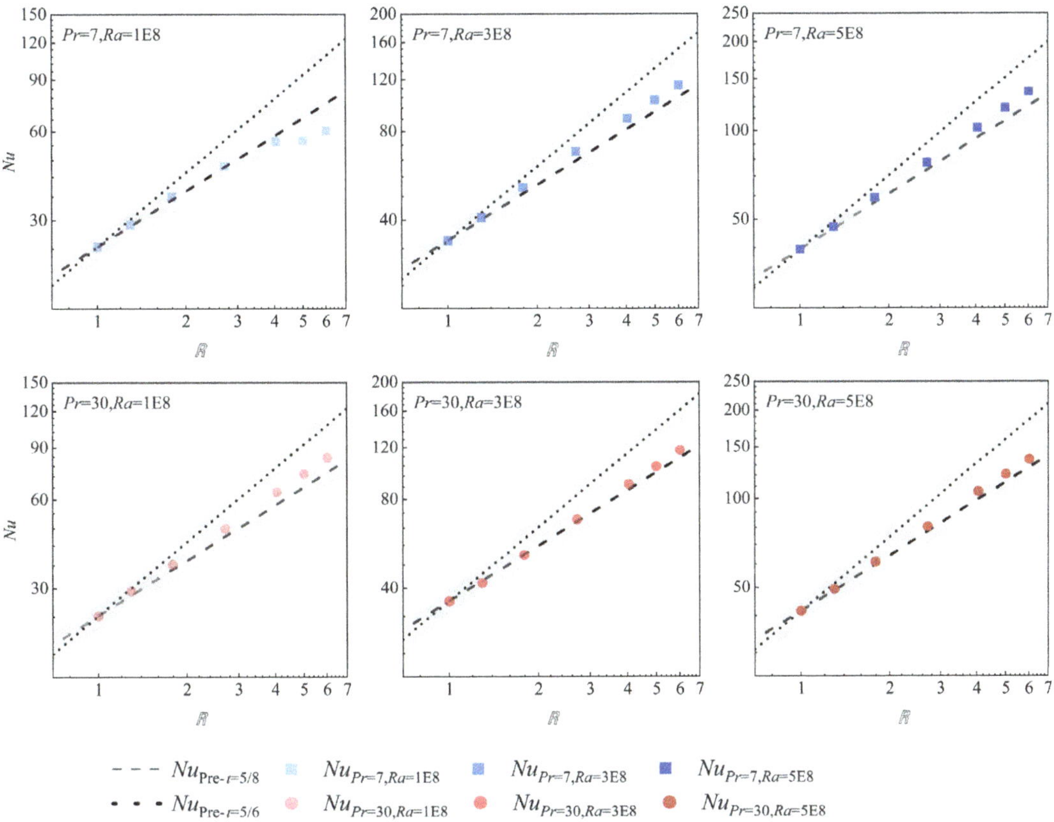

Figure 6. The comparison between DNS and prediction model: *Nu*. (The grey area represents a relative error of 10%).

According to the boundary division between the various regimes in the GL model [30–33], the parameter range in this paper is located in Regime I-b, where the exponent ι of $\mathbb{R}$ in the relation $Nu \sim \mathbb{R}^{\iota} Ra^{\alpha} Pr^{\gamma}$ is 5/6. However, as shown in Figure 6, the relative error between simulation results and the predicted results of the relation $Nu \sim \mathbb{R}^{5/8}$ in Regime I-a is relatively small (about 10%). This phenomenon may be attributed to the increase of δ_T, which arises due to the suppression of convection by thermal radiation and the increase in fluid temperature uniformity. Actually, the increase in BL thickness was also observed by Hossain et al. [45].

Figure 7 depicts the exponent ι of $\mathbb{R}$ obtained from theoretical analysis and DNS. It can be seen that the deviation in between is about 10%. Besides, the deviation decreases with the increase of *Pr* and *Ra*. Similar conclusions for the range of $Pr = 1$ (usually gas) and $Ra = (3 \sim 4) \times 10^8$ can be obtained in our previous work [46]. Both of them indicate that the derived scaling law can effectively predict the *Nu* of radiative Rayleigh-Bénard convection by applying the Rosseland approximation.

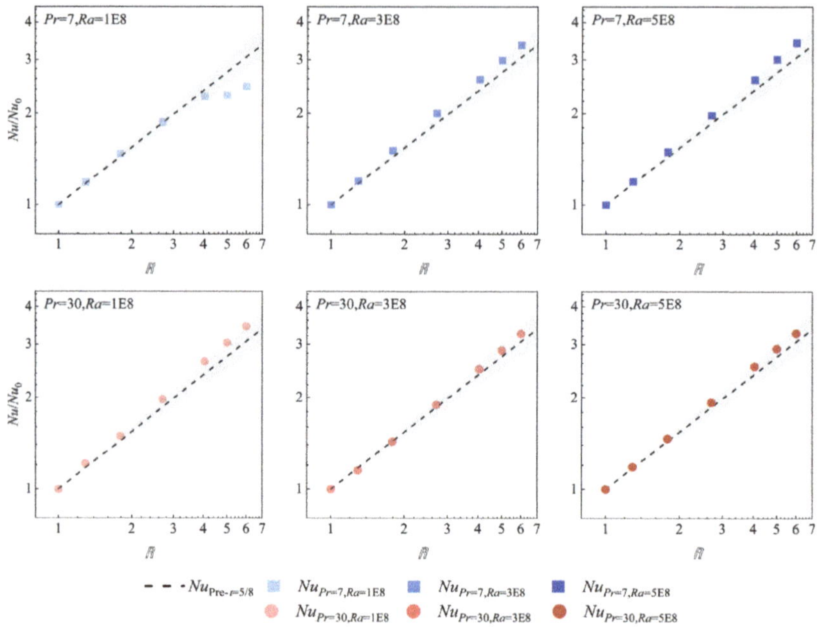

Figure 7. The comparison between DNS and prediction model: the exponent ι of $\mathbb{R}$. (The grey area represents a relative error of 10%).

5. Conclusions

In summary, the transport characteristics of radiative Rayleigh-Bénard convection in optically thick media with large Pr have been investigated by both theorical analysis and DNS. Main conclusions can be summarized as follows:

(1) Based on the turbulent dissipation theory and Rosseland approximation, the scaling laws in the four main regimes have been summarized within the scope of numerical simulation research presented in this paper ($Pr = (7, 30)$, $Ra = (1, 3, 5) \times 10^8$, $\mathbb{R} = 1.0 \sim 6.0$), $Nu = \mathbb{R}^{\frac{5}{8}} Ra^{\frac{1}{4}} Pr^{\frac{1}{8}}$.

(2) It has been found that due to the presence of thermal radiation, the Nu scaling law in the small Pr range of the GL model is more suitable for predicting the transport characteristics of optical thick media with large Pr. The deviation between the results of DNS and prediction model is approximately 10%.

(3) The determination of reference temperature is discussed, and there is almost no difference between the arithmetic mean temperature and radiative mean temperature, especially within the scope of application of Boussinesq approximation.

(4) The influence of thermal radiation results in a significant enhancement of heat transfer.

(5) When $Ra > 1 \times 10^8$, the LSC can been formed for weak radiation effect. The effect of radiation can stabilize flow, specifically manifested by a collapse of LSC as $\mathbb{R}$ increases. However, the radiation effect weakens for larger Pr.

These conclusions offer a theoretical underpinning for industrial designs related to energy storage, power generation, and thermal storage technology. Future research endeavors can focus on determining the pre-factors of scaling laws to further refine and enhance our understanding of these processes.

Author Contributions: Conceptualization, J.S.; methodology, J.S.; software, J.S., P.L. and L.C.; validation, J.S., Y.Z. and F.T.; formal analysis, J.S.; investigation, J.S.; resources, B.L.; data curation, J.S.; writing—original draft preparation, J.S.; writing—review and editing, J.S., P.L., L.C., Y.Z., F.T. and B.L.; visualization, J.S.; supervision, J.S. and B.L.; project administration, J.S. and B.L.; funding acquisition, J.S. and B.L. All authors have read and agreed to the published version of the manuscript.

Funding: This research was funded by the Central Guiding Local Science and Technology Development Fund Projects (No. 2023JH6/100100020), Liaoning Province International Science and Technology Cooperation Plan (No. 2023JH2/10700001), Dalian Science and Technology Plan Project (No. 2023JJ12SN028) and the Natural Science Foundation of China (No. 51976021).

Data Availability Statement: The data presented in this study are available on request from the corresponding author, since the data are not publicly available due to privacy and ethical restrictions.

Conflicts of Interest: Author Lu Chen was employed by the company, MCC Capital Engineering & Research Inc., Ltd. The remaining authors declare that the research was conducted in the absence of any commercial or financial relationships that could be construed as a potential conflict of interest.

Nomenclature

g	gravitational acceleration	$m \cdot s^{-2}$
h	length of the cavity	m
Ha	Hartmann number	-
I, I_b	radiation intensity, radiation intensity of black body	$W \cdot m^{-2}$
Nu	Nusselt number	-
p, P	pressure, dimensionless pressure	$kg \cdot m^{-1} \cdot s^{-1}$, -
Pl	Planck number	-
Pr	Prandtl number	-
q^R	radiative heat flux	$W \cdot m^{-2}$
Ra	Rayleigh number	-
Re	Reynold number	-
s	position vector	m
t	time	s
T	temperature	K
T_{ref}	reference temperature	K
$u = \{u, v, w\}$	velocity	$m \cdot s^{-1}$
u	dimensionless velocity vector	-
u_f	free-fall velocity	$m \cdot s^{-1}$
$\overline{u}$	the mean large-scale velocity	$m \cdot s^{-1}$
x, y, z	Cartesian coordinates	m
$\mathbb{R}$	radiation parameter	-
$\langle \cdot \rangle_A$	the average over any horizonal plane	-
$\langle \cdot \rangle_V$	the volume average	-
Δ	temperature difference	K
Greek symbols		
β	thermal expansion coefficient	K^{-1}
β_r	spectral extinction or attenuation coefficient	m^{-1}
Γ	aspect ratio	-
δ	the thickness of BL	m
ε	energy dissipation	-
ε_e	emissivity	-
θ	dimensionless temperature	-
κ	thermal diffusivity	$m^2 \cdot s^{-1}$
$\kappa_{a,\lambda}$	scattering coefficient	m^{-1}
λ	thermal conductivity	$W \cdot m^{-1} \cdot K^{-1}$
σ	Stefan-Boltzmann constant	$\sigma = 5.67 \times 10^{-8} W \cdot m^{-2} \cdot K^{-4}$
$\sigma_{s,\lambda}$	absorption coefficient	m^{-1}
τ	dimensionless time	-

τ_r	optical thickness	-
ν	kinematic viscosity	$m^2 \cdot s^{-1}$
Φ	scattering phase function	-
Ω, Ω'	solid angle, incident solid angle	sr
Subscripts		
h	hot plate	
c	cold plate	
u	kinetic related quantities	
T	thermal related quantities	
BL	contribution from BL	
bulk	contribution from bulk	
λ	spectrally dependent	

References

1. Arnaoutakis, G.E.; Katsaprakakis, D.A. Energy yield of spectral splitting concentrated solar power photovoltaic systems. *Energies* **2024**, *17*, 556. [CrossRef]
2. Zhou, W.; Yang, Z.; Lin, L.; Feng, Y. Molecular insights into thermal conductivity enhancement and interfacial heat transfer of molten salt/porous ceramic skeleton composite phase change materials. *Int. J. Heat Mass Transf.* **2024**, *232*, 125934. [CrossRef]
3. Yuan, H.M.; Liu, S.T.; Li, T.H.; Yang, L.Y.; Li, D.H.; Bai, H.; Wang, X.D. Review on thermal properties with influence factors of solid-liquid organic phase-change micro/nanocapsules. *Energies* **2024**, *17*, 604. [CrossRef]
4. Sheikholeslami, M.; Rokni, H.B. Numerical simulation for impact of Coulomb force on nanofluid heat transfer in a porous enclosure in presence of thermal radiation. *Int. J. Heat Mass Transf.* **2018**, *118*, 823–831. [CrossRef]
5. Qasem, S.A.; Sivasankaran, S.; Siri, Z.; Mothman, W.A. Effect of thermal radiation on natural convection of a nanofluid in a square cavity with a solid body. *Therm. Sci.* **2021**, *25*, 1949–1961. [CrossRef]
6. Alnaqi, A.A.; Aghakhani, S.; Pordanjani, A.H.; Bakhtiari, R.; Asadi, A.; Minh-Duc, T. Effects of magnetic field on the convective heat transfer rate and entropy generation of a nanofluid in an inclined square cavity equipped with a conductor fin: Considering the radiation effect. *Int. J. Heat Mass Transf.* **2019**, *133*, 256–267. [CrossRef]
7. Roy, N.C.; Saha, L.K.; Siddiqa, S. Electrohydrodynamics and thermal radiation effects on natural convection flow in an enclosed domain. *Int. Commun. Heat Mass Transf.* **2021**, *126*, 105437. [CrossRef]
8. Shah, F.; Asadullah; Ali, S.; Khan, S.U.; Akermi, M.; Nasr, S.; Hassani, R.; Jaballah, N.S. Thermal analysis for SWCNTs and MWCTs due to coaxially stretchable disks with applications of viscous dissipation and thermal radiation: Artificial neural network analysis. *Int. Commun. Heat Mass Transf.* **2024**, *155*, 107452. [CrossRef]
9. Ali, M.Y.; Rahman, M.; Ali, M.M.; Ahmmed, S.F.; Haque, S. Data analysis of non-linear radiative electro-periodic MHD flow past a stretching sheet with activation energy impact. *Int. Commun. Heat Mass Transf.* **2024**, *155*, 107572.
10. Kolsi, L.; Abidi, A.; Maatki, C.; Borjini, N.M.; Aissia, B.H. Combined radiation-natural convection in three-dimensional verticals cavities. *Therm. Sci.* **2011**, *15*, S327–S339. [CrossRef]
11. Leporini, M.; Corvaro, F.; Marchetti, B.; Polonara, F.; Benucci, M. Experimental and numerical investigation of natural convection in tilted square cavity filled with air. *Exp. Therm. Fluid Sci.* **2018**, *99*, 572–583. [CrossRef]
12. El Ayachi, R.; Raji, A.; Hasnaoui, M.; Bahlaoui, A. Combined effect of radiation and natural convection in a square cavity differentially heated with a periodic temperature. *Numer. Heat Transf. Part A-Appl.* **2008**, *53*, 1339–1356. [CrossRef]
13. Wang, Y.; Sergent, A.; Saury, D.; Lemonnier, D.; Joubert, P. Numerical study of an unsteady confined thermal plume under the influence of gas radiation. *Int. J. Therm. Sci.* **2020**, *156*, 106474. [CrossRef]
14. Martyushev, S.G.; Sheremet, M.A. Conjugate natural convection combined with surface thermal radiation in a three-dimensional enclosure with a heat source. *Int. J. Heat Mass Transf.* **2014**, *73*, 340–353. [CrossRef]
15. Sivaraj, C.; Miroshnichenko, I.V.; Sheremet, M.A. Influence of thermal radiation on thermogravitational convection in a tilted chamber having heat-producing solid body. *Int. Commun. Heat Mass Transf.* **2020**, *115*, 104611. [CrossRef]
16. Miroshnichenko, I.V.; Gibanov, N.S.; Sheremet, M.A. Numerical analysis of heat source surface emissivity impact on heat transfer performance in a rectangular enclosure at high Rayleigh numbers. *Int. J. Comput. Methods Eng. Sci. Mech.* **2020**, *21*, 205–214. [CrossRef]
17. Miroshnichenko, I.V.; Toilibayev, A.A.; Sheremet, M.A. Simulation of thermal radiation and turbulent free convection in an enclosure with a glass wall and a local heater. *Fluids* **2021**, *6*, 91. [CrossRef]
18. Qi, X.; Ma, W.; Liu, L. Effect of internal radiation on heat transfer during Ti:Sapphire crystal growth process by heat exchanger method. *Int. J. Heat Mass Transf.* **2021**, *170*, 121000. [CrossRef]
19. Parsaee, S.; Payan, S.; Payan, A. Semi-transient thermal analysis from MHD-Participating fluid into a square cavity with variable optical thickness. *Int. J. Therm. Sci.* **2021**, *169*, 107072. [CrossRef]
20. Gad, M.A.; Balaji, C. Effect of surface radiation on RBC in cavities heated from below. *Int. Commun. Heat Mass Transf.* **2010**, *37*, 1459–1464. [CrossRef]
21. Soucasse, L.; Podvin, B.; Riviere, P.; Soufiani, A. Reduced-order modelling of radiative transfer effects on Rayleigh-Benard convection in a cubic cell. *J. Fluid Mech.* **2020**, *898*, A2. [CrossRef]

22. Soucasse, L.; Podvin, B.; Riviere, P.; Soufiani, A. Low-order models for predicting radiative transfer effects on Rayleigh-Benard convection in a cubic cell at different Rayleigh numbers. *J. Fluid Mech.* **2021**, *917*, A5. [CrossRef]
23. Song, J.J.; Li, P.X.; Chen, L.; Li, C.H.; Li, B.W.; Huang, L.Y. A review on Rayleigh-Benard convection influenced by the complicating factors. *Int. Commun. Heat Mass Transf.* **2023**, *144*, 106784. [CrossRef]
24. Malkus, W.V.R. The Heat Transport and Spectrum of Thermal Turbulence. *Proc. R. Soc. Lond. Ser. A-Math. Phys. Sci.* **1954**, *225*, 196–212.
25. Heslot, F.; Castaing, B.; Libchaber, A. Transitions to turbulence in helium gas. *Phys. Rev. A* **1987**, *36*, 5870–5873. [CrossRef]
26. Castaing, B.; Gunaratne, G.; Heslot, F.; Kadanoff, L.; Libchaber, A.; Thomae, S.; Wu, X.Z.; Zaleski, S.; Zanetti, G. Scaling of hard thermal turbulence in Rayleigh-Benard convection. *J. Fluid Mech.* **1989**, *204*, 1–30. [CrossRef]
27. Sano, M.; Wu, X.Z.; Libchaber, A. Turbulence in helium-gas free-convection. *Phys. Rev. A* **1989**, *40*, 6421–6430. [CrossRef]
28. Shraiman, B.I.; Siggia, E.D. Heat-transport in high-Rayleigh-number convection. *Phys. Rev. A* **1990**, *42*, 3650–3653. [CrossRef]
29. Siggia, E.D. High Rayleigh number convection. *Annu. Rev. Fluid Mech.* **1994**, *26*, 137–168. [CrossRef]
30. Grossmann, S.; Lohse, D. Scaling in thermal convection: A unifying theory. *J. Fluid Mech.* **2000**, *407*, 27–56. [CrossRef]
31. Grossmann, S.; Lohse, D. Thermal convection for large Prandtl numbers. *Phys. Rev. Lett.* **2001**, *86*, 3316–3319. [CrossRef] [PubMed]
32. Grossmann, S.; Lohse, D. Prandtl and Rayleigh number dependence of the Reynolds number in turbulent thermal convection. *Phys. Rev. E* **2002**, *66*, 016305. [CrossRef] [PubMed]
33. Grossmann, S.; Lohse, D. Fluctuations in turbulent Rayleigh-Benard convection: The role of plumes. *Phys. Fluids* **2004**, *16*, 4462–4472. [CrossRef]
34. Chakraborty, S. On scaling laws in turbulent magnetohydrodynamic Rayleigh-Benard convection. *Phys. D-Nonlinear Phenom.* **2008**, *237*, 3233–3236. [CrossRef]
35. Zuerner, T.; Liu, W.; Krasnov, D.; Schumacher, J. Heat and momentum transfer for magnetoconvection in a vertical external magnetic field. *Phys. Rev. E* **2016**, *94*, 043108. [CrossRef]
36. Zurner, T. Refined mean field model of heat and momentum transfer in magnetoconvection. *Phys. Fluids* **2020**, *32*, 107101. [CrossRef]
37. Yan, M.; Tobias, S.M.; Calkins, M.A. Scaling behaviour of small-scale dynamos driven by Rayleigh-Benard convection. *J. Fluid Mech.* **2021**, *915*, A15. [CrossRef]
38. Urban, P.; Králík, T.; Hanzelka, P.; Musilova, V.; Vezník, T.; Schmoranzer, D.; Skrbek, L. Thermal radiation in Rayleigh-Benard convection experiments. *Phys. Rev. E* **2020**, *101*, 043106. [CrossRef] [PubMed]
39. Wu, X.Z.; Libchaber, A. Scaling relations in thermal turbulence—The aspect-ratio dependence. *Phys. Rev. A* **1992**, *45*, 842–845. [CrossRef]
40. Li, P.X.; Luo, X.H.; Chen, L.; Song, J.J.; Li, B.W.; Karcher, C. Numerical research for the effect of magnetic field on convective transport process of molten salt in Rayleigh-Bénard system. *Int. J. Therm. Sci.* **2024**, *195*, 108605. [CrossRef]
41. Zhang, S.; Sun, X. Convective and radiative heat transfer in molten salts. *Nucl. Technol.* **2020**, *206*, 1721–1739. [CrossRef]
42. Xie, M.; Zhu, Y.; Liu, Y.; Yuan, Y.; Tan, H. Measurement of spectral radiative characteristics of molten salt at high temperature using emission method. *Appl. Therm. Eng.* **2019**, *149*, 151–164. [CrossRef]
43. Huang, M.; Wang, Y.; Bao, Y.; He, X. Heat transport and temperature boundary-layer profiles in closed turbulent Rayleigh–Bénard convection with slippery conducting surfaces. *J. Fluid Mech.* **2022**, *943*, A2. [CrossRef]
44. Yang, J.L.; Zhang, Y.Z.; Jin, T.C.; Dong, Y.H.; Wang, B.F.; Zhou, Q. The Pr-dependence of the critical roughness height in two-dimensional turbulent Rayleigh–Bénard convection. *J. Fluid Mech.* **2021**, *911*, A52. [CrossRef]
45. Hossain, M.A.; Takhar, H.S. Thermal radiation effects on the natural convection flow over an isothermal horizontal plate. *Heat Mass Transf.* **1999**, *35*, 321–326. [CrossRef]
46. Song, J.; Li, P.; Chen, L.; Zhao, Y.; Tian, F.; Li, B. Correlations for the transport characteristics of Radiative Rayleigh-Bénard convection in optically thick media. In Proceedings of the 21st International Conference on Sustainable Energy Technologies, Shanghai, China, 12–14 August 2024.

energies

MDPI

Article

Towards Sustainable Energy Solutions: Evaluating the Impact of Floating PV Systems in Reducing Water Evaporation and Enhancing Energy Production in Northern Cyprus

Youssef Kassem [1,2,3,4,*], Hüseyin Gökçekuş [2,3] and Rifat Gökçekuş [2]

1 Department of Mechanical Engineering, Engineering Faculty, Near East University, via Mersin 10, Nicosia 99138, Cyprus
2 Department of Civil Engineering, Civil and Environmental Engineering Faculty, Near East University, via Mersin 10, Nicosia 99138, Cyprus; huseyin.gokcekus@neu.edu.tr (H.G.); rifatgokcekus@hotmail.com (R.G.)
3 Energy, Environment, and Water Research Center, Near East University, via Mersin 10, Nicosia 99138, Cyprus
4 Research Center for Science, Technology, and Engineering (BILTEM), Near East University, via Mersin 10, Nicosia 99138, Cyprus
* Correspondence: yousseuf.kassem@neu.edu.tr or youssef.kassem1986@hotmail.com; Tel.: +90-(392)-2236-464

Abstract: Floating photovoltaic systems (FPVSs) are gaining popularity, especially in countries with high population density and abundant solar energy resources. FPVSs provide a variety of advantages, particularly in situations where land is limited. Therefore, the main objective of the study is to evaluate the solar energy potential and investigate the techno-economic perspective of FPVSs at 15 water reservoirs in Northern Cyprus for the first time. Due to the solar radiation variations, solar power generation is uncertain; therefore, precise characterization is required to manage the grid effectively. In this paper, four distribution functions (Johnson SB, pert, Phased Bi-Weibull, and Kumaraswamy) are newly introduced to analyze the characteristics of solar irradiation, expressed by global horizontal irradiation (GHI), at the selected sites. These distribution functions are compared with common distribution functions to assess their suitability. The results demonstrated that the proposed distribution functions, with the exception of Phased Bi-Weibull, outperform the common distribution regarding fitting GHI distribution. Moreover, this work aims to evaluate the effects of floating photovoltaic systems on water evaporation rates at 15 reservoirs. To this aim, five methods were used to estimate the rate of water evaporation based on weather data. Different scenarios of covering the reservoir's surface with an FPVS were studied and discussed. The findings showed that annual savings at 100% coverage can reach 6.21×10^5 m^3 compared to 0 m^3 without PV panels. Finally, technical and economic assessment of FPVSs with various scales, floating assemblies, and PV technologies was conducted to determine the optimal system. The results revealed that a floating structure (North orientation-tilt 6°) and bifacial panels produced the maximum performance for the proposed FPVSs at the selected sites. Consequently, it is observed that the percentage of reduction in electricity production from fossil fuel can be varied from 10.19% to 47.21% at 75% FPV occupancy.

Keywords: Northern Cyprus; floating PV system; solar radiation distribution; water evaporation rates; floating structure; PV technologies

1. Introduction

Carbon emissions and the global need for energy are rising, causing environmental degradation [1]. By 2030, Sustainable Development Goal 7 (SDG 7) seeks to address ecological deterioration by achieving the goal of supplying "affordable, reliable, sustainable, and modern energy for all" [2]. Consequently, scientific researchers have been exploring alternative sources of renewable energy to replace fossil fuel-based generation systems [3].

Solar-powered photovoltaic (PV) systems play a critical role in achieving sustainable energy solutions for global growth [4]. In general, traditional PV systems are typically installed on land or rooftops. To address land scarcity, researchers propose installing solar-powered plants on artificial water bodies, known as floating photovoltaic (FPV) systems [5]. FPVSs increase solar panel power production through sunlight reflection and also conserve water and improve water quality [6,7]. According to Charles Lawrence Kamuyu [8], solar panels installed on water can produce up to 10% more electricity than those on land. Also, Choi et al. [9] found that floating solar PV systems have a higher generation efficiency of over 10% compared to ground-mounted systems. FPV systems can reduce water evaporation and save water [10]. Moreover, synergies can be developed by combining pumped-storage power systems with integrated FPV modules, which can minimize energy imbalance [11]. Additionally, Pringle et al. [12] proposed using water areas for floating solar energy and aquaculture to efficiently utilize water for food and energy generation. Furthermore, FPVSs can be combined with solar water heating systems for mining applications [13]. Also, FPV power plant pontoons can be used for compressed air energy storage [14]. In addition, combining FPV with wind and wave energy has been shown to improve power systems by smoothing power output [15]. Furthermore, several recent studies have examined various aspects of floating solar power systems, including total costs, environmental impacts, and energy generation efficiency. For instance, Durkovic and Djurisic [16] studied a large-scale FFV system in Montenegro. They found that there was a significant reduction in CO_2 emissions of about 83.42 ktons/year as well as water evaporation savings of about 5.41 million m^3/year. Sunny et al. [17] explored the potential installation of FPV systems in Bangladesh. The results indicated that the value of energy yield was 1564 kWh/kW.

Furthermore, implementing FPV technologies requires meticulous consideration of economic viability and environmental sustainability according to the previous studies. For instance, Ranjbaran et al. [18] studied various aspects of FPV systems, including design, power generation, evaporation reduction, economic assessment, and CO_2 emission reduction. The authors concluded that FPV is a promising solution for generating clean energy, reducing water issues, and reducing CO_2 emissions. Micheli et al. [19] estimated the economic viability of FPVSs and compared them with land-based PV systems in Europe. The results showed that the energy yields from FPVs can be up to 2% higher than land-based PV systems, particularly in the Mediterranean region. Islam et al. [20] investigated the techno-economic feasibility of a 10 MW FPV in Malaysia. The results indicated that the proposed system was economically feasible and significantly reduced carbon emissions by 11.135 ktons/year. Islam et al. [21] studied the feasibility of the FPV system at Hatirjheel Lake in Dhaka, Bangladesh. The results indicated that the development of FPVs is economically feasible, as it provides about 12.5% of the total domestic energy demand in addition to reducing environmental pollution by 6685 tons per year. Fereshtehpour et al. [22] evaluated the FPV systems' performance in Iran. The results showed that installing an FPV system produces between 194 GWh and 257 GWh annually, as well as significantly reducing CO_2 emissions, with a value of 118 ktons/year.

1.1. Energy Situation in Northern Cyprus

The demand for energy in Northern Cyprus is increasing due to the growth of educational institutions and the growth of the economy across various sectors. Generally, the rising need for energy in Northern Cyprus and the absence of local resources to meet the demands of the country represent two significant challenges for the energy sector. According to Akçaba and Eminer [23], insufficient generation and high demand contribute to shortages in Northern Cyprus, particularly in the summer. Approximately 30% and 20% of energy is used by the commercial and residential sectors, respectively [23]. Moreover, according to the Cyprus Turkish Electricity Authority (KIB-TEK), 94% of the power generated in Cyprus is generated from fossil fuels, with the remaining 6% sourced from renewable sources, as shown in Figure 1 [24].

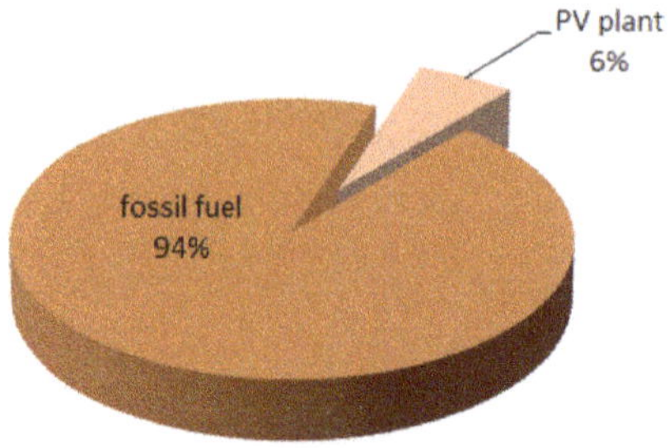

Figure 1. Structure of primary energy source for electricity generation in Northern Cyprus in 2023 [24].

Given the rise of commercial, educational, and industrial sectors, as well as the increasing cost of imported fuel and the resulting greenhouse gas emissions, the country's decision makers have made it their primary goal to provide clean, affordable, and sustainable energy. The country's main priorities now include implementing energy efficiency measures as well as pursuing the development of renewable energy sources, particularly solar energy.

Northern Cyprus is blessed to have an average of 320 sunny days per year and 5.6 to 6.13 $kWh/m^2/day$ of direct solar radiation according to the Global Solar Atlas [25]. The annual global solar irradiation on a horizontal surface varies between 5 $kW/m^2/day$ and 5.38 kW/m^2.

Moreover, based on the annual value of global horizontal irradiation and direct irradiation, the solar resources in Northern Cyprus can be categorized as good and excellent according to Prăvălie et al. [26]. The electricity generation capacity from solar plants can be within the range of 4.481–4.826 kWh/kWp/day.

Generally, most of the residences use solar water heaters in Cyprus. According to Refs. [27,28], solar water heating systems are installed on the rooftops of more than 90% of residences. Moreover, five solar PV plants can be found in Northern Cyprus: (1) a 1.3 MW solar PV power plant installation in 2011 in the Serhatköy region [29]; (2) a 1 MW PV power plant installation in 2015 at Middle East Technical University Northern Cyprus Campus; (3) a 1.3 MW PV power plant installation in 2016 at Cyprus International University; (4) a 50 kW PV system installation in 2017 at KKTCell Main Building; and (5) a 120 kW PV system installation in 2018 at Levent College [30,31]. Furthermore, KIB-TEK allows customers to produce and export excess energy from the PV system to the utility grid using a net metering system [32]. However, this mechanism has faced challenges with the increase in PV installations because the grid system remains isolated in Northern Cyprus [32]. According to the renewable energy board (Yek-Kurulu), single-phase grid-connected customers can currently install up to 5 kW of PV systems, while three-phase grid-connected customers can install up to 8 kW of PV systems [32]. Additionally, based on estimates from Kib-Tek and Yek-Kurulu, the total power generated by photovoltaic power plants in Northern Cyprus is around 74.3 MW [24,33].

Further, several researchers [34–62] have examined the solar energy potential and the technological-economic viability of photovoltaic systems in various Northern Cyprus regions, as tabulated in Table S1 of the Supplementary Materials. The following can be concluded:

- Installing a solar power plant could solve the country's energy crisis and reduce the dependency on fossil fuel.
- Using a photovoltaic system can achieve sustainable development goals by generating useful renewable energy without harming the land and the environment.
- According to the authors' review, only one study has compared the technical and economic aspects of a 1 MW grid-connected land-based PV plant and a floating PV system.

1.2. Water Resources in Northern Cyprus

Cyprus's water resources can be divided into four categories [63]: groundwater resources (75.5%), semi-perennial low-discharge springs (0.3%), various surface-water reservoirs (20.4%), and a limited number of small-scale desalination plants (3.8%). Moreover, Yildiz and Çakmak [64] reported that the total capacity of surface and groundwater in the region is approximately 173.18 MCM (million m^3). Additionally, there are three small private desalination plants located in Bafra, Eastern Mediterranean University, and Girne, with capacities of 730,000, 365,000, and 545,000 m^3/year, respectively [65]. The Güzelyurt basin's groundwater, situated in the western part of Northern Cyprus, is found under unconfined aquifer conditions [66–68]. The major aquifer in the region consists of quaternary deposits, while the pre-quaternary formations serve as the basement and impermeable or semi-permeable aquifer boundaries [61,62]. Previous research indicates a decline in groundwater levels by 45–50 m below sea level [69–71]. The Güzelyurt aquifer supplies water for irrigation as well as other purposes in the region and hosts some 920 mm^3 as its water reserve [72]. Typically, the economy of Northern Cyprus benefits considerably from the major agricultural activities primarily located in the Güzelyurt district [67–69]. The general agricultural production in the region is affected by the uneven distribution of rainfall and the water resource situation, including water deficit for agriculture [73]. In addition, 75% of all water withdrawals are for crop irrigation, and since the 1970s there has been a problem with drought and salinization [68–70,72,74]. Moreover, the region of Güzelyurt, which has plenty of farming areas, is now experiencing acute water problems, because of which water supplies have dropped drastically and produce quality has been reduced [75–77]. Furthermore, agriculture is divided into rain-fed and irrigated systems, producing various crops, with almond the most-grown crop [70,76].

1.3. Literature Review Related to Floating PV System (FPVS)

Due to a decrease in the cost of producing photovoltaic (PV) modules, solar energy has grown dramatically over the last decade [78]. PV modules are now being installed directly on water using FPVSs, a developing application compared to the conventional rooftop and ground-mounted alternatives [79,80]. According to Skoploki and Palyvos [81], the primary benefit of PV is the cooling impact of water on solar cells, which enhances energy conversion efficiency. In fact, according to Nisar et al. [82], Choi et al. [83], and Boduch et al. [84], solar water panels can produce up to 11% more electricity compared to those on the ground. Previous studies have indicated that the temperatures of floating PV modules were 2–10 °C lower than those of ground PV modules [84–87]. The capacity to scale the system from microwatts to megawatts, the abundance of water for cleaning the modules, the decrease in the development of algae caused by the modules' shade, and the decrease in natural evaporation in water reservoirs are additional benefits of this renewable energy technology [80].

The major components of FPVSs essentially consist of PV panels with their supporting electrical components and body construction that involves floats, mooring systems, waterproof materials, and buoyant force [88–90]. Figure 2 depicts a typical standalone FPV system.

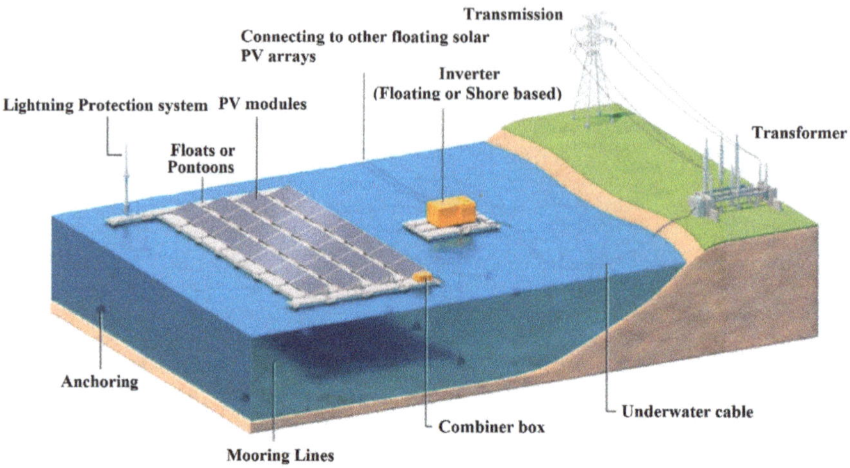

Figure 2. Typical standalone FPV system [91,92].

PV panels typically stand on an array of connected floats that are moored and fastened. These floats are usually made of lightweight materials such as Polyethylene (mostly HDPE), Hydro-Elastic Membranes, Polystyrene foam, Ferrocements, or Fiber-Reinforced Plastics with Aluminum or Steel rafts [90,93,94]. Meanwhile, electrical wires wrapped in thermoset rubber or polyurethane are commonly employed in FPVSs because of their water resistance [95]. Rubber matting protects the structure in the end [90]. Moreover, based on the previous studies [90,95,96], Figure 3 displays five common FPVS structure categories.

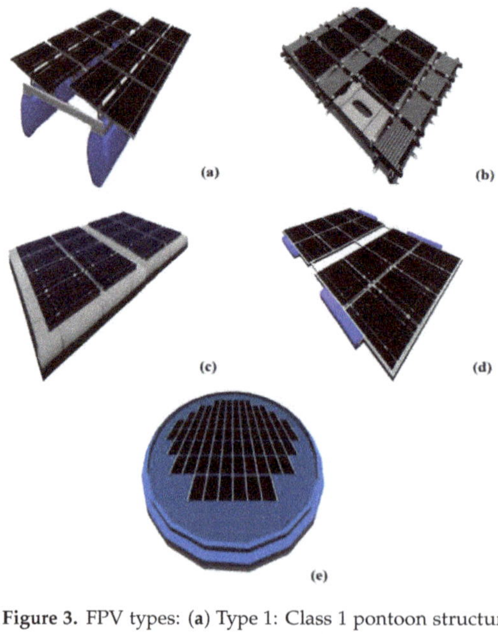

Figure 3. FPV types: (**a**) Type 1: Class 1 pontoon structure, (**b**) Type 2: Class 2 pontoon structure, (**c**) Type 3: Class 3 pontoon structure (**d**) Type 4: Superficial rigid structure, and (**e**) Type 5: Superficial flexible structure [97–99].

Generally, the geological and climatic conditions of the site should be taken into account during installation according to Ramanan et al. [90]. Other factors to consider during installation include array size, tilt angle, and solar panel type [90]. According to Ravichandran et al. [100], the surface area and depth of the water body affect the size of the matrix, which affects the ability of the floating structure to anchor and moor as well as its production capacity. The type of solar panels used must be carefully studied, as they have an important impact on the economic viability of FPVSs [101–104].

Furthermore, concrete block anchors were placed on the bottom of the water body and connected to the areas surrounding the PV array by mooring lines [105,106]. In addition, an effective anchoring system should be implemented to keep the floating structure stable in the presence of wind currents [100]. According to Ravichandran et al. [100], the floating structure is moored based on the reservoir's soil type and water level. Furthermore, based on the position of the FPV array, anchoring on the reservoir embankment or a nearby land area can be accomplished [100,104,105]. The mooring lines allow the FPV system to continue to adapt and stabilize in the presence of strong winds and waves. The flexible mooring lines are used to increase the flexibility of the FPV hull during water level deviation during monsoon and empty reservoir conditions [106]. Generally, underwater cables are employed to transport the electrical energy generated by the PV array mounted on the floating structure to the substation. In order to prevent resistance losses, the inverter station is either on the ground or on a separate floating platform following the PV array, depending on the distance between the substation and the FPV array [107].

1.4. Research Gap, Motivations, and Contributions

Previous research indicates that FPV devices can help lower surface evaporation in water bodies and are a viable alternative to conventional energy generation techniques. There is a lack of academic literature on FPV uses in Northern Cyprus; hence, further research is desperately needed in this field.

The present investigation provides a useful framework for assessing the various benefits of utilizing FPV systems, considering their unique features. To achieve this aim, the main contributions of this study are outlined as follows.

- An extensive and comprehensive analysis of solar energy potential is necessary as solar energy is intended as Northern Cyprus's main source of electricity. In this work, the distribution of solar radiation is analyzed using several distribution functions. In this paper, Johnson SB distribution, pert distribution, Phased Bi-Weibull distribution, and Kumaraswamy distribution are newly introduced to fit the solar radiation data, expressed by global solar radiation, and the probability distribution of Northern Cyprus sites. The selected distribution models are compared with the common distribution functions, including Normal, Gamma, Generalized Extreme Value, Log-normal, Weibull, and Logistics. Parameter estimation for distributions is conducted with the maximum likelihood estimation method.
- Investigating the utilization of floating photovoltaic systems for producing clean energy and minimizing water evaporation loss from selected water reservoirs is the main objective of the current paper. In this work, evaporation was estimated using Priestley–Taylor, deBruin–Keijman, Makkink, Penman–Monteith, Mass transfer, and Hamon methods. Various scenarios for using an FPVS to cover the reservoir's surface were investigated.
- The design process of various scales of floating photovoltaic systems and their technical and economic assessment were carried out in this paper, considering that there is no analysis addressing the application of this technology under the conditions of Northern Cyprus. Various floating assemblies and PV technologies were examined in order to determine the most suitable option.

The results will assist future FPV project developers and supporters in estimating the energy generated and identifying the most suitable places to implement this innovative technology to reduce electricity and water shortages in Northern Cyprus.

2. Materials and Methods

2.1. Case Study of Floating Solar PV Plant in Northern Cyprus

Cyprus, with a 9251 km^2 area, is the third largest island in the Mediterranean. It is situated in the Mediterranean's eastern region. Cyprus has a typical Mediterranean climate, meaning there are distinct seasons and substantial variations in temperature, precipitation, and overall weather patterns. The weather is hot and dry in summer, from mid-May to mid-September, and wet and volatile from early November to mid-March, divided by two brief seasons, spring and fall. During the summer, the island experiences high temperatures caused by the Great Continent passing over Central Asia. It is regarded as a season of high temperatures and nearly clear skies. In winter, Cyprus is subject to a few successive depressions that cross the Mediterranean Sea from west to east between Eurasia and the African belt. These depressions cause atmospheric disturbances, leading to the most rainfall of the year taking place in this season. The average annual rainfall is about 500 mm, with the maximum and minimum values of 759 mm and 182 mm recorded in 1968–1969 and 1972–1973, respectively. Generally, dams are constructed across rivers to recharge aquifers, store surface runoff for use during future dry seasons, and occasionally minimize hazardous flooding. Northern Cyprus experienced a shortage of water in 1960, which prompted the construction of dams. After realizing the importance of dams and ponds as renewable water resources, Turkey's State Hydraulic Works (DSI) rapidly established 41 reservoirs with varying capacities. Of these, eighteen were constructed for irrigation and twenty-three were constructed to prevent temporary streams from going straight into the Mediterranean Sea. Consequently, these infrastructures have accomplished the tasks for which they were designed. Over 27 million cubic meters (MCM) of water are taken annually from 28 streams in Northern Cyprus. About 43 MCM of water is also released annually by ten seasonal streams that rise from the Troodos Mountains in Southern Cyprus. However, these rivers' upstream portions have since been dammed in the southern part of the island. In this study, 15 reservoirs/dams have been considered to assess the potential of floating photovoltaic installations. Table 1 tabulates the selected reservoirs/dams, and Figure 4 shows the locations of the selected reservoirs/dams. Moreover, Figure S1 in the Supplementary Materials presents the long-term averages of solar resources, including Global Horizontal Irradiation (GHI), Global Tilted Irradiation (GTI), Direct Normal Irradiation (DNI), Air temperature (AT), and Global Tilted Irradiation (GTI) as generated by Global Solar Atlas, the principal phenomena of climate that determine solar power generation.

Table 1. Selected reservoirs/dams in Northern Cyprus.

Location No.	Dam/Reservoir/ Lake	Dam/Reservoir Location	Year of Construction	Latitude [°N]	Longitude [°E]	Capacity [m^3]	GHI [kWh/m^2]	DNI [kWh/m^2]	AT [°C]
D#1	Akdeniz	Güzelyurt	1988	35.26	32.98	1.47×10^6	1963.6	2203.5	19.9
D#2	Arapköy Ayanidere	Girne	1990	35.32	33.43	6.09×10^5	1894.4	2038.3	20.4
D#3	Arapköy Uzundere	Girne	1990	35.32	33.44	4.44×10^5	1885.3	2000.7	20.1
D#4	Dağyolu Üçparmakdere	Girne	1994	35.27	33.2	3.92×10^5	1944.7	2194.9	19.2
D#5	Değirmenlik Çataldere	Lefkoşa	1990	35.25	33.46	2.97×10^5	1898.7	2067	19.9
D#6	Ergazi Sayadere	Gazimağusa	1989	35.37	33.94	4.05×10^5	1856	1964.4	20.3
D#7	Geçitkale Eğridere	Gazimağusa	1989	35.29	33.72	1.36×10^6	1875.8	2028.5	20.4
D#8	Gemikonağı Madendere	Güzelyurt	1988	35.11	32.83	4.12×10^6	1926	2111.3	20.5
D#9	Gönendere	Gazimağusa	1987	35.29	33.66	9.39×10^5	1878.2	2033.4	20.3
D#10	Gönyeli	Lefkoşa	1962	35.24	33.3	4.54×10^5	1932.1	2168.5	19.9
D#11	Hamitköy Baştanlıkdere	Lefkoşa	1992	35.24	33.36	5.29×10^5	1920.6	2137.3	19.9
D#12	Haspolat	Lefkoşa	1964	35.23	33.41	1.17×10^5	1910.8	2117.2	19.9
D#13	Kanlıköy	Lefkoşa	1963	35.24	33.26	7.30×10^5	1939.9	2180.9	19.6
D#14	Serdarlı Ağillidere	Lefkoşa	1992	35.25	33.58	3.92×10^5	1881.7	2035.3	20.3
D#15	Yılmazköy Polatdere	Lefkoşa	1994	35.26	33.17	5.17×10^5	1952.3	2210.7	19.3

GHI: global horizontal irradiation; DNI: direct normal irradiation; AT: air temperature.

Figure 4. Locations of the selected reservoirs/dams.

2.2. Dataset

Generally, it is challenging to design solar energy systems in a location where the measurement data are not available. The number of stations and data is limited in Northern Cyprus. It is considered a major problem for working with the observed rainfall records. Thus, it is necessary to evaluate the satellite climate dataset products with locally measured data before utilizing these products as sources for evaluating the performance of solar energy systems at the specific location. In the literature, NASA, a leader in space technology, provides model-based and satellite products via its network of satellite systems, which are essential for addressing agricultural needs, increasing energy efficiency, and studying climate dynamics [108–110]. The NASA Earth Science Research Program's POWER (Prediction of Worldwide Energy Resources) Project made this possible. With a spatial resolution of $0.5° \times 0.5°$, this worldwide grid offers solar and meteorological data from 1981 onward and covers areas without surface meteorological observations [108–110]. The POWER data from NASA is freely available to users. Several studies evaluated the potential for solar energy in different regions using the NASA database [111–113]. Moreover, a robust association between observed data (GHI) and the NASA database was shown by Belkilani et al. [114], Gairaa and Bakelli [115], and Tahir et al. [116]. Moreover, solar radiation can be considered constant over large areas, ignoring the effect of cloud absorption, according to Bhatia [117].

Therefore, the daily meteorological data for the selected dams were collected between 1982 and 2023 using the NASA POWER dataset. For estimating water losses, the NASA power dataset was employed. A selection of reservoirs and dams' evaporation was determined using daily temperature, solar radiation, relative humidity, and wind speed data spanning from 1982 to 2023.

2.3. Solar Radiation

2.3.1. Characteristics of Global Solar Irradiation (GHI)

The most fundamental factor that determines the feasible locations for solar power plants and floating PV systems is GHI [118,119]. According to Castillo et al. [120], GHI can be defined as the solar energy (light) arriving at the surface of the Earth every year. Therefore, finding the location of SI is essential for evaluating the performance of the system [121]. According to Gökmener et al. [122], Suh and Brownson [123], and Wang [124], PV systems give better performances in sunnier regions.

Moreover, examining the characteristics of GHI data using probability distribution functions, such as the gamma, Weibull, and normal distribution functions, is crucial for evaluating PV systems. Several probability distribution functions (PDFs) have been utilized for analyzing the characteristics of SI, as shown in Table 2.

Table 2. List of PDFs used for analyzing the SI characteristics.

Ref.	Region/Country	Distribution Function Used
[125]	Kumasi/Ghana	Exponential Distribution, Normal Distribution, Weibull Distribution, Gamma Distribution, Lognormal Distribution, Beta Distribution
[126]	Ibadan, Nigeria	Gamma Distribution, Extreme Value Distribution, Generalized Extreme Value Distribution, Weibull Distribution, Lognormal Distribution, Logistic Distribution, Normal Distribution
[127]	Taipei, Taichung, Tainan, Kaohsiung, Hualien, Taitung	Weibull Distribution, Logistic Distribution, Normal Distribution, Lognormal Distribution
[128]	India	Normal distribution, Rayleigh distribution, Weibull distribution, Lognormal Distribution
[129]	Algeria	Beta Distribution
[130]	France areas	Weibull distribution
[131]	Yenagoa/Nigeria	Normal Distribution, Lognormal Distribution, Lognormal Distribution, Logistic Distribution, Extreme Value Distribution, Weibull Distribution
[132]	-	Weibull Distribution, Logistics Distribution, Lognormal Distribution, Beta Distribution, Gamma Distribution

In this study, Johnson SB distribution, pert distribution, Phased Bi-Weibull distribution, and Kumaraswamy distribution are proposed for the first time for solar irradiation modeling. The proposed PDF and common distribution functions for analyzing the characteristics of SI at the selected reservoir/dam in Northern Cyprus are listed in Table 3.

Table 3. Expression of the probability density function used to analyze the distribution of GHI data.

Distribution Function	Probability Density Function	Parameters	
Kumaraswamy (K)	$f(S) = abS^{a-1}(1 - S^a)^{b-1}$	a b	Shape parameters Shape parameters
Johnson SB (JSB)	$f(S) = \frac{\delta}{\lambda\sqrt{2\pi}} \frac{1}{\sqrt{1+\left(\frac{S-\xi}{\lambda}\right)^2}} exp\left(-\frac{1}{2}\left[\gamma + \delta sinh^{-1}\left(\frac{S-\xi}{\lambda}\right)\right]^2\right)$	δ λ γ ξ	Location parameter Scale parameter Shape parameter Shape parameter
Normal (N)	$f(S) = \frac{1}{\sqrt{2\pi\sigma^2}} exp\left(-\frac{S-\mu}{2\sigma^2}\right)$	σ μ	Standard deviation Mean parameter
Gamma (G)	$f(S) = \frac{S^{\alpha-1}}{\beta^\alpha \Gamma(\alpha)} exp\left(-\left(\frac{S}{\beta}\right)\right)$	β α	Shape parameter Scale parameter
Generalized Extreme Value (GEV)	$f(S) = \begin{cases} \frac{1}{\sigma}\left[\left(1+\zeta\left(\frac{S-\mu}{\sigma}\right)\right)^{-1/\zeta}\right]^{\zeta+1} exp\left(-\left(1+\zeta\left(\frac{S-\mu}{\sigma}\right)\right)^{-1/\zeta}\right) & k \neq 0 \\ \frac{1}{\sigma}\left[exp\left(-\left(\frac{S-\mu}{\sigma}\right)\right)\right]^{\zeta+1} exp\left(exp\left(-\left(\frac{S-\mu}{\sigma}\right)\right)\right) & k = 0 \end{cases}$	α ζ	Scale parameter Shape parameter
Log-normal (LN)	$f(S) = \frac{1}{S\sigma\sqrt{2\pi}} exp\left[-\frac{1}{2}\left(\frac{In(S)-\mu}{\sigma}\right)^2\right]$	σ μ	Shape parameter Scale parameter
Weibull (W)	$f(S) = \left(\frac{\alpha}{\sigma}\right)\left(\frac{S}{\sigma}\right)^{\alpha-1} exp\left(-\left(\frac{S}{\sigma}\right)^\alpha\right)$	α σ	Shape parameter Scale parameter

Table 3. *Cont.*

Distribution Function	Probability Density Function	Parameters	
Logistic (L)	$f(S) = \frac{exp\left(-\frac{S-\mu}{\sigma}\right)}{\sigma\left\{1+exp\left(-\frac{S-\mu}{\sigma}\right)\right\}^2}$	μ	Location parameter
		σ	Scale parameter
Phased Bi-Weibull (PBW)	$f(S) = \begin{cases} \frac{\beta_1}{\eta_1}\left(\frac{S}{\eta_1}\right)^{\beta_1-1} exp\left[-\left(\frac{S}{\eta_1}\right)^{\beta_1}\right] & 0 \le S < \tau \\ \frac{\beta_2}{\eta_2}\left(\frac{S-\tau}{\eta_2}\right)^{\beta_1-1} exp\left[-\left(\frac{S-\tau}{\eta_2}\right)^{\beta_2}\right] & S \ge \tau \end{cases}$	β_1	Shape parameter of the first Weibull distribution phase
		η_1	Scale parameter of the first Weibull distribution phase
		β_2	Shape parameter of the second Weibull distribution phase
		η_2	Scale parameter of the second Weibull distribution phase
		τ	Transition point
Pert (P)	$f(S) = \frac{(S-a)^{\left(1+4\frac{b-a}{c-a}\right)}(c-S)^{\left(1+4\frac{c-b}{c-a}\right)}}{B\left(\left[1+4\frac{b-a}{c-a}\right]\left[1+4\frac{c-b}{c-a}\right]\right)(c-a)^{\left\{\left[1+4\frac{b-a}{c-a}\right]+\left[1+4\frac{c-b}{c-a}\right]-1\right\}}}$	a	Minimum
		b	Most likely
		c	Maximum

S: global solar irradiation.

2.3.2. Solar Radiation Incident on Tilted Surface

Solar radiation generally consists of radiant radiation as well as diffuse radiation and radiation reflected from the Earth [133,134]. Because few measurement tools and methods are available, it is rare to find data recorded on an inclined surface [133,135]. In general, there are three aspects to the incident global solar irradiance (GSI_i) on an inclined surface [133,134]:

(i) The beam component from direct sunlight on the tilted surface (G_b);
(ii) The diffuse component (G_d);
(iii) A reflected component that measures the radiation reflected from the ground to the tilted surface (G_r).

It can be calculated using the following equation [134].

$$GSI_i = G_b + G_d + G_r \tag{1}$$

The incidence angle between the sun's rays and the surface normal can be utilized for calculating the G_b. It can be expressed as Equation (2) [134].

$$G_b = \frac{G_{DH}}{cos\theta_z} cos\theta_i \tag{2}$$

where G_{DH} is the direct horizontal solar irradiance, θ_z is the solar zenital angle (Equation (3)), and θ_i is the incidence angle of the beam radiation on the tilted surface (Equation (4)).

$$cos\theta_z = sin\phi \cdot sin\delta + cos\phi \cdot cos\delta \cdot cos\omega \tag{3}$$

$$\begin{aligned} cos\theta_i = {} & sin\delta \cdot sin\phi \cdot cos\beta - sin\delta \cdot cos\phi \cdot sin\beta \cdot cos\alpha + cos\delta \cdot cos\phi \cdot cos\beta \\ & \cdot cos\omega + cos\delta \cdot cos\phi \cdot sin\beta \cdot cos\alpha \cdot cos\omega + cos\delta \cdot sin\beta \cdot sin\alpha \\ & \cdot sin\omega \end{aligned} \tag{4}$$

where δ is the solar declination angle, ϕ is the location's latitude, ω is the hour angle, β is the surface tilt angle concerning the horizontal plane, and α is the surface azimuth angle.

The ω is computed using the local apparent time (LAT) [133,136]. Two modifications can be made to the normal time shown on a clock to accomplish this: The location's longitude and the meridian that serves as the basis for standard time disagree, which leads to the first correction. For every degree of longitude variation, the adjustment has a magnitude of 4 min. The Earth's orbit and rotational speed are prone to minute fluctuations, which is why the second correction is known as the equation of time (*EOT*). Equation (5) is used to calculate the ω [133].

$$\omega = 15(12 - LAT) \tag{5}$$

$$\begin{aligned} LAT = {} & standard\ time\ (clock\ time) \\ & \pm 4(standard\ time\ longitude - longitude\ of\ location) \\ & + EOT \end{aligned} \tag{6}$$

$$\begin{aligned} EOT = 229.18(& 0.000075 + 0.001868 \cdot cos(B) - 0.032077 \cdot sin(2 \cdot B) \\ & - 0.014615 \cdot cos(2 \cdot B) - 0.04089 \cdot sin(2 \cdot B)) \end{aligned} \tag{7}$$

$$B = \frac{360 \cdot (N - 1)}{365} \tag{8}$$

where N is the day of the year.

It should be noted that the first correction for LAT applies to the eastern hemisphere in the negative sign, while the western hemisphere applies to the positive sign.

2.4. Electricity Generated by PV System

Energy generation represents one of the most important factors in the assessment of the floating PV system (FPV). Equation (9) is used to estimate the energy generation (EG) by the FPV system [80].

$$EG = \sum_{i=1}^{n} \eta_{PV} P_{STC} \left(\frac{G}{G_{STC}} \right) [1 - \alpha_p (T_C - T_{STC})] N \Delta t_i \tag{9}$$

where η_{PV} is the individual PV module derating factor, which accounts for wiring losses, inverter inefficiency, component failures, soiling, and aging, among other effects (η_{PV} = 0.85 here); P_{STC} is the nominal power of an individual PV module in terms of power output under Standard Test Conditions (STCs); G is the effective, or plane-of-array irradiance, i.e., the incident irradiance less self-shading losses; G_{STC} is the reference plane of-array (POA) irradiance under STC = 1 kW/m^2; α_p is the PV panel temperature coefficient of power; T_C is the operating cell temperature; T_{STC} is the STC operating cell temperature = 25 °C; N is the number of installed PV modules; and Δt_i is the duration of the n time steps considered.

Generally, the amount of EG is a function of the solar radiation, the system's efficiency, and the total area of the panels. Moreover, the energy output is not much affected by variations in water level according to Ref. [137]. Moreover. According to Fereshtehpour et al. [22], the optimal angle of floating solar panels mounted on the water's surface can be significantly affected by variations in water level. This kind of situation is avoided by anchoring floating solar panels using mooring systems. This technique allows the system to be kept facing the Sun while adjusting for variations in the water level. Furthermore, the minimum water depth in the range of water level changes should not be less than a suitable value for the seamless installation and functioning of the floating PV.

2.5. Water Depth

According to Gökmener et al. [122], the reservoir/dam must have enough water depth to support a floating photovoltaic system, as the water depth directly impacts the floating panels on the water's surface. The setup and maintenance of the panels on the water's surface are also significantly impacted by the depth of the water. Generally, the dead water level of each reservoir is subtracted from the water level to determine the water depth. The fluctuation of water levels, seasonal variations, and the inflow of debris floating around are all related to water depth. The recommended water depth for a reservoir can be 8–10 cm [138], 2 m [139], 1–5 m [140], and 2–4 m [141], based on previous scientific studies.

2.6. Local Average Wind Speed

In general, FPV system performance may be positively or negatively impacted by the average wind speed in the region. According to previous studies, wind speed is one of the most important variables that should be taken into account when designing floating photovoltaic systems [142,143]. Furthermore, Golroodbari et al. [143] concluded that wind speed can lower a PV module's temperature, indicating that wind speed can significantly affect the production of power. Furthermore, high wind speeds have been shown to destabilize and harm floating photovoltaic (PV) systems [144,145]. Furthermore, Habib et al. [145] found that wind speed contributes to the removal of dust deposition from PV cell surfaces.

2.7. Evaporation Estimation

In general, the floating PV system protects the entire reservoir/dam's surface from evaporation, not simply the area covered after installation. The decline in water evaporation can be attributed to two main reasons. The first reason, which directly affects the decrease in evaporation, is the reduction in air and water contact from the covered area [119]. The second reason is that the power plant's construction changes the lake's heat balance, resulting in a decrease in total evaporation across the lake's surface and a colder temperature [119,146].

Thus, precise evaporation estimation is essential. Generally, air saturation deficit above the surface, wind speed, solar radiation reaching the water's surface, air pressure, and the chemical makeup of the water are all significant factors that influence the process of estimating water evaporation, according to Fereshtehpour et al. [22].

In the literature, various methods are utilized to estimate water evaporation, such as Penman's method and the bulk aerodynamic method. Table 4 lists the methods that can be used for estimating the evaporation rate (E).

Table 4. Methods used for estimating E [mm/day].

Method Name	Equation	Reference
Priestley–Taylor	$E = \alpha \cdot \frac{\Delta}{\Delta+\gamma}\left(\frac{R_n - \Delta Q}{\lambda \rho_w}\right) \cdot D$	[146]
deBruin–Keijman	$E = \frac{\Delta}{0.85 \cdot \Delta + 0.63 \cdot \gamma}\left(\frac{R_n - \Delta Q}{\lambda \rho_w}\right) \cdot D$	[146]
Makkink	$E = 0.61 \cdot \frac{\Delta}{\Delta+\gamma} \cdot \frac{R_n}{\lambda} - 0.012$	[147]
Penman Monteith	$E = \frac{0.047 \cdot \Delta \cdot R_n + \gamma \cdot \frac{900}{T+273} \cdot U_2 (e_s - e_a)}{\Delta + \gamma \cdot (1 + 0.34 \cdot U_2)}$	[148]
Mass transfer	$E = 10 \cdot [NU_2(e_s - e_a)]$	[149]
Hamon	$E = \frac{35.8 \cdot K \cdot N \cdot e_s}{T + 273.3}$	[150]

α: Priestley–Taylor empirically derived constant, dimensionless ($\alpha = 1.26$). ρ_w: density of water (1000 kg m^{-3}). R_n: the net radiation [W/m^2]. ΔQ: the heat storage change in the water body [W/m^2]. γ: Psychrometric "constant" (depends on temperature and atmospheric pressure) [Pa °C^{-1}]; $\gamma = 0.665 \times 10^{-3} \cdot P$. P: Atmospheric pressure [kPa]. e_s: saturated vapor pressure at temperature of the air [kPa]; $e_s = \frac{1}{2} \cdot \left[0.6108 \cdot exp\left(\frac{17.27 T_{max}}{T_{max}+237.3}\right) + 0.6108 \cdot exp\left(\frac{17.27 T_{min}}{T_{min}+237.3}\right)\right]$. e_a: vapor pressure at temperature and relative humidity of the air [kPa]; $e_a = \frac{RH}{100} \cdot e_s$. Δ: the slope of saturated vapor pressure–air temperature curve [kPa/°C]. λ: the latent heat of vaporization (2.45×10^6 J/kg). RH: Relative humidity [%].

2.8. Water Volume Lost by Evaporation Estimation

The daily average of the evaporation rate (E_{daily}) was used to compute the volume of water lost due to evaporation every day. Equation (10) is utilized to calculate the daily volume of water lost (V_{Daily}) [149,151].

$$V_{Daily} = E_{daily} \times A \tag{10}$$

where A is the surface area of the reservoir/dam in km^2.

2.9. Relation between Evaporation and Water Depth

According to Abd-Elhamid et al. [149] and Wong et al. [152], Equation (11) shows the relationship between water loss ($\frac{E}{V}$) and water depth (d). It is noticed that the evaporation rate increases exponentially at lower depths.

$$\frac{E}{V} = 10^{-0.8-0.48d} \quad (11)$$

2.10. Reduction in Evaporation

The floating photovoltaic system is one of the potential techniques that may be utilized to mitigate evaporation from open water surfaces [149,153]. In this research, a floating photovoltaic system was implemented to reduce the evaporation rate. This study investigated four different covering ratios of the reservoir/dam's surface: 25%, 50%, 75%, and 100%. The surface area of the reservoir/dam is calculated with Google Maps. The water savings from installing a floating PV system are estimated using Equation (12) [149,154].

$$Annual\ water\ saving = Annual\ evaporation \times Percentage\ of\ insulated\ area \times 0.70 \quad (12)$$

2.11. Techno-Economic Analysis

In the literature, several methods and techniques are used for evaluating the performance of floating PV (FPV) systems. For instance, Islam et al. [21] evaluated the performance of a floating PV system in Bangladesh using PVsyst software. Febrian et al. [153] evaluated the performance of floating PV systems with various types of solar panels in Indonesia using RETScreen software. Yashas et al. [155] analyzed floating PV system technology to harness sustainable energy in India using RETScreen, PVWatts, and PVGIS. Semeskandeh et al. [156] utilized RETScreen software to evaluate the techno-economic-environmental feasibility performance of a 5 kW floating PV system and ground PV power plant in Iran. Mittal et al. [157] evaluated the feasibility performance of installing a floating PV system in India utilizing RETScreen software. Alhassan et al. [158] used RETScreen software to analyze the feasibility of a floating PV-Hydro hybrid plant system. Nagababu et al. [159] and Taye et al. [160] used mathematical modeling to calculate the performance of a floating PV system in India.

Based on the above, RETScreen, developed by Natural Resources Canada, has been a popular tool for assessing the viability of various renewable energy systems [57]. Its goal is to calculate the risk, costs, emissions reduction, financial viability, and energy production and savings of different renewable energy solutions. The National Aeronautics and Space Administration (NASA) database's long-term monthly average meteorological data are used by the RETScreen to obtain meteorological data for a particular location. It is a quick and effective method for assessing the energy efficiency and environmental impact reduction of the renewable energy system. RETScreen software was utilized in this study to assess the grid-connected PV development's techno-economic viability. The impact of the orientation angles on the output PV and capacity factor power (Equation (13)) was investigated. Furthermore, the economic feasibility indicators, including net present value (Equation (14)), Levelized cost of energy (Equation (15)), simple payback (Equation (16)), equity payback (Equation (17)), annual life cycle savings (Equation (18)), annual greenhouse gas (GHG) emissions (Equation (19)), GHG emission reduction cost (GHG-E-RC) (Equation (20)), and Benefit–cost ratio (B–C) (Equation (21)) were calculated using RETScreen software. It should be mentioned that the predicted values of these economic feasibility indicators were used to evaluate the floating PV project's profitability and return on investment. Furthermore, RETScreen can estimate these indicators.

Capacity factor (CF)

$$CF = \frac{P_{out}}{P \times 8760} \quad (13)$$

Net present value (NPV)

$$NPV = \sum_{n=0}^{N} \frac{C_n}{(1+r)^n} \tag{14}$$

Levelized cost of energy (LCOE)

$$LCOE = \frac{sum\ of\ cost\ over\ lifetime}{s\ of\ electricity\ generated\ over\ the\ lifetime} \tag{15}$$

Simple payback (SP)

$$SP = \frac{C - IG}{(C_{ener} + C_{capa} + C_{RE} + C_{GHG}) - \left(C_{o\&M} + C_{fuel}\right)} \tag{16}$$

Equity payback (EP)

$$EP = \sum_{n=0}^{N} C_n \tag{17}$$

Annual life cycle savings (ALCS)

$$ALCS = \frac{NPV}{\frac{1}{r}\left(1 - \frac{1}{(1+r)^N}\right)} \tag{18}$$

Annual GHG emission reduction (A-GHG)

$$\begin{aligned} A - GHG = [& (Base\ case\ GHG\ emission\ factor) \\ & -(Proposed\ case\ GHG\ emission\ factor)] \\ & \times End\ use\ energy\ delivered \end{aligned} \tag{19}$$

GHG emission reduction cost (GHG-E-RC)

$$\text{GHG} - \text{E} - \text{RC} = \frac{ALCS}{\Delta_{GHG}} \tag{20}$$

Benefit–cost ratio (B–C)

$$B - C = \frac{NPV + (1 - f_d)C}{(1 - f_d)C} \tag{21}$$

where P_{out} is the energy generated per year, P is installed capacity, N is the project life in years, C_n is the after-tax cash flow in year n, and r is the discount rate. C is the total initial cost of the project, f_d is the debt ratio, B is the total benefit of the project, IG is the incentives and grants, and C_{ener} is the annual energy savings or income. C_{capa} is the annual capacity savings or income, C_{RE} is the annual renewable energy (RE) production credit income, C_{GHG} is the GHG reduction income, and $C_{o\&M}$ is the yearly operation and maintenance costs incurred by the clean energy project. C_{fuel} is the annual cost of fuel, which is zero for renewable projects, and Δ_{GHG} is the annual GHG emission reduction.

3. Results

3.1. Solar Energy Assessment

The average value of GHI, DNI, and AT of the chosen dams/reservoirs are within the range of 1875.8–1963.6 kWh/m^2, 1964.4–2210.7 kWh/m^2, and 19.2–20.5 °C, respectively, as shown by Table 1 and Figure S1 in the Supplementary Materials. Furthermore, Bhatia [117] claimed that solar radiation can be considered constant over large areas by ignoring the effect of cloud absorption. Figure 5 illustrates the monthly variation of GHI for all selected dams/reservoirs located in Northern Cyprus. It is observed that the amount of solar energy received remains constant across the specified areas, with slight deviations. In considering this, the solar radiation analysis of particular selected dams/reservoirs (Gemikonağı Madendere, Hamitköy Baştanlıkdere, and Ergazi Sayadere) was made using different distribution functions. Figure 6 shows the mean daily solar radiation (SR) for Gemikonağı Madendere, Hamitköy Baştanlıkdere, and Ergazi Sayadere. It was found that the maximum and minimum average daily season values of SR are recorded in summer in Ergazi Sayadere and in winter in Gemikonağı Madendere, as 9.80 kWh/m^2/day and 3.06 kWh/m^2/day, respectively (see Figure 6a–c). As shown in Figure 6d, the highest and lowest value of mean daily variation SR is recorded at Hamitköy Baştanlıkdere, with the value of 7.06 kWh/m^2/day and 6.28 kWh/m^2/day, respectively.

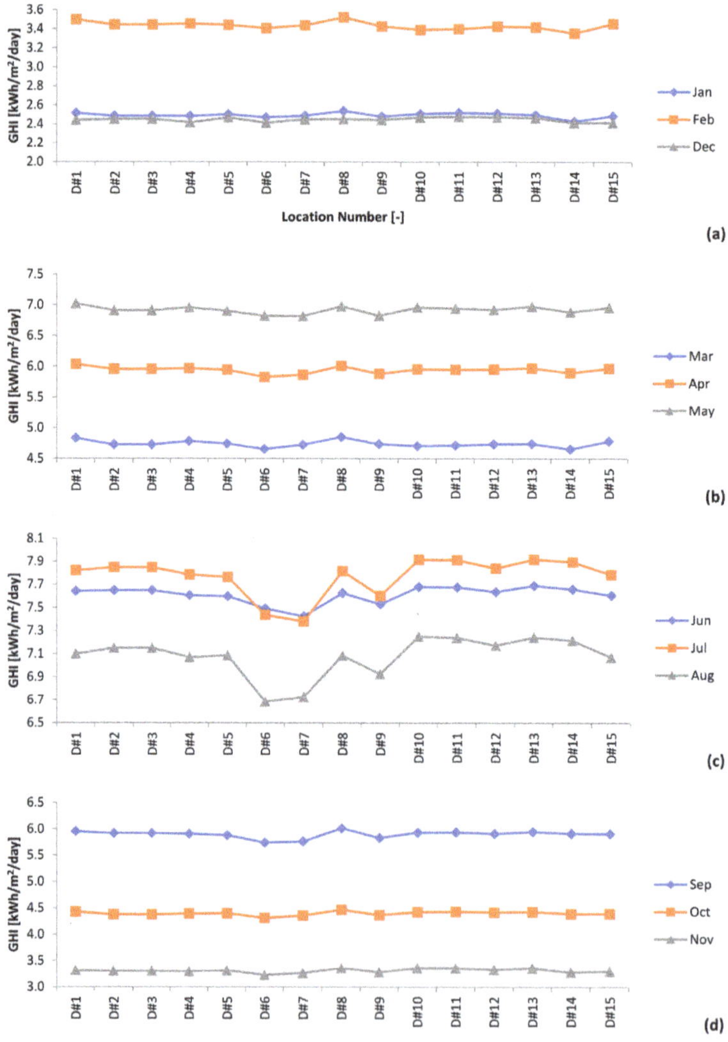

Figure 5. Mean monthly variation of GHI at selected dams during the studied periods: (**a**) winter (Jan: January; Feb: February; and Dec: December), (**b**) spring (Mar: March; Apr: April; and May), (**c**) summer (Jun: June; Jul: July; and Aug: August) and (**d**) autumn (Sep: September; Oct: October; and Nov: November).

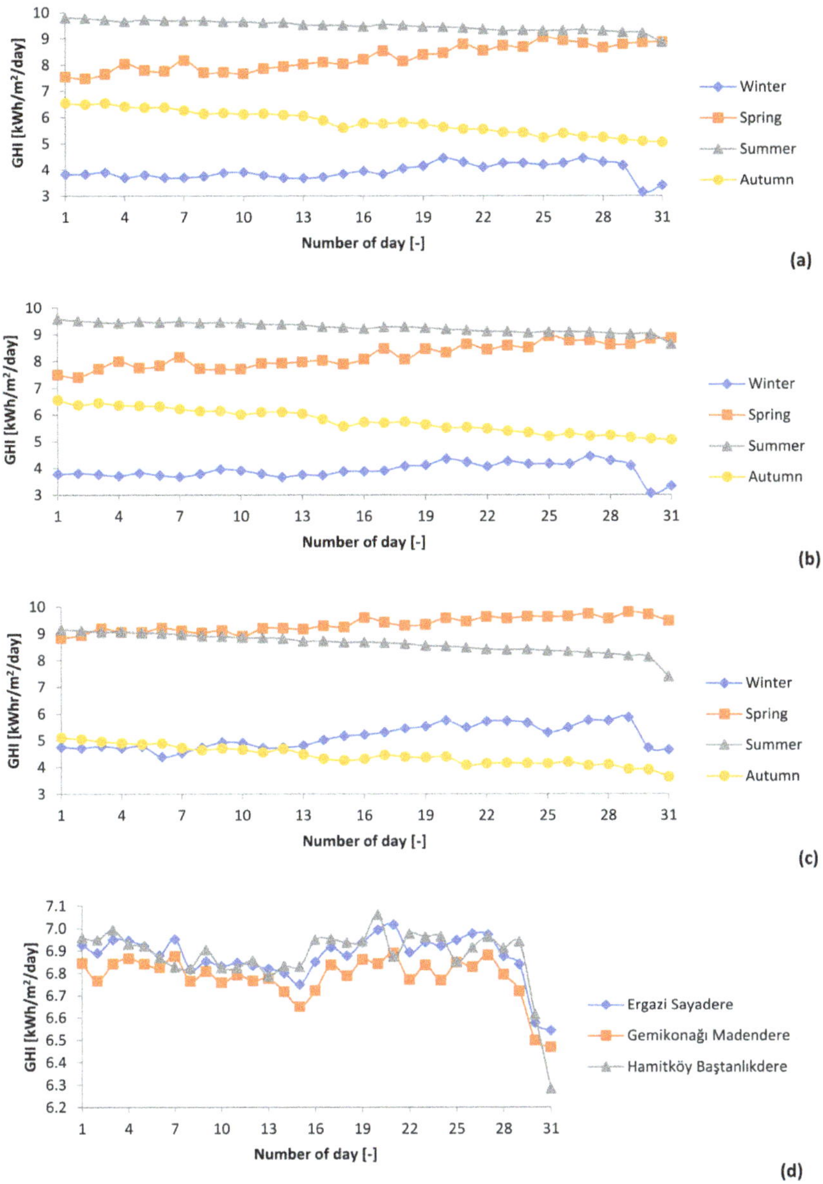

Figure 6. Daily variation of GHI at selected dams during the studied periods: (**a**) seasonal variation at Ergazi Sayadere, (**b**) seasonal variation at Gemikonağı Madendere, (**c**) seasonal variation at Hamitköy Baştanlıkdere. and (**d**) mean daily variation for all locations.

Furthermore, all locations' solar resources are categorized as excellent (class 5) according to the assessment of solar potential based on annual GHI and DNI [26]. Consequently, all locations are suitable for installing PV/flat-plate systems.

To determine the appropriate statistical distribution that best represents the GHI characteristics associated with the dams and reservoirs, we use several distribution functions. In this study, the efficiency of the Johnson SB distribution, pert distribution, Phased Bi-Weibull distribution, and Kumaraswamy distribution are employed to analyze the GHI data.

Tables S2–S4 in the Supplementary Materials include comprehensive details on the distribution functions that are most effective when taking into account aspects related to shape, scale, and location. The fitted probability density function (PDF) models for Gemikonağı Madendere and Ergazi Sayadere are presented as examples in Figures S2 and S3 in the Supplementary Materials, respectively. It was noticed that when the values of the Anderson–Darling (AD) and Kolmogorov–Smirnov (K–S) tests are near zero, the distribution function more closely resembles the real data. Based on the K–S and A–D tests, Table 5 provides the distribution models' ranking for the selected dams/reservoirs. Therefore, JSB has the lowest value according to the K–S test, and this distribution function has been considered to be the most appropriate for analyzing solar radiation at all dams/reservoirs in spring and autumn.

Table 5. Ranking of distribution functions based on the K–S and A–D tests.

Season	Location	Rank of the Distribution Function Based on K–S									
		1	2	3	4	5	6	7	8	9	10
Winter	Gemikonağı Madendere	PBW	GEV	K	JSB	P	L	N	G	W	LN
	Hamitköy Baştanlıkdere	K	L	JSB	GEV	P	N	PBW	G	W	LN
	Ergazi Sayadere	PBW	K	GEV	JSB	L	P	N	G	W	LN
Spring	Gemikonağı Madendere	JSB	K	GEV	W	N	P	G	LN	L	*
	Hamitköy Baştanlıkdere	JSB	GEV	K	G	N	LN	P	L	W	PBW
	Ergazi Sayadere	JSB	K	P	GEV	W	N	G	LN	L	*
Summer	Gemikonağı Madendere	L	JSB	K	N	G	LN	W	GEV	P	PBW
	Hamitköy Baştanlıkdere	GEV	JSB	K	W	N	L	G	LN	P	PBW
	Ergazi Sayadere	K	JSB	GEV	W	P	L	N	G	LN	PBW
Autumn	Gemikonağı Madendere	JSB	K	N	GEV	W	G	LN	L	P	PBW
	Hamitköy Baştanlıkdere	JSB	GEV	G	LN	P	N	L	W	K	PBW
	Ergazi Sayadere	JSB	GEV	N	G	K	LN	P	L	W	PBW
Yearly	Gemikonağı Madendere	JSB	GEV	P	W	PBW	N	G	LN	K	L
	Hamitköy Baştanlıkdere	GEV	P	JSB	W	N	G	LN	L	K	PBW
	Ergazi Sayadere	JSB	P	W	GEV	K	N	G	LN	L	PBW
Season	**Location**	**Rank of the Distribution function based on A–D**									
		1	2	3	4	5	6	7	8	9	10
Winter	Gemikonağı Madendere	PBW	K	P	L	N	G	W	LN	GEV	JSB
	Hamitköy Baştanlıkdere	GEV	P	JSB	K	L	N	W	G	LN	PBW
	Ergazi Sayadere	K	P	L	PBW	N	G	W	LN	GEV	JSB
Spring	Gemikonağı Madendere	GEV	N	W	P	G	LN	L	JSB	K	*
	Hamitköy Baştanlıkdere	JSB	GEV	N	W	G	P	LN	PBW	L	K
	Ergazi Sayadere	GEV	W	P	N	G	LN	L	JSB	K	*
Summer	Gemikonağı Madendere	K	L	N	W	G	LN	P	GEV	JSB	PBW
	Hamitköy Baştanlıkdere	GEV	JSB	W	L	N	G	LN	P	K	PBW
	Ergazi Sayadere	K	JSB	W	L	N	P	G	LN	GEV	PBW
Autumn	Gemikonağı Madendere	JSB	GEV	G	N	LN	P	W	L	K	PBW
	Hamitköy Baştanlıkdere	GEV	JSB	LN	G	N	P	L	W	K	PBW
	Ergazi Sayadere	JSB	GEV	G	N	LN	P	W	L	PBW	K
Yearly	Gemikonağı Madendere	GEV	P	W	JSB	N	G	LN	L	K	PBW
	Hamitköy Baştanlıkdere	JSB	GEV	W	N	P	G	LN	L	K	PBW
	Ergazi Sayadere	GEV	K	JSB	N	G	LN	P	L	W	PBW

* PBW = No fit.

Moreover, it is found that JSB, GEV, and P are identified as the most efficient and secondarily successful overall among all distributions for all dams/reservoirs when analyzing the average daily solar radiation. It is found that K is the best overall for the winter season in Hamitköy Baştanlıkdere, whereas PBW is listed as the best overall for the winter season in Gemikonağı Madendere and Ergazi Sayadere. Furthermore, the most suitable distribution function for the summer season is determined to be L, GEV, and K for Gemikonağı Madendere, Ergazi Sayadere, and Hamitköy Baştanlıkdere, respectively.

Based on AD tests, the GEV, JBS, and K distributions show higher compatibility with the observed data when compared with other researched distribution functions. Based on the categories of data, these distributions were determined to be the most accurate statistical distribution to predict the solar radiations of the locations in the study.

In conclusion, distribution models provide valuable insights into solar radiation patterns during specific seasons while having no direct impact on PV system performance. The size and orientation angles of the photovoltaic system can be intelligently adjusted to maximize its overall efficiency by positioning the solar panels to capture the most sunlight.

3.2. *Results of Average Monthly Evaporation*

Because the study location's particular characteristics may have significant impacts on the expected evaporation rates, it is crucial to select an approach that includes these variables. If the region experiences significant variations in wind and humidity, for example, the Penman methodology or the Mass transfer method might produce more accurate predictions. When radiation is the primary contributor, on the other hand, the Priestley–Taylor or Makkink techniques would be more suitable.

3.2.1. Monthly Variation of Climate Data

In this work, the monthly variations of evaporation from a chosen dam, reservoir, or lake from 2010 to 2023 were computed using mean monthly meteorological data, including temperature (maximum temperature and minimum temperature), relative humidity, wind speed, and solar radiation. The average monthly variation of meteorological data is illustrated in Figure 7. The following can be observed:

It is found that the mean maximum temperature is within the range of 15.71–38.11 °C, with an average value of 26.51 °C. The highest and lowest value of mean maximum temperature is recorded in August at D#13 (Kanlıköy), with a value of 38.11 °C, and in January at D#14 (Serdarlı Ağillidere), with a value of 15.71 °C.

The highest value of mean monthly minimum temperature is found to be 24.69 °C, which was recorded in August at D#7 (Geçitkale Eğridere), while the lowest value of 7.08 °C was recorded in February at D#12 (Haspolat).

The relative humidity value at D#12 (Haspolat) and D#13 (Kanlıköy) reaches a high value of 77.13% in January. The minimum mean relative humidity occurs at D#10 (Gönyeli) during July, at 45.87%.

The highest value of wind speed is recorded at D#2 (Arapköy Ayanidere) and D#3 (Arapköy Uzundere), with a value of 3.44 m/s in July. The minimum wind speed of 1.96 m/s is recorded at D#7 (Geçitkale Eğridere) in April.

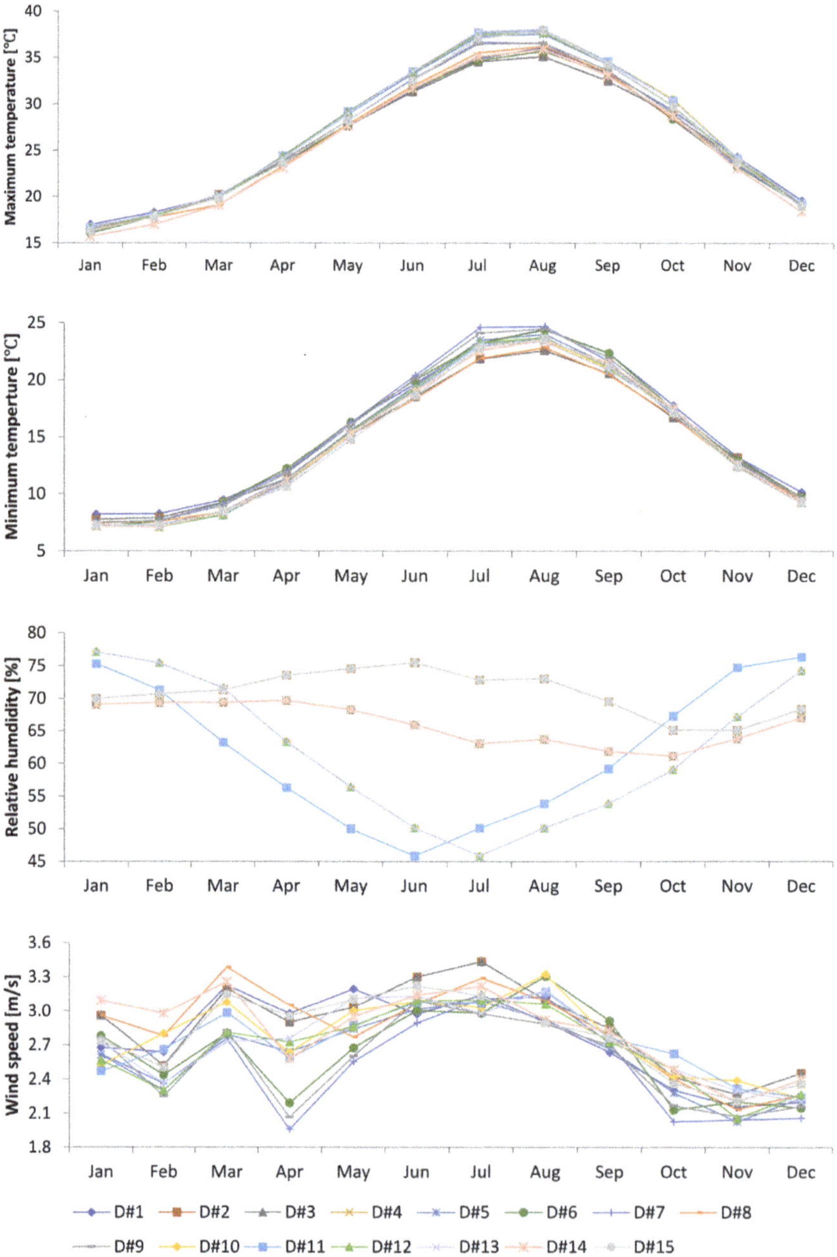

Figure 7. Monthly variation of meteorological data at selected dams/reservoirs/lakes during the studied period.

3.2.2. Monthly Variation of Evaporation

Table S5 in the Supplementary Materials displays the average monthly evaporation for all selected sites using six different estimation methods (Priestley–Taylor, Penman, Makkink, deBruin–Keijman, Hamon, and Mass transfer). Furthermore, Figure 8 shows the monthly variation of E as an example for some of the selected locations.

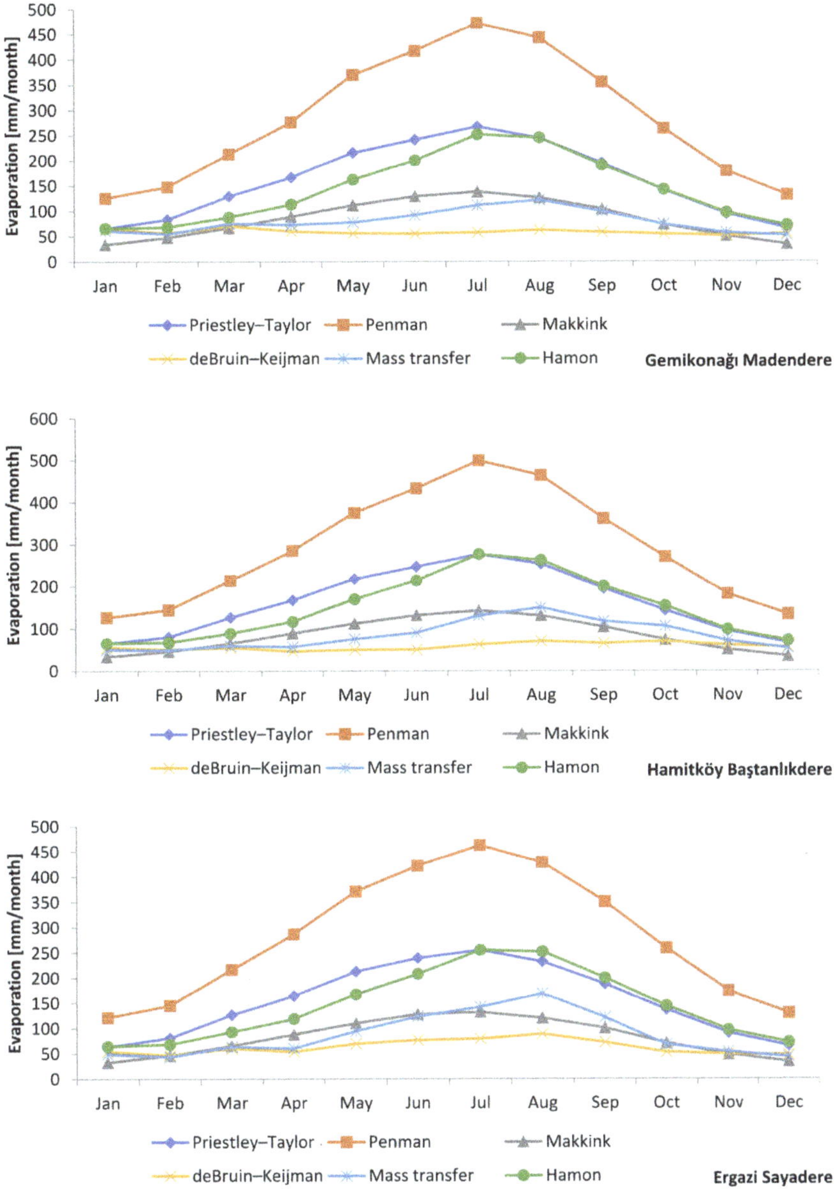

Figure 8. Monthly variation of E at some selected sites during the studied periods.

It is found that all approaches show higher evaporation rates in the warmer months (May to August) and the colder months (December to February). The various methods' evaporation rates (E) varied significantly. According to the Priestley–Taylor method, E is highest in July at D#10 (Gönyeli), with a value of 276.61 mm/month, and the lowest value of E (64.70 mm/month) is in December at D#14 (Serdarlı Ağillidere). Based on the Penman technique, the highest and lowest values are reported at D#11 (Hamitköy Baştanlıkdere) and D#15 (Yılmazköy Polatdere), with a value of 498.70 mm/month in July and 125.04 mm/month in December, respectively. According to the de Bruin–Keijman and

Makkink methods, July has the maximum value of E at D#11 (Hamitköy Baştanlıkdere) and D#13 (Kanlıköy), with a value of 142.60 mm/month and 142.84 mm/month, respectively. Additionally, the Mass Transfer method indicates fluctuation; D#15 (Yılmazköy Polatdere) had the highest E (201.94 mm/month) in July. Based on the Hamon method, the highest E for July is at D#10 (276.40 mm/month).

Moreover, various scenarios of the reservoir's covered area were examined, including 25%, 50%, 75%, and 100% covered areas. It should be noted that the Google Earth engine was used to determine the surface area of the lake, reservoir, and dams (see Figure S4 in the Supplementary Materials). The annual water saving is estimated using Equation (7). Figure S5 in the Supplementary Materials illustrates the annual water losses from the selected water reservoirs with different percentages of the reservoir's covered area. The system's effectiveness in reducing evaporation is demonstrated by the fact that the water savings increase significantly as coverage rises from 0% to 100%. The findings showed that, depending on the reservoir's surface area, the annual water losses ranged from 2.47×10^4 m^3 to 1.11×10^6 m^3, as shown in Figure S5 in the Supplementary Materials, while the annual water savings for the 0% covered area were estimated to be 0 m^3. For instance, annual savings at 100% coverage can reach 6.21×10^5 m^3 (Location D#13) compared to 0 m^3 in the absence of PV panels, as shown in Figure 9.

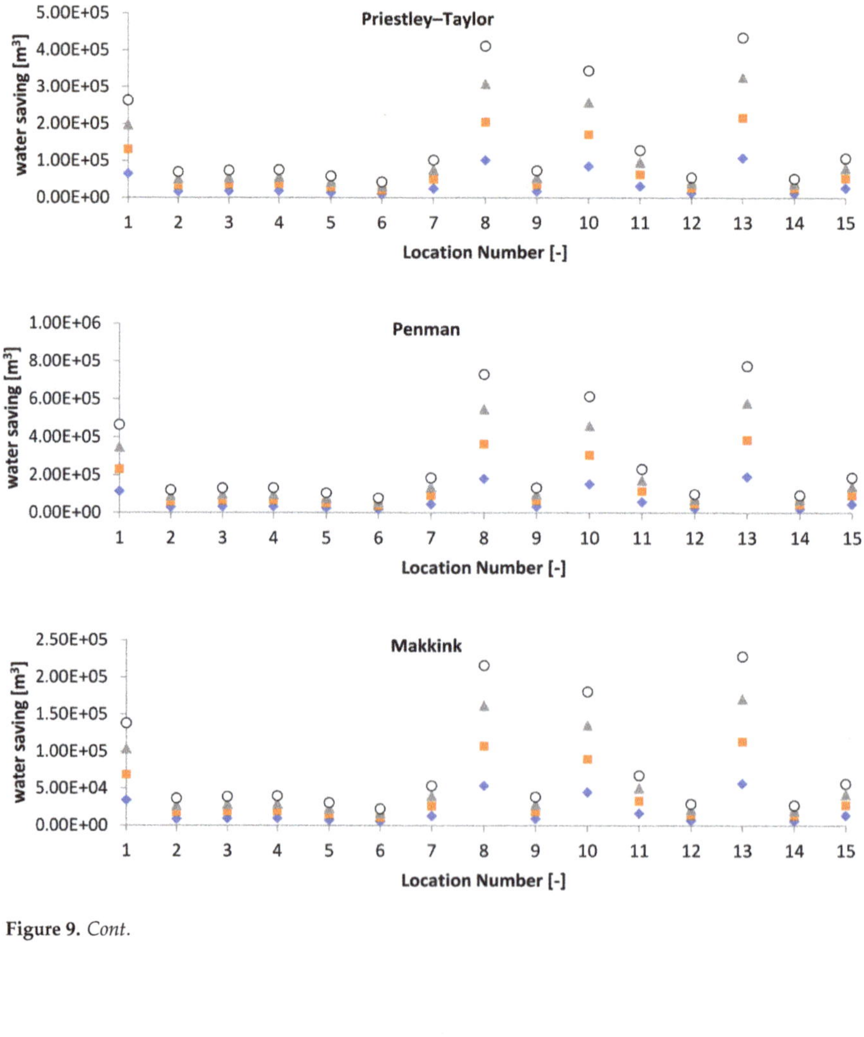

Figure 9. *Cont.*

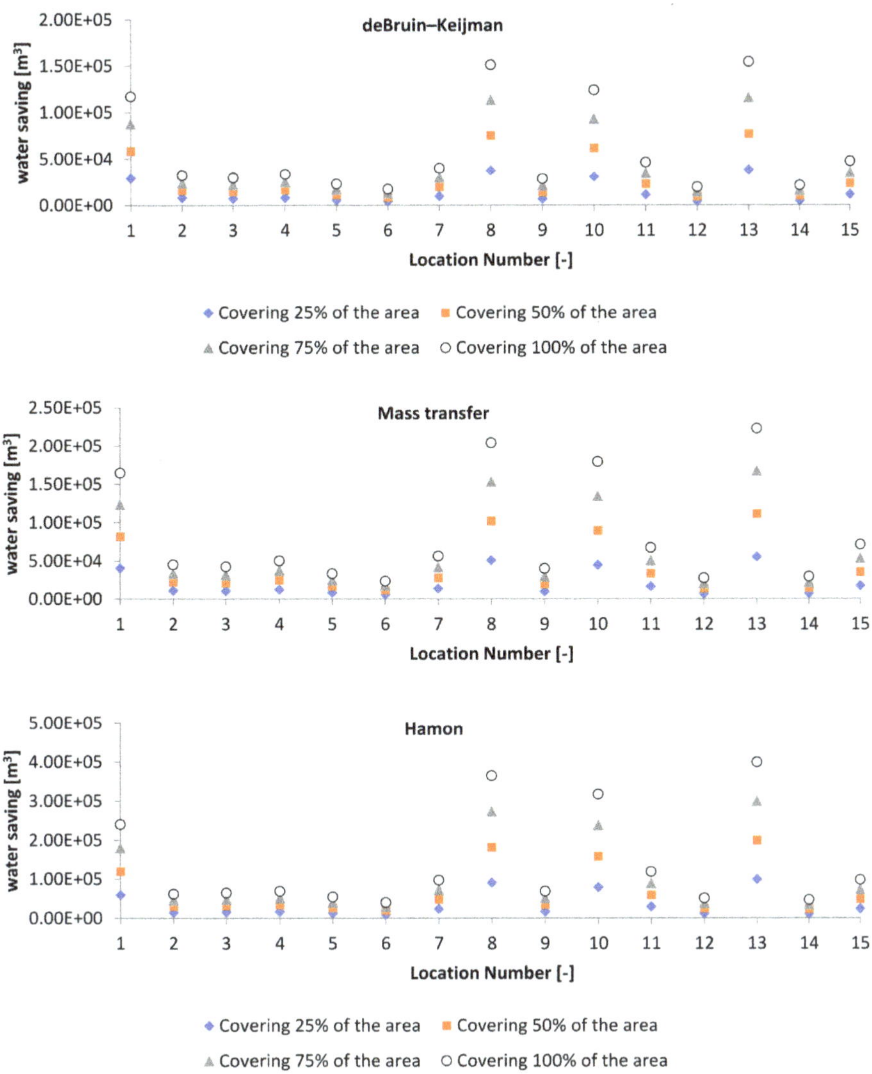

Figure 9. Annual water savings at the selected locations.

This trend highlights the clear correlation between evaporation reduction and covering area. Controlling water resources in reservoirs and dams can benefit from floating photovoltaic systems because of their potential as an achievable option for water conservation, as seen by the notable savings that increase with higher coverage.

3.3. Best Floating Structure

The economic viability of developing different scale grid-connected FPV systems was examined in this paper by mathematical modeling, taking into account technical, financial, and environmental points of view.

3.3.1. Proposed FPV System

Considering that the N-type Mono-crystalline module from Jinko Solar Company is one of the most efficient PV modules on the market currently, it was selected for the recommended FPV systems. The detailed specifications of the selected module are presented in Table S6 in the Supplementary Materials. Additionally, Central 850CP XT central inverters with a 954 kW capacity were used. Table S7 in the Supplementary Materials provides the used inverter's specifications. The cost of the inverter and PV panel was found to be USD 155,647 and USD 173, respectively. Moreover, four floating structural variations were taken into consideration in this study, as shown in Table 6.

Table 6. Different types of floating structures considered in this study.

Company	Type	Characteristics
	Hydrelio® Floating Solar	North orientation—tilt 0°
Ciel&Terre	C&T S12	South orientation—tilt 12°
	C&T EW12	East/west orientation—tilt 12°
Solaris Synergy	SolSyn S25	South orientation—tilt 25°
Swimsol	Swimsol S10	South orientation—tilt 10°
Isigenere	Isifloating EW15	East/west orientation—tilt 15°
Ocean Sun	OceanSun S6	North/south orientation—tilt 6°
Seaflex	Seaflex S5	South orientation—tilt 5°
Eco Wave Power	EWP EW20	East/west orientation—tilt 20°

3.3.2. Economic Model

As mentioned previously, the study employed mathematical modeling to assess the techno-economic feasibility of the proposed floating photovoltaic plant. In this study, different important economic measures were used to evaluate the performance of the developed FPV system. Table 7 lists the assumptions used in this study.

Table 7. List of assumptions taken into account in this study.

Assumption	Unit	Value
PV's lifetime	Year	25
Discount rate	%	6
Inflation rate	%	8
Engineering cost	%	2
Civil work	%	5
Installation cost	%	8
Cost of transport	%	2
Cost of the electrical connection	%	7
Miscellaneous costs	%	1
Cost of a unit of inverter	USD	155,647
Electricity price	USD	0.10
Cost of FPV	USD/Watt	1.2

3.3.3. Technical Viability

According to Owolabi et al. [113], the annual energy production (AEP) by solar photovoltaic plants and its capacity factor (CF) are both influenced by the solar radiation of the location and the number of clear sunny days. In this study, three different scenarios of covering ratios of the reservoir/dam's surface (25%, 50%, and 75%) were considered for technical feasibility. Figure 10 illustrates the annual energy production from the proposed FPVS at various orientations angles for the first scenario (covered area 25%). It is found that the AEP varies between 2.25 GWh and 30.73 GWh. The maximum and minimum values of AEP are recorded at D#13 (Kanlıköy) and D#6 (Ergazi Sayadere), respectively. Further, it was found that the floating structure (North orientation—tilt 6°) produced the maximum AEP for all sites. Figures S6 and S7 in the Supplementary Materials show the annual energy production from the proposed FPV at various orientations angles for the second (covered area 50%) and third (covered area 75%) scenarios, respectively. It was observed that the covered area of an FPV increases with power generation. This is because more solar panels can be fitted due to the greater covered area, which instantly increases the production and capacity of the system. Moreover, it is found that the CF value is within the range of 13.87–19.81%, as shown in Table 8.

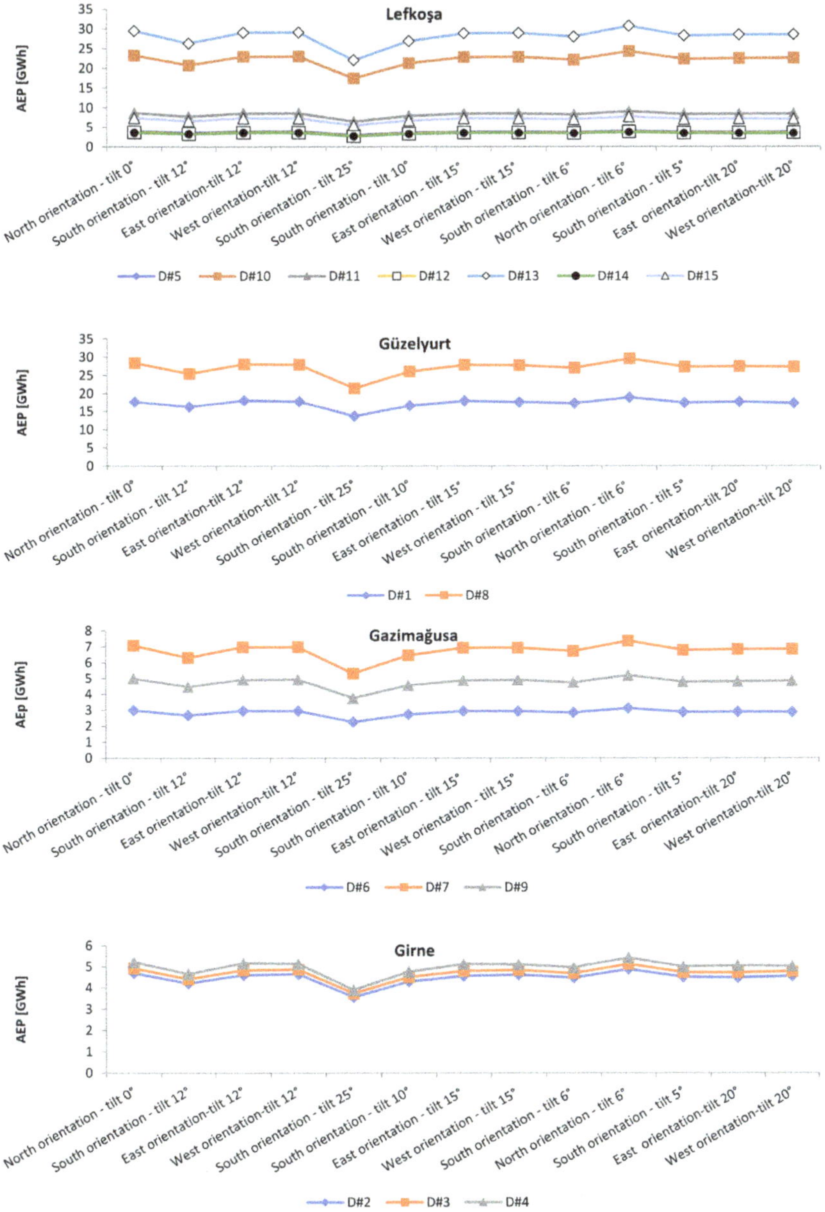

Figure 10. Annual energy production with various floating PV structures at different locations for the first scenario (covered area of 25%).

Table 8. Annual capacity factor [%] with various floating PV structures at different locations.

Site	S#1	S#2	S#3	S#4	S#5	S#6	S#7	S#8	S#9	S#10	S#11	S#12	S#13
D#1	18.48	17.04	18.89	18.62	14.31	17.41	18.78	18.45	18.11	19.80	18.27	18.54	18.14
D#2	18.31	16.42	17.94	18.14	13.87	16.77	17.79	18.02	17.43	19.04	17.59	17.51	17.79
D#3	18.33	16.44	18.00	18.13	13.88	16.79	17.86	18.01	17.45	19.07	17.60	17.60	17.78
D#4	18.92	16.88	18.72	18.64	14.15	17.25	18.60	18.51	17.97	19.73	18.13	18.38	18.27
D#5	18.63	16.62	18.33	18.44	13.94	16.99	18.19	18.33	17.69	19.43	17.86	17.93	18.11
D#6	18.57	16.61	18.36	18.29	13.98	16.97	18.23	18.16	17.66	19.34	17.82	18.01	17.92
D#7	18.53	16.56	18.29	18.29	13.92	16.92	18.16	18.17	17.61	19.31	17.77	17.93	17.95
D#8	18.91	16.93	18.66	18.59	14.24	17.30	18.52	18.45	17.99	19.67	18.15	18.26	18.18
D#9	18.47	16.50	18.19	18.28	13.87	16.86	18.05	18.17	17.54	19.25	17.71	17.80	17.97
D#10	18.77	16.74	18.51	18.54	14.04	17.12	18.37	18.42	17.82	19.57	17.99	18.13	18.20
D#11	18.74	16.71	18.47	18.53	14.01	17.08	18.34	18.41	17.79	19.54	17.95	18.10	18.20
D#12	18.64	16.62	18.36	18.41	13.94	16.99	18.22	18.29	17.69	19.43	17.86	17.97	18.07
D#13	18.84	16.80	18.58	18.60	14.09	17.18	18.44	18.48	17.89	19.64	18.06	18.20	18.25
D#14	18.84	16.80	18.58	18.60	14.09	17.18	18.44	18.48	17.89	19.64	18.06	18.20	18.25
D#15	19.00	16.95	18.82	18.69	14.20	17.32	18.70	18.55	18.04	19.81	18.21	18.49	18.29

S: Structure of FPV; S#1: North orientation—tilt 0°; S#2: South orientation—tilt 12°; S#3: East orientation—tilt 12°; S#4: West orientation—tilt 12°; S#5: South orientation—tilt 25°; S#6: South orientation—tilt 10°; S#7: East orientation—tilt 15°; S#8: West orientation—tilt 15°; S#9: South orientation—tilt 6°; S#10: North orientation—tilt 6°; S#11: South orientation—tilt 5°; S#12: East orientation—tilt 20°; S#13:West orientation—tilt 20°.

When utilizing the OceanSun S6 floating PV system structure, which is specified with a north orientation and a tilt angle of 6°, the maximum value of CF is observed at D#1 (Akdeniz). The SolSyn S25 floating PV system structure, which is designed with a south orientation and a tilt angle of 25°, records the highest value of CF at D#2 (Arapköy Ayanidere), as shown in Table 8. It should be noted that the CF of a solar power system is simply an assessment of the system's efficiency in producing electricity relative to its maximum output potential during a specific duration. CF is dependent on the overall performance of the system, shade, system orientation, and solar radiation availability. The CF is independent of the surface area covered by PV panels. The surface area changes the overall amount of electricity produced but does not affect the value of CF.

3.4. Performance Assessment of FPV with Various PV Technologies

After finding the best floating PV system structure (OceanSun S6), which is specified with a north orientation and a tilt angle of 6°, the performance of various scaled FPV systems with four PV technologies in terms of AEP, CF, SPP, energy production cost, and annual GHG emission reduction was evaluated using mathematical modeling. Table 9 lists the PV technologies used in this study.

Table 9. Specification of various PV technologies.

PV Technologies	Model	Power [W]	Efficiency [%]	Frame Area [m^2]	Cost [USD/W]
Monocrystalline Silicon	JKM445N-54HL4R-V	445	22.27	1.998	0.39
Polycrystalline Silicon	AS-P72 310-345W	345	17.78	1.940	0.43
Cadmium Telluride	FS-6445/FS-6445A	445	18.00	2.457	0.50
Bifacial (Bifacial Gain: 30%)	CS3W-445MB-AG	445	25.9	2.234	0.70
Bifacial (Bifacial Gain: 20%)	CS3W-445MB-AG	445	23.9	2.234	0.65
Bifacial (Bifacial Gain: 10%)	CS3W-445MB-AG	445	21.9	2.234	0.60
Bifacial (Bifacial Gain: 5%)	CS3W-445MB-AG	445	20.9	2.234	0.55

The suggested systems' annual electricity generation (AEP) is displayed in Figure 11. The range of AEP values is found to be 2.5–38.7 GWh for a 25% covered area, 5.0–77.3 GWh for a 50% covered area, and 7.5–116.0 GWh for a 75% covered area. Moreover, it is observed that the highest AEP is recorded when the bifacial panel (Bifacial Gain: 30%) is employed. Additionally, using polycrystalline silicon panels results in the lowest AEP.

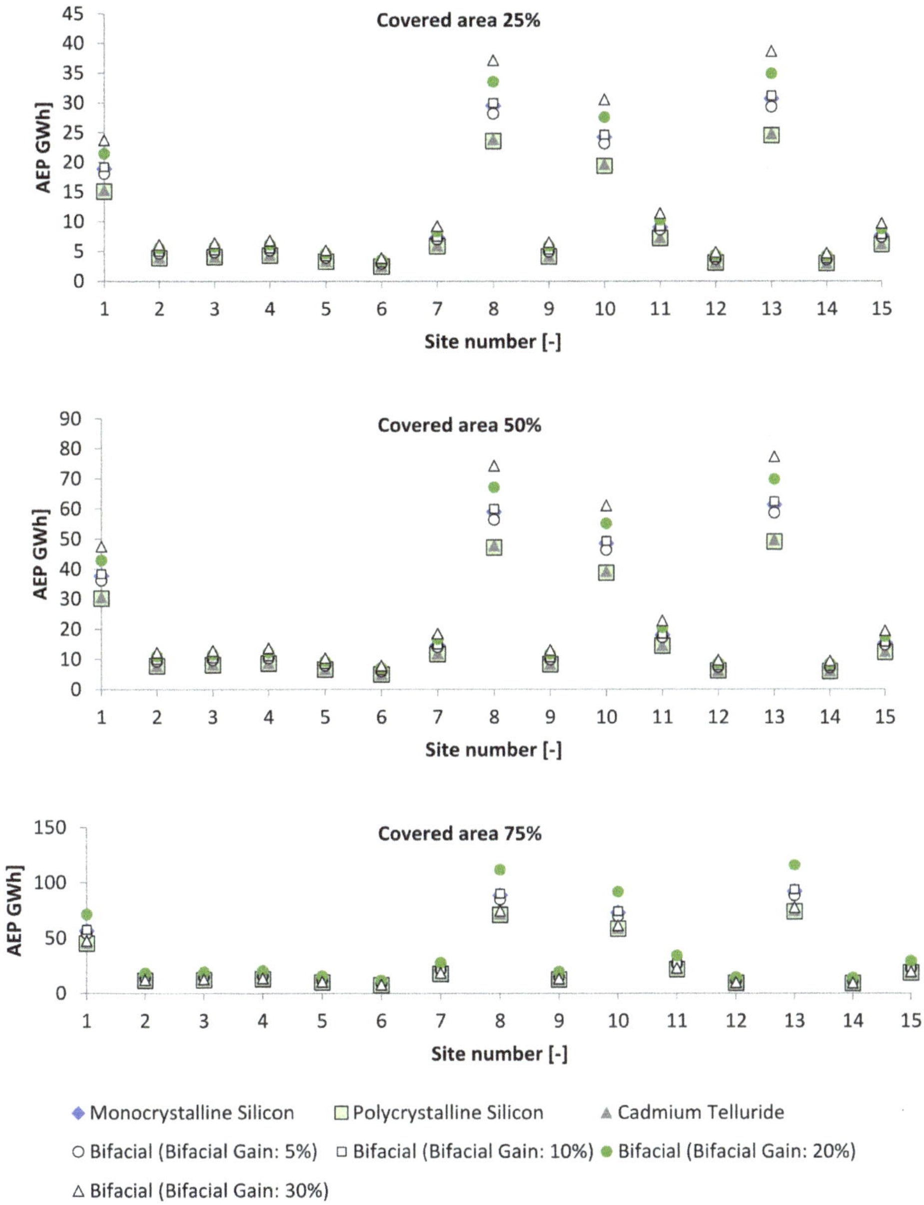

Figure 11. Annual energy production with various PV technologies and covered areas.

Furthermore, it is noticed that bifacial panels have a clear advantage over traditional panel technologies, particularly those with significant bifacial gains. Making effective utilization of reflected light increases the total quantity of energy produced. In terms of performance, high-bifacial-gain panels outperform monocrystalline silicon panels. Polycrystalline silicon and cadmium telluride panels have lower AEPs and are less effective at maximizing energy production, despite being considerably less costly to manufacture.

As previously stated, the percentage of surface area covered by PV panels has no impact on the CF. It does change the overall quantity of the power produced. Table 10 summarizes the annual value of CF for all proposed floating PV systems. It is found that the CF values of FPV systems varied from 18.60% (D#2: Arapköy Uzundere) to 27.37% (D#15: Yılmazköy Polatdere). Additionally, it is found that Cadmium telluride panels are more inefficient than monocrystalline, polycrystalline, and bifacial panels, as shown in Table 10. By using reflected light to maximize energy output, bifacial panels significantly raise capacity factors, with greater bifacial gains.

Table 10. Annual value of CF [%] for all selected sites with various PV technologies.

PV Technology	D#1	D#2	D#3	D#4	D#5	D#6	D#7	D#8	D#9	D#10	D#11	D#12	D#13	D#14	D#15
Monocrystalline	19.46	18.71	18.74	19.38	19.09	19.00	18.97	19.33	18.91	19.22	19.20	19.09	19.30	19.30	19.46
Polycrystalline	19.46	18.71	18.73	19.38	19.09	19.00	18.97	19.33	18.91	19.22	19.19	19.09	19.29	19.29	19.46
Cadmium Telluride	19.34	18.60	18.62	19.27	18.97	18.89	18.86	19.21	18.80	19.11	19.08	18.98	19.18	19.18	19.34
Bifacial (5%)	20.76	19.97	19.99	20.68	20.37	20.28	20.24	20.63	20.18	20.51	20.48	20.37	20.59	20.59	20.77
Bifacial (10%)	22.08	21.24	21.26	22.00	21.66	21.57	21.53	21.94	21.46	21.82	21.78	21.67	21.90	21.90	22.09
Bifacial (20%)	24.72	23.78	23.80	24.63	24.25	24.14	24.11	24.56	24.03	24.42	24.39	24.26	24.52	24.52	24.73
Bifacial (30%)	27.37	26.31	26.35	27.26	26.84	26.72	26.68	27.18	26.60	27.03	26.99	26.85	27.13	27.13	27.37

Furthermore, the main results regarding the economic performance of the proposed floating PV systems are tabulated in Table 11. It is found that the maximum (6.24 years) and minimum values (4.07 years) of SPP are recorded in D#2 (Arapköy Uzundere) and D#1 (Akdeniz), respectively, as shown in Table 11. Furthermore, when a Cadmium Telluride panel is employed in D#2 (Arapköy Uzundere), the maximum value of 0.025 USD/kWh is found, while the lowest value of LCOE is found in D#1 (Akdeniz) at 0.0163 USD/kWh when the Monocrystalline Silicon panel is used.

Table 11. Values of simple payback period and energy production costs for all selected sites with various PV technologies.

PV Technology	D#1	D#2	D#3	D#4	D#5	D#6	D#7	D#8	D#9	D#10	D#11	D#12	D#13	D#14	D#15
Simple payback period [Year]															
Monocrystalline	4.07	4.23	4.23	4.09	4.15	4.17	4.18	4.10	4.19	4.12	4.13	4.15	4.11	4.11	4.07
Polycrystalline	4.37	4.54	4.53	4.38	4.45	4.47	4.48	4.39	4.49	4.42	4.43	4.45	4.40	4.40	4.37
Cadmium Telluride	6.00	6.24	6.23	6.02	6.12	6.14	6.15	6.04	6.17	6.07	6.08	6.12	6.05	6.05	6.00
Bifacial (5%)	5.44	5.66	5.66	5.47	5.55	5.58	5.58	5.48	5.60	5.51	5.52	5.55	5.49	5.49	5.44
Bifacial (10%)	5.46	5.68	5.67	5.48	5.57	5.59	5.60	5.50	5.62	5.53	5.54	5.57	5.51	5.51	5.46
Bifacial (20%)	5.18	5.39	5.38	5.20	5.28	5.31	5.31	5.22	5.33	5.24	5.25	5.28	5.23	5.23	5.18
Bifacial (30%)	4.96	5.15	5.15	4.98	5.05	5.08	5.08	4.99	5.10	5.02	5.02	5.05	5.00	5.00	4.96
Energy production cost [USD/kWh]															
Monocrystalline	0.0163	0.0169	0.0169	0.0164	0.0166	0.0167	0.0167	0.0164	0.0168	0.0165	0.0165	0.0166	0.0164	0.0164	0.0163
Polycrystalline	0.0175	0.0182	0.0181	0.0175	0.0178	0.0179	0.0179	0.0176	0.0180	0.0177	0.0177	0.0178	0.0176	0.0176	0.0175
Cadmium Telluride	0.0240	0.0250	0.0249	0.0241	0.0245	0.0246	0.0246	0.0242	0.0247	0.0243	0.0243	0.0245	0.0242	0.0242	0.0240
Bifacial (5%)	0.0218	0.0226	0.0226	0.0219	0.0222	0.0223	0.0223	0.0219	0.0224	0.0220	0.0221	0.0222	0.0220	0.0220	0.0218
Bifacial (10%)	0.0218	0.0227	0.0227	0.0219	0.0223	0.0224	0.0224	0.0220	0.0225	0.0221	0.0221	0.0223	0.0220	0.0220	0.0218
Bifacial (20%)	0.0207	0.0216	0.0215	0.0208	0.0211	0.0212	0.0213	0.0209	0.0213	0.0210	0.0210	0.0211	0.0209	0.0209	0.0207
Bifacial (30%)	0.0198	0.0206	0.0206	0.0199	0.0202	0.0203	0.0203	0.0200	0.0204	0.0201	0.0201	0.0202	0.0200	0.0200	0.0198

Figure 12 illustrates the total annual reduction of GHG emissions for each site. The floating PV system in D#13 (Kanlıköy) was found to have the highest overall GHG emission reduction, with reductions of 14804.63, 29609.27, and 44413.90 tCO2/year for covered areas of 25%, 50%, and 75%, respectively. This is followed by the D#8 (Gemikonağı Madendere) site project, with the lowest GHG reduction resulting from the D#6 (Ergazi Sayadere) site project.

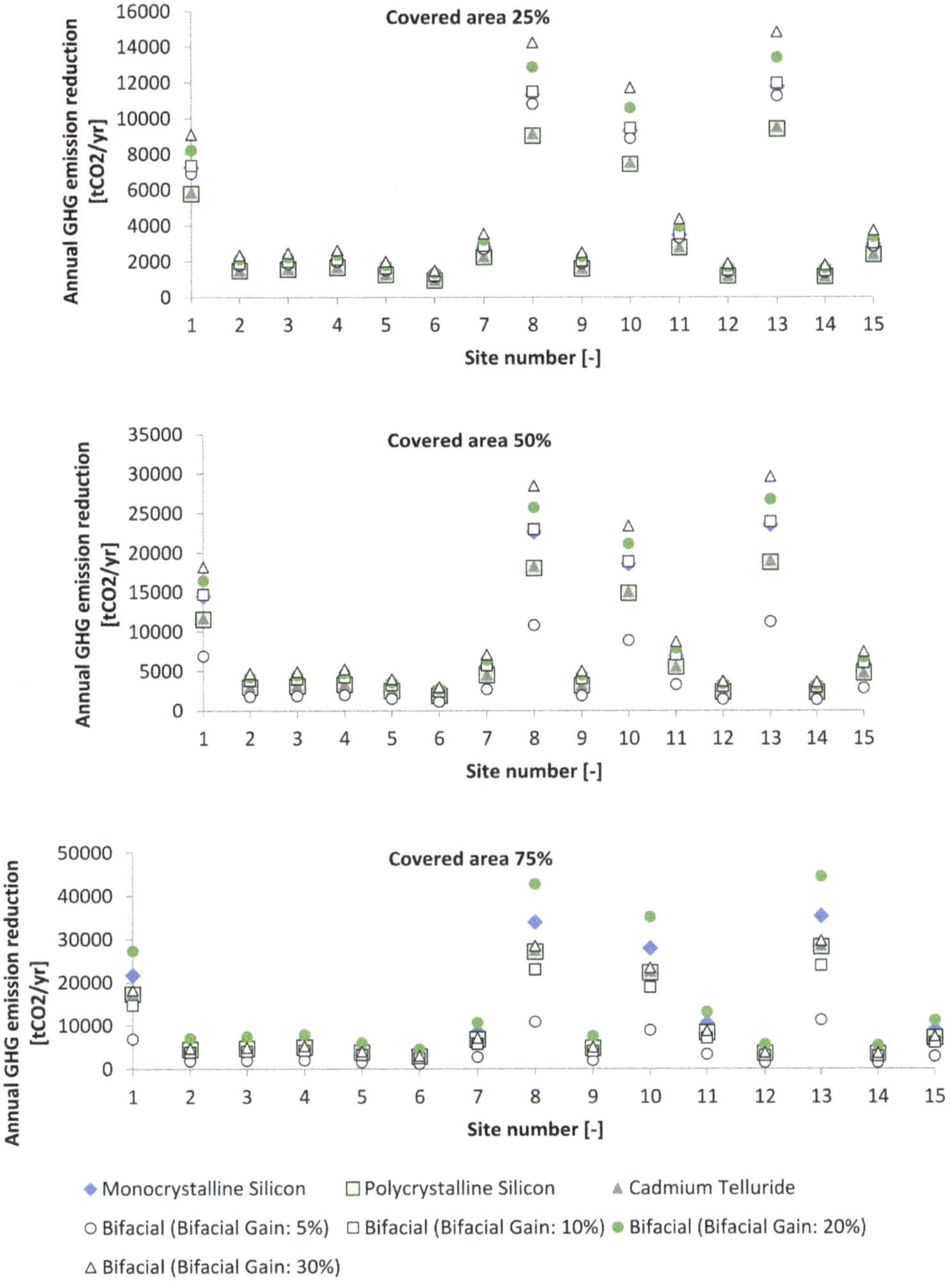

Figure 12. Annual GHG emission reduction with various PV technologies and covered area.

4. Discussion and Conclusions

In general, the growth of higher education in Northern Cyprus has resulted in a significant increase in energy consumption and the need for infrastructure to support this development [55,161]. Therefore, a solar energy system such as the FPV system can be an acceptable solution to produce green and clean energy that covers energy demand and achieves sustainable development goals. In the literature, understanding the distribution of solar radiation is crucial for effectively implementing solar power systems at particular locations. In general, A floating/fixed-tilt PV plant's performance is greatly influenced by the quantity and distribution of solar radiation at a particular location. Thus, an accurate comprehension of these variables facilitates the estimation of energy yield and the optimization of PV system design. According to Kebede [162], global horizontal irradiation (GHI) is the essential parameter to assess the energy generation for flat-plate photovoltaic systems. Frequency analysis can be utilized to determine the probability that different quantities of solar radiation will occur [163].

Therefore, the Johnson SB distribution, Pert distribution, Phased Bi-Weibull distribution, and Kumaraswamy distribution are added to the pool of parametric distributions for fitting the solar radiation data for the first time in this study, which was inspired by the list of PDFs used to analyze the characteristics of solar radiation (Table 2). These distribution functions are compared with common distribution functions (Normal, Gamma, Generalized Extreme Value, Log-normal, Weibull, and Logistic) to assess their suitability. The best-fit result of each site was taken as the distribution with the lowest value from the Kolmogorov–Smirnov (K–S) test and the Anderson–Darling (A–D) test. The results showed that the Johnson SB distribution, Burt distribution, and Coomaraswamy distribution performed better than the common distribution in terms of fitting performance and modeling of solar radiation statistics. Langat et al. [164] indicated that the frequency regime of the data series influences whether the distribution function is optimal. Therefore, the authors believe that the current results encourage further research to take these distributions into account when fitting global solar radiation data.

Moreover, the floating photovoltaic system for reservoirs is an emerging and novel technology that has the potential to significantly reduce evaporation, according to prior studies. Several models have been used in the literature to study the evaporation rate on the water's surface. Hence, Priestley–Taylor, deBruin–Keijman, Makkink, Penman–Monteith, mass transfer, and Hamon methods were employed to estimate the rate of evaporation at the selected reservoirs. The evaporation rate was estimated using the following data: temperature, humidity, wind speed, and solar radiation. It is crucial to note that the site's conditions play a role in identifying the best approach. It is found that the Priestley–Taylor and Hamon methods are expected to yield similar results because they are simple and highly sensitive to temperature and radiation. In contrast, the Penman method considers wind speed and vapor pressure deficits, resulting in higher evaporation estimates. The Penman method integrates mass transfer and energy balance approaches and is widely used for estimating evaporation, especially in the summer, producing greater estimates [165,166]. On the other hand, the Priestley–Taylor method assumes constant humidity, resulting in lower estimations, particularly in aerodynamically significant situations [167]. The Makkink and de Bruin–Keijman approaches also produce lower estimations due to their potential inability to consider humidity and wind speed variations [168,169]. The Hamon method, due to its temperature-based methodology, gives little consideration to meteorological factors like wind speed, humidity, or radiation, leading to different evaporation patterns than other methods [170]. These findings suggest that future studies should compare model data with measured data to evaluate the model's accuracy. A deeper comprehension of the data's applicability and dependability for assessing the evaporation rate in Northern Cyprus can be attained through this comparison analysis, which can also help reveal the constraints and uncertainties surrounding the data.

Furthermore, different scenarios of covering the reservoir's surface with an FPVS were studied and discussed. It was found that water savings grow dramatically as coverage increases from 0% to 100%, indicating that the system is successful in minimizing evaporation. For example, annual savings with 100% coverage can be as high as 6.21×10^5 m^3, as opposed to 0 m^3 in the absence of photovoltaic panels. This pattern makes the obvious relationship between covering area and evaporation decrease evident. Floating PV systems can be a useful tool for managing water resources in reservoirs and dams, since they represent a feasible choice for water conservation, as demonstrated by the significant savings that rise with increased coverage. These results can be supported by previous studies. For instance, according to Abd-Elhamid et al. [149], areas of the lake that are covered by 25%, 50%, 75%, and 100% can save around 2.1, 4.2, 6.3, 7.0, and 8.4×10^9 m^3/year. Ilgen et al. [171] estimated water savings of up to 5.9 billion m^3/year and a 49.7% evaporation reduction at 90% FPV occupancy.

Moreover, a technical and financial assessment was conducted on the installation of FPV systems with various PV technologies, floating structures, and capacity, taking into account that the covering area (25, 50, and 75 percent) of the reservoir determines the FV system's capacity. The results indicated that the floating PV system structure (OceanSun S6), which is specified with a north orientation and a tilt angle of 6°, produced the maximum performance compared to other floating PV system structures. It is found that the AEP values are within the range of 2.5–38.7 GWh for a 25% covered area, 5.0–77.3 GWh for a 50% covered area, and 7.5–116.0 GWh for a 75% covered area. Also, it is found that the FPV systems' annual CF values range from 18.50% to 27.37%. These findings can be confirmed by several scientific researchers, who examined the techno-economic performance of FPV systems in terms of CF. For instance, López et al. [141] found the CF value for the proposed FPV system in mainland Spain varied between 13.64% and 21.97%. Lopes et al. [172] found that the annual CF value for proposed FPV systems in Brazil is within the range of 18.8–20.0%. Al Saadi and Ghosh [173] estimated that the value of CF for the proposed FPV system in Oman was 23.3%. Consequently, based on the technical viability of the solar system, it is technically feasible to construct a floating solar photovoltaic system in each of the 15 sites in Northern Cyprus, since the values of AEP and CF there are compatible with the acceptable values. Based on the analysis, the percentage of reduction in electricity production from fossil fuels due to solar energy production from the proposed FPV system is illustrated in Figure 13. It is observed that the percentage of reduction in electricity production from fossil fuels can vary from 10.19% to 47.21% at 75% FPV occupancy. It should be noted that the monthly electrical consumption for the period of January–July 2024 is estimated to be 169.52, 153.49, 144.87, 128.27, 141.31, 214.33, and 247.70 GWh, respectively, according to the Cyprus Turkish Electricity Authority (KIB-TEK).

Furthermore, the feasibility of floating PV systems in Northern Cyprus was studied. It is found that, depending on the PV technology employed, the SPP value ranged between 4.07 and 6.24 years, while the value of energy production cost varied between 0.0163 and 0.025 USD/kWh. The annual reduction in GHG was estimated at 44,413.90 tons at 75% FPV occupancy.

In summary, this study is the first attempt to illustrate the importance of implementing an FPV system on the reservoir surfaces of Northern Cyprus by focusing on the most significant factors. The FPV system ultimately creates a new investment avenue, where several variables contribute to cost savings and environmental improvement.

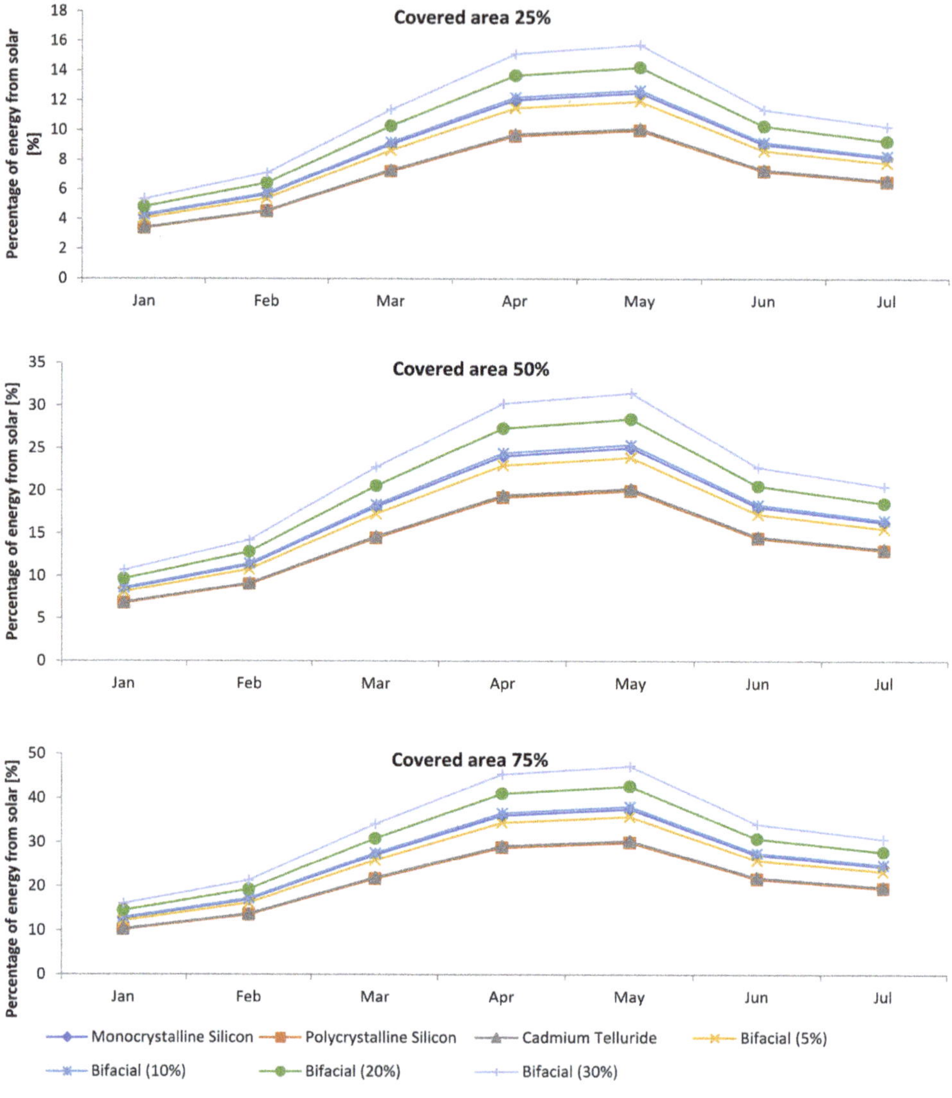

Figure 13. Monthly percentage of total energy production from FPV systems with various PV technologies and covered areas.

Supplementary Materials: The following supporting information can be downloaded at: https://www.mdpi.com/article/10.3390/en17215300/s1, Figure S1. Long-term averages of solar resource [25]; Figure S2. Probability density function for solar radiation data at Gemikonağı Madendere; Figure S3. Probability density function for solar radiation data at Ergazi Sayadere; Figure S4. Surface area of the selected lake, reservoir, and dams; Figure S5. Annual water losses at the selected locations; Figure S6. Annual energy production with various floating PV structures at different locations for first scenario (covered area 50%); Figure S7. Annual energy production with various floating PV structures at different locations for first scenario (covered area 75%); Table S1. Summary of previous studies focusing on solar energy potential in Northern Cyprus; Table S2. The most efficient distribution functions (DF) fitting the average daily GHI data at Gemikonağı Madendere; Table S3. The most efficient distribution functions (DF) fitting the average daily GHI data at Hamitköy Baştanlıkdere; Table S4.

The most efficient distribution functions (DF) fitting the average daily GHI data at Ergazi Sayadere; Table S5. Monthly average evaporation [mm/month] at various water body storage durign the period of 2010–2023; Table S6. Main specifications of the selected PV modules; Table S7. Main specifications of the selected inverter.

Author Contributions: Conceptualization, Y.K. and R.G.; methodology, Y.K. and R.G.; software, Y.K.; validation, Y.K. and R.G.; investigation, Y.K. and H.G.; writing—original draft preparation, Y.K. and R.G.; writing—review and editing, Y.K. and H.G.; supervision, Y.K. and H.G. All authors have read and agreed to the published version of the manuscript.

Funding: This research received no external funding.

Data Availability Statement: The original contributions presented in the study are included in the article/Supplementary Material, further inquiries can be directed to the corresponding author.

Acknowledgments: The authors would like to thank the Civil and Environmental Engineering Faculty at Near East University for their support and encouragement.

Conflicts of Interest: The authors declare no conflicts of interest.

References

1. Jawad, A.; Hasan, M.S.; Faruqui, M.F.I. Small-scale floating photovoltaic systems in university campus: A pathway to achieving SDG 7 goals in Bangladesh. *Energy Convers. Manag.* **2023**, *297*, 117722. [CrossRef]
2. Sánchez, A.; Zhang, Q.; Martín, M.; Vega, P. Towards a new renewable power system using energy storage: An economic and social analysis. *Energy Convers. Manag.* **2022**, *252*, 115056. [CrossRef]
3. Holechek, J.L.; Geli, H.M.; Sawalhah, M.N.; Valdez, R. A global assessment: Can renewable energy replace fossil fuels by 2050? *Sustainability* **2022**, *14*, 4792. [CrossRef]
4. Obaideen, K.; Olabi, A.G.; Al Swailmeen, Y.; Shehata, N.; Abdelkareem, M.A.; Alami, A.H.; Sayed, E.T. Solar energy: Applications, trends analysis, bibliometric analysis and research contribution to sustainable development goals (SDGs). *Sustainability* **2023**, *15*, 1418. [CrossRef]
5. Vivar, M.; Sharon, H.; Fuentes, M. Photovoltaic system adoption in water related technologies—A review. *Renew. Sustain. Energy Rev.* **2024**, *189*, 114004. [CrossRef]
6. Du, S.; Liang, C.; Sun, H.; Wang, K.; Wang, J.; Li, H.; Zhang, W. Evaluating the potential benefits of float solar photovoltaics through the water footprint recovery period. *J. Clean. Prod.* **2024**, *446*, 141399. [CrossRef]
7. Jin, Y.; Hu, S.; Ziegler, A.D.; Gibson, L.; Campbell, J.E.; Xu, R.; Zeng, Z. Energy production and water savings from floating solar photovoltaics on global reservoirs. *Nat. Sustain.* **2023**, *6*, 865–874. [CrossRef]
8. Charles Lawrence Kamuyu, W.; Lim, J.R.; Won, C.S.; Ahn, H.K. Prediction model of photovoltaic module temperature for power performance of floating PVs. *Energies* **2018**, *11*, 447. [CrossRef]
9. Choi, Y.K.; Lee, Y.G. A study on development of rotary structure for tracking-type floating photovoltaic system. *Int. J. Precis. Eng. Manuf.* **2014**, *15*, 2453–2460. [CrossRef]
10. Abdelgaied, M.; Kabeel, A.E.; Zeleňáková, M.; Abd-Elhamid, H.F. Floating Photovoltaic Plants as an Effective Option to Reduce Water Evaporation in Water-Stressed Regions and Produce Electricity: A Case Study of Lake Nasser, Egypt. *Water* **2023**, *15*, 635. [CrossRef]
11. Liu, L.; Sun, Q.; Li, H.; Yin, H.; Ren, X.; Wennersten, R. Evaluating the benefits of integrating floating photovoltaic and pumped storage power system. *Energy Convers. Manag.* **2019**, *194*, 173–185. [CrossRef]
12. Pringle, A.M.; Handler, R.M.; Pearce, J.M. Aquavoltaics: Synergies for dual use of water area for solar photovoltaic electricity generation and aquaculture. *Renew. Sustain. Energy Rev.* **2017**, *80*, 572–584. [CrossRef]
13. Taboada, M.E.; Cáceres, L.; Graber, T.A.; Galleguillos, H.R.; Cabeza, L.F.; Rojas, R. Solar water heating system and photovoltaic floating cover to reduce evaporation: Experimental results and modeling. *Renew. Energy* **2017**, *105*, 601–615. [CrossRef]
14. Cazzaniga, R.; Cicu, M.; Rosa-Clot, M.; Rosa-Clot, P.; Tina, G.M.; Ventura, C. Compressed air energy storage integrated with floating photovoltaic plant. *J. Energy Storage* **2017**, *13*, 48–57. [CrossRef]
15. Oliveira-Pinto, S.; Stokkermans, J. Assessment of the potential of different floating solar technologies–Overview and analysis of different case studies. *Energy Convers. Manag.* **2020**, *211*, 112747. [CrossRef]
16. Durković, V.; Đurišić, Ž. Analysis of the potential for use of floating PV power plant on the Skadar Lake for electricity supply of aluminium plant in Montenegro. *Energies* **2017**, *10*, 1505. [CrossRef]
17. Sunny, M.R.; Kabir, M.A.; Sarker, M.S.Z.; Aghaloo, K.; Ali, T. Feasibility study of floating solar photovoltaic systems using techno-economic assessment and multi-criteria decision-making method: A case study of Bangladesh. *Energy* **2024**, *310*, 133202. [CrossRef]
18. Ranjbaran, P.; Yousefi, H.; Gharehpetian, G.B.; Astaraei, F.R. A review on floating photovoltaic (FPV) power generation units. *Renew. Sustain. Energy Rev.* **2019**, *110*, 332–347. [CrossRef]

19. Micheli, L.; Talavera, D.L.; Tina, G.M.; Almonacid, F.; Fernández, E.F. Techno-economic potential and perspectives of floating photovoltaics in Europe. *Sol. Energy* **2022**, *243*, 203–214. [CrossRef]
20. Islam, M.I.; Jadin, M.S.; Mansur, A.A.; Kamari, N.A.M.; Jamal, T.; Hossain Lipu, M.S.; Shihavuddin, A.S.M. Techno-economic and carbon emission assessment of a large-scale floating solar Pv system for sustainable energy generation in support of Malaysia's renewable energy roadmap. *Energies* **2023**, *16*, 4034. [CrossRef]
21. Islam, M.I.; Maruf, M.H.; Al Mansur, A.; Ashique, R.H.; ul Haq, M.A.; Shihavuddin, A.S.M.; Jadin, M.S. Feasibility analysis of floating photovoltaic power plant in Bangladesh: A case study in Hatirjheel Lake, Dhaka. *Sustain. Energy Technol. Assess.* **2023**, *55*, 102994. [CrossRef]
22. Fereshtehpour, M.; Sabbaghian, R.J.; Farrokhi, A.; Jovein, E.B.; Sarindizaj, E.E. Evaluation of factors governing the use of floating solar system: A study on Iran's important water infrastructures. *Renew. Energy* **2021**, *171*, 1171–1187. [CrossRef]
23. Akçaba, S.; Eminer, F. Sustainable energy planning for the aspiration to transition from fossil energy to renewable energy in Northern Cyprus. *Heliyon* **2022**, *8*, e09813. [CrossRef] [PubMed]
24. Kibris Türk Elektrik Kurumu. Available online: https://www.kibtek.com/ (accessed on 5 August 2024).
25. Solargis. Global Solar Atlas. Available online: https://globalsolaratlas.info/ (accessed on 5 June 2024).
26. Prăvălie, R.; Patriche, C.; Bandoc, G. Spatial assessment of solar energy potential at global scale. A geographical approach. *J. Clean. Prod.* **2019**, *209*, 692–721. [CrossRef]
27. Agathokleous, R.A.; Kalogirou, S.A. PV roofs as the first step towards 100% RES electricity production for Mediterranean islands: The case of Cyprus. *Smart Energy* **2021**, *4*, 100053. [CrossRef]
28. Shemshaki, M.S. Efficiency of Solar Domestic Hot Water System: The Case of Famagusta Northern Cyprus. Master's Thesis, Eastern Mediterranean University (EMU)-Doğu Akdeniz Üniversitesi (DAÜ), Gazimağusa, Turkey, 2016.
29. Maltini, F.; Minder, R. The Serhatköy photovoltaic power plant and the future of renewable energy on the Turkish Republic of Northern Cyprus: Integrating Solar Photovoltaic and Wind Farms into Electricity Transmission and Distribution Networks. In *Eco-Friendly Innovation in Electricity Transmission and Distribution Networks*; Woodhead Publishing: Cambridge, MA, USA, 2015.
30. Egelioglu, F.; Mohamad, A.; Guven, H. Economic variables and electricity consumption in Northern Cyprus. *Energy* **2001**, *26*, 355–362. [CrossRef]
31. Al-Ghussain, L.; Taylan, O.; Samu, R.; Fahrioglu, M. Techno-economic analysis of photovoltaic-hydrogen fuel cell/pumped hydro storage system for micro grid applications: Case study in Cyprus. In Proceedings of the 2018 International Conference on Photovoltaic Science and Technologies (PVCon), Ankara, Turkey, 4–6 July 2018; pp. 1–6.
32. Dakyen, M.M.; Dagbasi, M.; Özdenefe, M. Energy models for cost-optimal analysis: Development and calibration of residential reference building models for Northern Cyprus. *Indoor Built Environ.* **2022**, *31*, 657–681. [CrossRef]
33. Northern Cyprus Renewable Energy Board (YEKKurulu). Announcements. Available online: http://yek.gov.ct.tr (accessed on 2 May 2024).
34. Abbasoglu, S. Techno-economic and environmental analysis of PV power plants in Northern Cyprus. *Energy Educ. Sci. Technol. Part A* **2011**, *28*, 357–368.
35. Yenen, M.; Ercan, F.; Fahrioglu, M. Solar thermal system analysis of Northern Cyprus. In Proceedings of the EECS'12—7th International Symposium on Electrical and Computer Systems, Lefke, Northern Cyprus, 29–30 November 2012.
36. Okoye, C.O.; Atikol, U. A parametric study on the feasibility of solar chimney power plants in North Cyprus conditions. *Energy Convers. Manag.* **2014**, *80*, 178–187. [CrossRef]
37. Maltini, F.; Minder, R. The Serhatköy photovoltaic power plant and the future of renewable energy on the Turkish Republic of Northern Cyprus. In *Eco-Friendly Innovation in Electricity Transmission and Distribution Networks*; Woodhead Publishing: Cambridge, UK, 2015; pp. 377–402.
38. Ozerdem, O.C.; Tackie, S.; Biricik, S. Performance evaluation of Serhatkoy (1.2 MW) PV power plant. In Proceedings of the 9th International Conference on Electrical and Electronics Engineering (ELECO 2015), Bursa, Turkey, 26–28 November 2015.
39. Kamali, S. Feasibility analysis of standalone photovoltaic electrification system in a residential building in Cyprus. *Renew. Sustain. Energy Rev.* **2016**, *65*, 1279–1284. [CrossRef]
40. Dagbasi, M.; Bamisile, O.; Adii, C. The techno-economic comparison of solar power generation methods for Turkish Republic of North Cyprus. In Proceedings of the 2016 HONET-ICT, Nicosia, Cyprus, 13–14 October 2016; pp. 17–23.
41. Cabacaba, N.; Abbasoğlu, S. Evaluation of Wind–Solar Hybrid System for a Household in Northern Cyprus. In *Towards 100% Renewable Energy*; Springer: Cham, Switzerland, 2017; pp. 313–321.
42. Kassem, Y.; Gökçekuş, H.; Çamur, H. Economic assessment of renewable power generation based on wind speed and solar radiation in urban regions. *Glob. J. Environ. Sci. Manag.* **2018**, *4*, 465–482.
43. Kassem, Y.; Gökçekuş, H. GHG Emissions and Energy Performance of 1MW Grid-Connected Solar PV Plant at Lefke in Northern Cyprus: Case Study. *Disaster Sci. Eng.* **2018**, *4*, 90–98.
44. Ouria, M.; Sevinc, H. Evaluation of the potential of solar energy utilization in Famagusta, Cyprus. *Sust. Cities Society* **2018**, *37*, 189–202. [CrossRef]
45. Al-Ghussain, L.; Abujubbeh, M.; Fahrioglu, M. Assessment of PV investments in Northern Cyprus. In Proceedings of the 16th International Conference on Clean Energy, Famagusta, Cyprus, 9–11 May 2018; pp. 9–11.
46. Hastunç, M.; Tekbıyık-Ersoy, N. Optimizing Residential Renewable Energy Utilization in North Cyprus: Case Study of Solar Energy. In Proceedings of the 16th International Conference on Clean Energy, Famagusta, Cyprus, 9–11 May 2018.

47. Kassem, Y.; Al Zoubi, R.; Gökçekuş, H. The possibility of generating electricity using small-scale wind turbines and solar photovoltaic systems for households in Northern Cyprus: A comparative study. *Environments* **2019**, *6*, 47. [CrossRef]
48. Kassem, Y.; Gokcekus, H.; Alsayas, S.M. Freestanding PV solar system—Example of Lefke town in Northern Cyprus. *Int. J. Appl. Eng. Res.* **2019**, *14*, 2522–2526.
49. Ogbeba, J.; Hoskara, E. The Evaluation of Single-Family Detached Housing Units in terms of Integrated Photovoltaic Shading Devices: The Case of Northern Cyprus. *Sustainability* **2019**, *11*, 593. [CrossRef]
50. Kassem, Y.; Gökçekuş, H.; Güvensoy, A. Solar Potential assessment in Near East University, Northern Cyprus. *Int. J. Eng. Res. Technol.* **2019**, *12*, 3061–3069.
51. Gökçekuş, H.; Kassem, Y.; Abdi, S. Simulation and performance analysis of 110 KWP grid-connected photovoltaic (PV) systems for residential building in Northern Cyprus. *Int. J. Innov. Technol. Explor. Eng.* **2019**, *8*, 3082–3091.
52. Oner, H. Economic Feasibility Assessment of Solar Powered Seawater Desalination Plants: Unconventional Fresh Water Supply for Guzelyurt, Northern Cyprus. Master's Thesis, Middle East Technical University, Nicosia, Cyprus, 2019.
53. Al-Ghussain, L.; Taylan, O. Sizing methodology of a PV/wind hybrid system: Case study in cyprus. *Environ. Prog. Sustain. Energy* **2019**, *38*, e13052. [CrossRef]
54. Kassem, Y.; Gokcekus, H.; Filitoglu, Ü.B. Performance Characteristics of Building Integrated and Freestanding Photovoltaic System with Various PV Technologies and Angles: A Case Study in NEU Grand Library, North Nicosia. *J. Eng. Appl. Sci.* **2020**, *15*, 1027–1042. [CrossRef]
55. Kassem, Y.; Çamur, H.; Alhuoti, S.M.A. Solar energy technology for Northern Cyprus: Assessment, statistical analysis, and feasibility study. *Energies* **2020**, *13*, 940. [CrossRef]
56. Al-Turjman, F.; Qadir, Z.; Abujubbeh, M.; Batunlu, C. Feasibility analysis of solar photovoltaic-wind hybrid energy system for household applications. *Comput. Electr. Eng.* **2020**, *86*, 106743. [CrossRef]
57. Kassem, Y.; Gökçekuş, H.; Güvensoy, A. Techno-economic feasibility of grid-connected solar PV system at Near East University hospital, Northern Cyprus. *Energies* **2021**, *14*, 7627. [CrossRef]
58. Kassem, Y.; Gökçekuş, H.; Gökçekuş, R. Economic Feasibility of Large-Scale Renewable Energy Projects in Mountain Location, Northern Cyprus. In *International Conference on Natural Resources and Sustainable Environmental Management*; Springer International Publishing: Cham, Switzerland, 2021; pp. 66–71.
59. Kassem, Y.; Gökçekuş, H.; Iravanian, A.; Gökçekuş, R. Predictive suitability of renewable energy for desalination plants: The case of güzelyurt region in northern Cyprus. *Model. Earth Syst. Environ.* **2022**, *8*, 1–21. [CrossRef]
60. Kassem, Y.; Gokcekus, H.; Aljatlawe, A. Utilization of Solar Energy for Electric Vehicle Charging and the Energy Consumption of Residential Buildings in Northern Cyprus: A Case Study. *Eng. Technol. Appl. Sci. Res.* **2023**, *13*, 11598–11607. [CrossRef]
61. Kassem, Y.; Gokcekus, H.; Hamad, O.A.M.; Fayid, F.M.B. Economic viability of a 6.5 kW off-grid solar PV with various sun-tracking systems in Northern Cyprus: A case study. *Eng. Technol. Appl. Sci. Res.* **2023**, *13*, 10608–10621. [CrossRef]
62. Ünlükuş, U.A. Economic Feasiblity of Floating Solar pv and Land-Based Solar pv in Northern Cyprus. Master's Thesis, Middle East Technical University, Ankara, Türkiye, 2023.
63. Elkiran, G.; Aslanova, F.; Hiziroglu, S. Effluent water reuse possibilities in Northern Cyprus. *Water* **2019**, *11*, 191. [CrossRef]
64. Yildiz, D.; Çakmak, C. *Peace River to Cyprus from Turkey*; Hydropolitics Academy: Ankara, Turkey, 2014; pp. 1–34.
65. Gökçekuş, H.; Olgun, E. Cenozoic stratigraphy of the Güzelyurt Basin, Turkish Republic of Northern Cyprus. *TAPG Bull.* **1993**, *5*, 57–68.
66. Gökçekuş, H.; Doyuran, V. Evaluation of hydrochemical data from Güzelyurt Aquifer (Turkish Republic of Northern Cyprus) through Cluster Analysis. *Turk. J. Earth Sci.* **1995**, *4*, 1–10.
67. Gökçekuş, H.; Doyuran, V. Groundwater budget for the Güzelyurt basin (North Cyprus). *Proc. Int. Conf. Water Probl. Mediterr. Ctries.* **1997**, *11*, 779–787.
68. Gökçekuş, H.; Nourani, V. Lessons from Groundwater Quantity and Quality Problems in the Güzelyurt Coastal Aquifer, North Cyprus. In *Advances in Environmental Research*; Nova Science Publishers, Inc.: New York, NY, USA, 2018; Volume 64.
69. Gökçekuş, H.; Doyuran, V. Evaluation of domestic and agricultural quality of the groundwater in the Güzelyurt Basin (North Cyprus). *Proc. Int. Conf. Water Probl. Mediterr. Ctries.* **1997**, *11*, 1305–1314.
70. Arslan, B.; Akün, E. Management, contamination, and quality evaluation of groundwater in North Cyprus. *Agric. Water Manag.* **2019**, *222*, 1–11. [CrossRef]
71. Gökçekuş, H.; Doyuran, V. Groundwater budget for the Güzelyurt Aquifer: An example for an overdraft. In *Proceedings of the International Conference on Water Problems in the Mediterranean Countries*; NEU: Lefkoşa, Cyprus, 1998; Volume 5, pp. 779–787.
72. Gökçekuş, H.; Türker, U.; Sözen, S.; Orhon, D. Water management difficulties with limited and contaminated water resources. In Proceedings of the International Conference on the Environmental Problem of the Mediterranean Region, Nicosia, Cyprus, 12–15 April 2002; pp. 1241–1249.
73. Payab, A.H.; Türker, U. Comparison of standardized meteorological indices for drought monitoring at northern part of Cyprus. *Environ. Earth Sci.* **2019**, *78*, 309. [CrossRef]
74. Park, E.J. Strategy of Water Distribution for Sustainable Community: Who Owns Water in Divided Cyprus? *Sustainability* **2020**, *12*, 8978. [CrossRef]
75. Kahramanoğlu, I.; Usanmaz, S.; Alas, T. Water footprint and irrigation use efficiency of important crops in Northern Cyprus from an environmental, economic and dietary perspective. *Saudi J. Biol. Sci.* **2020**, *27*, 134–141. [CrossRef]

76. Elkiran, G.; Ongul, Z. Implications of excessive water withdrawals to the environment of Northern Cyprus. *Water Environ. J.* **2009**, *23*, 245. [CrossRef]
77. Sofroniou, A.; Bishop, S. Water Scarcity in Cyprus: A Review and Call for Integrated Policy. *Water* **2014**, *6*, 2898–2928. [CrossRef]
78. Feldman, D.; Barbose, G.; Margolis, R.; Bolinger, M.; Chung, D.; Fu, R.; Wiser, R. Photovoltaic System Pricing Trends: Historical, Recent, and Near-Term Projections 2015 Edition. 2015. Available online: https://escholarship.org/content/qt9pc3x32t/qt9pc3x32t.pdf (accessed on 15 October 2024).
79. El Hammoumi, A.; Chalh, A.; Allouhi, A.; Motahhir, S.; El Ghzizal, A.; Derouich, A. Design and construction of a test bench to investigate the potential of floating PV systems. *J. Clean. Prod.* **2021**, *278*, 123917. [CrossRef]
80. Marion, W.F. *Overview of the PV Module Model in PVWatts*; National Renewable Energy Laboratory: Golden, CO, USA, 2010.
81. Skoplaki, E.; Palyvos, J.A. On the temperature dependence of photovoltaic module electrical performance: A review of efficiency/power correlations. *Sol. Energy* **2009**, *83*, 614–624. [CrossRef]
82. Nisar, H.; Janjua, A.K.; Hafeez, H.; Shahzad, N.; Waqas, A. Thermal and electrical performance of solar floating PV system compared to on-ground PV system-an experimental investigation. *Sol. Energy* **2018**, *241*, 231–247. [CrossRef]
83. Choi, Y.K.; Choi, W.S.; Lee, J.H. Empirical research on the efficiency of floating PV systems. *Sci. Adv. Mater.* **2016**, *8*, 681–685. [CrossRef]
84. Boduch, A.; Mik, K.; Castro, R.; Zawadzki, P. Technical and economic assessment of a 1 MWp floating photovoltaic system in Polish conditions. *Renew. Energy* **2022**, *196*, 983–994. [CrossRef]
85. Liu, H.; Krishna, V.; Lun Leung, J.; Reindl, T.; Zhao, L. Field experience and performance analysis of floating PV technologies in the tropics. *Prog. Photovolt. Res. Appl.* **2018**, *26*, 957–967. [CrossRef]
86. Goswami, A.; Sadhu, P.K. Degradation analysis and the impacts on feasibility study of floating solar photovoltaic systems. *Sustain. Energy Grids Netw.* **2021**, *26*, 100425. [CrossRef]
87. Junianto, B.; Dewi, T.; Sitompul, C.R. Development and Feasibility Analysis of Floating Solar Panel Application in Palembang, South Sumatra. In *Journal of Physics: Conference Series*; IOP Publishing: Cambridge, MA, USA, 2020; Volume 1500, p. 012016.
88. Essak, L.; Ghosh, A. Floating photovoltaics: A review. *Clean Technol.* **2022**, *4*, 752–769. [CrossRef]
89. Badhoutiya, A. Advancements in PV Technology-Floating Photovoltaics. In Proceedings of the 2023 5th International Conference on Smart Systems and Inventive Technology (ICSSIT), Tirunelveli, India, 23–25 January 2023; pp. 382–385.
90. Ramanan, C.J.; Lim, K.H.; Kurnia, J.C.; Roy, S.; Bora, B.J.; Medhi, B.J. Towards sustainable power generation: Recent advancements in floating photovoltaic technologies. *Renew. Sustain. Energy Rev.* **2024**, *194*, 114322.
91. Koondhar, M.A.; Albasha, L.; Mahariq, I.; Graba, B.B.; Touti, E. Reviewing floating photovoltaic (FPV) technology for solar energy generation. *Energy Strategy Rev.* **2024**, *54*, 101449. [CrossRef]
92. Ghigo, A.; Faraggiana, E.; Sirigu, M.; Mattiazzo, G.; Bracco, G. Design and analysis of a floating photovoltaic system for offshore installation: The case study of Lampedusa. *Energies* **2022**, *15*, 8804. [CrossRef]
93. Li, Z.; Chen, D.; Feng, X.; Chen, J.F. Hydroelastic analysis and structural design of a modular floating structure applying ultra-high performance fiber-reinforced concrete. *Ocean Eng.* **2023**, *277*, 114266. [CrossRef]
94. Ravichandran, N.; Ravichandran, N.; Panneerselvam, B. Review on the structural components of floating photovoltaic covering systems. In *Intelligent Manufacturing and Energy Sustainability: Proceedings of ICIMES 2021*; Springer: Singapore, 2021; pp. 125–133.
95. Rosa-Clot, M. *Floating PV Plants*; Academic Press: Cambridge, MA, USA, 2020.
96. Claus, R.; López, M. A methodology to assess the dynamic response and the structural performance of floating photovoltaic systems. *Sol. Energy* **2023**, *262*, 111826. [CrossRef]
97. Claus, R.; López, M. Key issues in the design of floating photovoltaic structures for the marine environment. *Renew. Sustain. Energy Rev.* **2022**, *164*, 112502. [CrossRef]
98. Ghosh, A. A comprehensive review of water based PV: Flotavoltaics, under water, offshore & canal top. *Ocean Eng.* **2023**, *281*, 115044.
99. Delacroix, S.; Bourdier, S.; Soulard, T.; Elzaabalawy, H.; Vasilenko, P. Experimental modelling of a floating solar power plant array under wave forcing. *Energies* **2023**, *16*, 5198. [CrossRef]
100. Ravichandran, N.; Fayek, H.H.; Rusu, E. Emerging floating photovoltaic system—Case studies high dam and Aswan reservoir in Egypt. *Processes* **2021**, *9*, 1005. [CrossRef]
101. Chiang, C.H.; Young, C.H. An engineering project for a flood detention pond surface-type floating photovoltaic power generation system with an installed capacity of 32,600.88 kWp. *Energy Rep.* **2022**, *8*, 2219–2232. [CrossRef]
102. Trapani, K.; Millar, D.L. Proposing offshore photovoltaic (PV) technology to the energy mix of the Maltese islands. *Energy Convers. Manag.* **2013**, *67*, 18–26. [CrossRef]
103. Kim, S.H.; Yoon, S.J.; Choi, W. Design and construction of 1 MW class floating PV generation structural system using FRP members. *Energies* **2017**, *10*, 1142. [CrossRef]
104. Sukarso, A.P.; Kim, K.N. Cooling effect on the floating solar PV: Performance and economic analysis on the case of west Java province in Indonesia. *Energies* **2020**, *13*, 2126. [CrossRef]
105. Cazzaniga, R.; Rosa-Clot, M.; Rosa-Clot, P.; Tina, G.M. Integration of PV floating with hydroelectric power plants. *Heliyon* **2019**, *5*, e01918. [CrossRef]
106. Huang, G.; Tang, Y.; Chen, X.; Chen, M.; Jiang, Y. A comprehensive review of floating solar plants and potentials for offshore applications. *J. Mar. Sci. Eng.* **2023**, *11*, 2064. [CrossRef]

107. Lopes, M.P.C.; de Andrade Neto, S.; Branco, D.A.C.; de Freitas, M.A.V.; da Silva Fidelis, N. Water-energy nexus: Floating photovoltaic systems promoting water security and energy generation in the semiarid region of Brazil. *J. Clean. Prod.* **2020**, *273*, 122010. [CrossRef]
108. Setiya, P.; Singh, M.; Nain, A.S. Evaluating the performance of Grid IMD, NASA POWER, and MarkSim timeseries weather dataset for Uttarakhand Climatic Condition. *Theor. Appl. Climatol.* **2024**, *155*, 2657–2668. [CrossRef]
109. Quansah, A.D.; Dogbey, F.; Asilevi, P.J.; Boakye, P.; Darkwah, L.; Oduro-Kwarteng, S.; Mensah, P. Assessment of solar radiation resource from the NASA-POWER reanalysis products for tropical climates in Ghana towards clean energy application. *Sci. Rep.* **2022**, *12*, 10684. [CrossRef] [PubMed]
110. Aboelkhair, H.; Morsy, M.; El Afandi, G. Assessment of agroclimatology NASA POWER reanalysis datasets for temperature types and relative humidity at 2 m against ground observations over Egypt. *Adv. Space Res.* **2019**, *64*, 129–142. [CrossRef]
111. Ferreira, M.M.; Santos, J.A.; da Silva, L.R.; Abrahao, R.; Gomes FDS, V.; Braz, H.D.M. A new index to evaluate renewable energy potential: A case study on solar, wind and hybrid generation in Northeast Brazil. *Renew. Energy* **2023**, *217*, 119182. [CrossRef]
112. Hamad, J.; Ahmad, M.; Zeeshan, M. Solar energy resource mapping, site suitability and techno-economic feasibility analysis for utility scale photovoltaic power plants in Afghanistan. *Energy Convers. Manag.* **2024**, *303*, 118188. [CrossRef]
113. Owolabi, A.B.; Nsafon, B.E.K.; Roh, J.W.; Suh, D.; Huh, J.S. Validating the techno-economic and environmental sustainability of solar PV technology in Nigeria using RETScreen Experts to assess its viability. *Sustain. Energy Technol. Assess.* **2019**, *36*, 100542. [CrossRef]
114. Belkilani, K.; Ben Othman, A.; Besbes, M. Assessment of global solar radiation to examine the best locations to install a PV system in Tunisia. *Appl. Phys. A* **2018**, *124*, 122. [CrossRef]
115. Gairaa, K.; Bakelli, Y. Solar energy potential assessment in the Algerian south area: Case of Ghardaïa region. *J. Renew. Energy* **2013**, *2013*, 496348. [CrossRef]
116. Tahir, Z.R.; Asim, M. Surface measured solar radiation data and solar energy resource assessment of Pakistan: A review. *Renew. Sustain. Energy Rev.* **2018**, *81*, 2839–2861. [CrossRef]
117. Bhatia, S.C. *Advanced Renewable Energy Systems*; Woodhead: New Delhi, India, 2014.
118. Jiang, Z.; Li, R.; Li, A.; Ji, C. Runoff forecast uncertainty considered load adjustment model of cascade hydropower stations and its application. *Energy* **2018**, *158*, 693–708. [CrossRef]
119. Karipoğlu, F.; Koca, K.; İlbahar, E. Convenient Site Selection of a Floating PV Power Plant in Türkiye by using GIS-Fuzzy Analytical Hierarchy Process. *Environ. Sci. Pollut. Res.* **2024**, *31*, 23193–23210. [CrossRef] [PubMed]
120. Yilmaz, O.S.; Ateş, A.M.; Gülgen, F. A novel approach suggestion for assessing the impact of topographic shading on the estimation of the floating photovoltaic technical potential. *Energy* **2023**, *283*, 128479. [CrossRef]
121. Noorollahi, E.; Fadai, D.; Akbarpour Shirazi, M.; Ghodsipour, S.H. Land suitability analysis for solar farms exploitation using GIS and fuzzy analytic hierarchy process (FAHP)—A case study of Iran. *Energies* **2016**, *9*, 643. [CrossRef]
122. Gökmener, S.; Oğuz, E.; Deveci, M.; Göllü, K. Site selection for floating photovoltaic system on dam reservoirs using sine trigonometric decision making model. *Ocean Eng.* **2023**, *281*, 114820. [CrossRef]
123. Suh, J.; Brownson, J.R. Solar farm suitability using geographic information system fuzzy sets and analytic hierarchy processes: Case study of Ulleung Island, Korea. *Energies* **2016**, *9*, 648. [CrossRef]
124. Wang, Y. Application of TOPSIS and AHP in the Multi-Objective Decision-Making Problems. In *MATEC Web of Conferences*; EDP Sciences: Les Ulis, France, 2018; Volume 228, p. 05002.
125. Arthur, Y.D.; Gyamfi, K.B.; Appiah, S.K. Probability distributional analysis of hourly solar irradiation in Kumasi-Ghana. *Int. J. Bus. Soc. Res. (IJBSR)* **2013**, *3*, 63–75.
126. Ayodele, T.R. Determination of probability distribution function for modelling global solar radiation: Case study of Ibadan, Nigeria. *Int. J. Appl. Sci. Eng.* **2015**, *13*, 233–245.
127. Chang, T.P. Investigation on frequency distribution of global radiation using different probability density functions. *Int. J. Appl. Sci. Eng.* **2010**, *8*, 99–107.
128. Dukkipati, S.; Sankar, V.; Varma, P.S. Forecasting of solar irradiance using probability distributions for a PV System: A case study. *Int. J. Renew. Energy Res. (IJRER)* **2019**, *9*, 741–748.
129. Ettoumi, F.Y.; Mefti, A.; Adane, A.; Bouroubi, M.Y. Statistical analysis of solar measurements in Algeria using beta distributions. *Renew. Energy* **2002**, *26*, 47–67. [CrossRef]
130. Kam, O.M.; Noël, S.; Ramenah, H.; Kasser, P.; Tanougast, C. Comparative Weibull distribution methods for reliable global solar irradiance assessment in France areas. *Renew. Energy* **2021**, *165*, 194–210. [CrossRef]
131. Abiodun, C.I.; Anisiji, O.E. Performance Probability Distribution Function for Modelling Solar Radiation in South Southern Nigeria: A Case Study of Yenagoa. *J. Energy Res. Rev.* **2019**, *2*, 1–8. [CrossRef]
132. Abdulkarim, A.; Abdelkader, S.M.; John Morrow, D. Statistical analyses of wind and solar energy resources for the development of hybrid microgrid. In Proceedings of the 2nd International Congress on Energy Efficiency and Energy Related Materials (ENEFM2014) Proceedings, Oludeniz, Fethiye/Mugla, Turkey, 16–19 October 2015; pp. 9–14.
133. Shukla, K.N.; Rangnekar, S.; Sudhakar, K. Mathematical modelling of solar radiation incident on tilted surface for photovoltaic application at Bhopal, MP, India. *Int. J. Ambient Energy* **2016**, *37*, 579–588. [CrossRef]
134. Demain, C.; Journée, M.; Bertrand, C. Evaluation of different models to estimate the global solar radiation on inclined surfaces. *Renew. Energy* **2013**, *50*, 710–721. [CrossRef]

135. Othman, A.B.; Belkilani, K.; Besbes, M. Global solar radiation on tilted surfaces in Tunisia: Measurement, estimation and gained energy assessments. *Energy Rep.* **2018**, *4*, 101–109. [CrossRef]
136. Yahyaoui, I. Chapter 2-Modeling of the photovoltaic irrigation plant components. In *Specifications of Photovoltaic Pumping Systems in Agriculture*; Elsevier: Amsterdam, The Netherlands, 2017; pp. 15–57.
137. Castillo, C.P.; e Silva, F.B.; Lavalle, C. An assessment of the regional potential for solar power generation in EU-28. *Energy Policy* **2016**, *88*, 86–99. [CrossRef]
138. Spencer, R.S.; Macknick, J.; Aznar, A.; Warren, A.; Reese, M.O. Floating photovoltaic systems: Assessing the technical potential of photovoltaic systems on man-made water bodies in the continental United States. *Environ. Sci. Technol.* **2018**, *53*, 1680–1689. [CrossRef]
139. Kim, S.M.; Oh, M.; Park, H.D. Analysis and prioritization of the floating photovoltaic system potential for reservoirs in Korea. *Appl. Sci.* **2019**, *9*, 395. [CrossRef]
140. Nebey, A.H.; Taye, B.Z.; Workineh, T.G. GIS-Based Irrigation Dams Potential Assessment of Floating Solar PV System. *J. Energy* **2020**, *2020*, 1268493. [CrossRef]
141. López, M.; Soto, F.; Hernández, Z.A. Assessment of the potential of floating solar photovoltaic panels in bodies of water in mainland Spain. *J. Clean. Prod.* **2022**, *340*, 130752. [CrossRef]
142. Liu, J.; Xu, F.; Lin, S. Site selection of photovoltaic power plants in a value chain based on grey cumulative prospect theory for sustainability: A case study in Northwest China. *J. Clean. Prod.* **2017**, *148*, 386–397. [CrossRef]
143. Golroodbari, S.Z.; van Sark, W. Simulation of performance differences between offshore and land-based photovoltaic systems. *Prog. Photovolt. Res. Appl.* **2020**, *28*, 873–886. [CrossRef]
144. Deveci, M.; Pamucar, D.; Oguz, E. Floating photovoltaic site selection using fuzzy rough numbers based LAAW and RAFSI model. *Appl. Energy* **2022**, *324*, 119597. [CrossRef]
145. Habib, S.M.; Suliman AER, E.; Al Nahry, A.H.; Abd El Rahman, E.N. Spatial modeling for the optimum site selection of solar photovoltaics power plant in the northwest coast of Egypt. *Remote Sens. Appl. Soc. Environ.* **2020**, *18*, 100313. [CrossRef]
146. Meng, X.; Liu, H.; Du, Q.; Xu, L.; Liu, Y. Evaluation of the performance of different methods for estimating evaporation over a highland open freshwater lake in mountainous area. *Water* **2020**, *12*, 3491. [CrossRef]
147. Moazed, H.; Ghaemii, A.A.; Rafiee, M.R. Evaluation of Several Reference Evapotranspiration Methods: A Comparative Study of Greenhouse and Outdoor Conditions. *Trans. Civ. Eng.* **2014**, *38*, 421.
148. Dlouhá, D.; Dubovský, V.; Pospíšil, L. Optimal calibration of evaporation models against Penman–Monteith equation. *Water* **2021**, *13*, 1484. [CrossRef]
149. Abd-Elhamid, H.F.; Ahmed, A.; Zeleňáková, M.; Vranayová, Z.; Fathy, I. Reservoir management by reducing evaporation using floating photovoltaic system: A case study of Lake Nasser, Egypt. *Water* **2021**, *13*, 769. [CrossRef]
150. Ghumman, A.R.; Ghazaw, Y.M.; Alodah, A.; Rauf, A.U.; Shafiquzzaman, M.; Haider, H. Identification of parameters of evaporation equations using an optimization technique based on pan evaporation. *Water* **2020**, *12*, 228. [CrossRef]
151. Junzeng, X.U.; Qi, W.E.I.; Shizhang PEN, G.; Yanmei, Y.U. Error of saturation vapor pressure calculated by different formulas and its effect on calculation of reference evapotranspiration in high latitude cold region. *Procedia Eng.* **2012**, *28*, 43–48. [CrossRef]
152. Wong, C.P.; Jiang, B.; Bohn, T.J.; Lee, K.N.; Lettenmaier, D.P.; Ma, D.; Ouyang, Z. Lake and wetland ecosystem services measuring water storage and local climate regulation. *Water Resour. Res.* **2017**, *53*, 3197–3223. [CrossRef]
153. Febrian, H.G.; Supriyanto, A.; Purwanto, H. Calculating the energy capacity and capacity factor of floating photovoltaic (FPV) power plant in the cirata reservoir using different types of solar panels. In *Journal of Physics: Conference Series*; IOP Publishing: Bristol, UK, 2023; Volume 2498, p. 012007.
154. Sahu, A.; Yadav, N.; Sudhakar, K. Floating photovoltaic power plant: A review. *Renew. Sustain. Energy Rev.* **2016**, *66*, 815–824. [CrossRef]
155. Yashas, V.; Aman, B.; Dhanush, S. Feasibility study of floating solar panels over lakes in Bengaluru City, India. *Proc. Inst. Civ. Eng.-Smart Infrastruct. Constr.* **2021**, *174*, 1–10. [CrossRef]
156. Semeskandeh, S.; Hojjat, M.; Hosseini Abardeh, M. Techno–economic–environmental comparison of floating photovoltaic plant with conventional solar photovoltaic plant in northern Iran. *Clean Energy* **2022**, *6*, 353–361. [CrossRef]
157. Mittal, D.; Saxena, B.K.; Rao, K.V.S. Potential of floating photovoltaic system for energy generation and reduction of water evaporation at four different lakes in Rajasthan. In Proceedings of the 2017 International Conference on Smart Technologies for Smart Nation (SmartTechCon), Bengaluru, India, 17–19 August 2017; pp. 238–243.
158. Alhassan, M.O.; Opoku, R.; Uba, F.; Obeng, G.Y.; Sekyere, C.K.; Nyanor, P. Techno-economic and environmental estimation assessment of floating solar PV power generation on Akosombo dam reservoir in Ghana. *Energy Rep.* **2023**, *10*, 2740–2755. [CrossRef]
159. Nagababu, G.; Patil, P.; Bhatt, T.N.; Srinivas, B.A.; Puppala, H. Floating solar panels: A sustainable solution to meet energy demands and combat climate change in offshore regions. *J. Therm. Anal. Calorim.* **2024**, *13*, 1–8. [CrossRef]
160. Taye, B.Z.; Nebey, A.H.; Workineh, T.G. Design of floating solar PV system for typical household on Debre Mariam Island. *Cogent Eng.* **2020**, *7*, 1829275. [CrossRef]
161. Katircioğlu, S.T. Estimating higher education induced energy consumption: The case of Northern Cyprus. *Energy* **2014**, *66*, 831–838. [CrossRef]

162. Kebede, K.Y. Viability study of grid-connected solar PV system in Ethiopia. *Sustain. Energy Technol. Assess.* **2015**, *10*, 63–70. [CrossRef]
163. Khamees, A.K.; Abdelaziz, A.Y.; Eskaros, M.R.; Attia, M.A.; Badr, A.O. The mixture of probability distribution functions for wind and photovoltaic power systems using a metaheuristic method. *Processes* **2022**, *10*, 2446. [CrossRef]
164. Langat, P.K.; Kumar, L.; Koech, R. Identification of the most suitable probability distribution models for maximum, minimum, and mean streamflow. *Water* **2019**, *11*, 734. [CrossRef]
165. Sumner, D.M.; Jacobs, J.M. Utility of Penman–Monteith, Priestley–Taylor, reference evapotranspiration, and pan evaporation methods to estimate pasture evapotranspiration. *J. Hydrol.* **2005**, *308*, 81–104. [CrossRef]
166. McKenney, M.S.; Rosenberg, N.J. Sensitivity of some potential evapotranspiration estimation methods to climate change. *Agric. For. Meteorol.* **1993**, *64*, 81–110. [CrossRef]
167. Xiaoying, L.; Erda, L. Performance of the Priestley–Taylor equation in the semiarid climate of North China. *Agric. Water Manag.* **2005**, *71*, 1–17. [CrossRef]
168. Jansen, F.A.; Teuling, A.J. Evaporation from a large lowland reservoir–(dis) agreement between evaporation models from hourly to decadal timescales. *Hydrol. Earth Syst. Sci.* **2020**, *24*, 1055–1072. [CrossRef]
169. Liu, X.; Xu, C.; Zhong, X.; Li, Y.; Yuan, X.; Cao, J. Comparison of 16 models for reference crop evapotranspiration against weighing lysimeter measurement. *Agric. Water Manag.* **2017**, *184*, 145–155. [CrossRef]
170. Naganathini, R.; Nagavinothini, R. Investigation on floating photovoltaic covering system in rural Indian reservoir to minimize evaporation loss. *Int. J. Sustain. Energy* **2021**, *40*, 781–805. [CrossRef]
171. Ilgen, K.; Schindler, D.; Armbruster, A.; Ladwig, R.; Eppinger Ruiz de Zarate, I.; Lange, J. Evaporation reduction and energy generation potential using floating photovoltaic power plants on the Aswan High Dam Reservoir. *Hydrol. Sci. J.* **2024**, *47*, 709–720. [CrossRef]
172. Lopes, M.P.C.; Nogueira, T.; Santos, A.J.L.; Branco, D.C.; Pouran, H. Technical potential of floating photovoltaic systems on artificial water bodies in Brazil. *Renew. Energy* **2022**, *181*, 1023–1033. [CrossRef]
173. Al Saadi, K.; Ghosh, A. Investigating the integration of floating photovoltaics (FPV) technology with hydrogen (H_2) energy for electricity production for domestic application in Oman. *Int. J. Hydrogen Energy* **2024**, *80*, 1151–1163. [CrossRef]

energies

MDPI

Article

Non-Condensation Turbulence Models with Different Near-Wall Treatments and Solvers Comparative Research for Three-Dimensional Steam Ejectors

Yiqiao Li [1,2,*], Hao Huang [1], Dingli Duan [3], Shengqiang Shen [4], Dan Zhou [2] and Siyuan Liu [1]

1 Zhan Tianyou College, Dalian Jiaotong University, Dalian 116028, China; cqhuanghao@outlook.com (H.H.); liusiyuan020812@outlook.com (S.L.)
2 Bingshan Refrigeration & Heat Transfer Technologies Co., Ltd., Dalian 116630, China; 13700098947@139.com
3 Department of Thermal Energy and Power Engineering, Yantai University, Yantai 264005, China; duandingli@ytu.edu.cn
4 School of Energy and Power Engineering, Dalian University of Technology, Dalian 116024, China; zzbshen@dlut.edu.cn
* Correspondence: author: liyiqiao@djtu.edu.cn

Abstract: Steam ejectors are important energy-saving equipment for solar thermal energy storage; however, a numerical simulation research method has not been agreed upon. This study contributes to a comprehensive selection of turbulence models, near-wall treatments, geometrical modeling (2-D and 3-D), solvers, and models (condensation and ideal-gas) in the RANS equations approach for steam ejectors through validation with experiments globally and locally. The turbulence models studied are k-ε Standard, k-ε RNG, k-ε Realizable, k-ω Standard, k-ω SST, Transition SST, and linear Reynolds Stress. The near-wall treatments assessed are Standard Wall Functions, Non-equilibrium Wall Functions, and Enhanced Wall Treatment. The solvers compared are pressure-based and density-based solvers. The root causes of their distinctions in terms of simulation results, applicable conditions, convergence, and computational cost are explained and compared. The complex phenomena involving shock waves, choking, and vapor condensation captured by different models are discussed. The internal connections of their performance and flow phenomena are analyzed from the mechanism perspective. The originality of this study is that both condensation and 3-D asymmetric effects on the simulation results are considered. The results indicate that the k-ω SST non-equilibrium condensation model coupling the low-Re boundary conditions has the most accurate prediction results, best convergence, and fit for the widest range of working conditions. A 3-D asymmetric condensation model with a density-based solver is recommended for simulating steam ejectors accurately.

Keywords: non-equilibrium condensation; near-wall treatments; RANS; steam ejectors; three-dimensional effect; turbulence model; solvers

1. Introduction

Steam ejectors are important and widely used energy-saving equipment that can only utilize residual pressure by recovering low-pressure steam [1]. Although a steam ejector's structure is simple and reliable, the flow and mixing process of its internal fluid is very complicated [2]. Many kinds of flow phenomena that affect the ejector performance occur in the flow field, such as shock waves [3], choking [4], and condensation [5]. These flow phenomena cause the transonic phase transition and the sudden change of various flow parameters in the flow field. It is difficult to obtain accurate measurements in supersonic fluids by current experimental methods. Computational fluid dynamics can quantitatively describe the complex flow phenomena that make up for the shortage of experiments. Assessment of these dynamics is an effective method to study the relationship between flow phenomenon and ejector performance. Therefore, the method of simulated steam ejectors to get accurate results is worth investigating.

The RANS (Reynolds-Averaged Navier–Stokes) equations are the most frequently used equations in scientific research and engineering practice simulations. According to the different closures, the turbulence models of the RANS equations are divided into two categories: the Reynolds Stress Model and eddy-viscosity closures. The eddy-viscosity closures include the zero-, one-, and two-equation turbulence models [6]. Among these, the two-equation turbulence models that include the k-ε series and k-ω series models are adopted in ejector simulations.

The turbulence models' closures of the RANS equations in simulating ejectors have not reached a consensus [7,8]. For air ejectors, Hemidi et al. [9] compared two turbulence models. They observed that both k-ε Standard and k-ω SST predicted almost the same Er (entrainment ratio) but had completely different local flow characteristics under higher working pressures. Overall, the k-ε Standard is better. Gagan et al. [10] compared six turbulence models (k-ε series, k-ω series, and RSM) of flow visualization using the PIV technique. The supersonic flow shape and the subsonic region velocity are best obtained by the k-ε Standard. For N_2 ejectors, Zhu and Jiang [11] compared four turbulence models (k-ε series, k-ω SST) based on visualization experiments. They stated that the shock train configuration obtained by the k-ε RNG accorded best with the experimental data. For R134a ejectors, Croquer et al. [12] compared four turbulence models (k-ε series, k-ω SST) with the ideal-gas equation and real-gas equation. They found the k-ω SST obtains accurate Er only when the real-gas equation is used. For steam ejectors, scholars neglect condensation effects during the comparison of turbulence models. Ruangtrakoon [13] and Besagni [8] both recommended the k-ω SST. The former compared two models (k-ε Realizable, k-ω SST), and the latter compared seven models (k-ε Standard, k-ε RNG, k-ε Realizable k-ω standard, k-ω SST, Spalart-Allmaras, and RSM). Han et al. [14] and Varga et al. [15] obtained different comparison results. Han et al. [14] compared four models (k-ε series, k-ω SST) and stated that the pressure distribution of the k-ε Realizable model accorded best with experiments. Varga et al. [15] compared six models (k-ε series, k-ω series, Transition SST) and stated that the ejector performance of the Transition SST accorded best with experiments. However, they assumed that water vapor was an ideal gas and ignored condensation.

Apart from turbulence models, near-wall treatments should be cautiously adopted. Two commonly used near-wall treatments are the wall function and near-wall modeling. The former is to resolve the viscosity-influenced region by wall functions containing semi-empirical formulas, and its turbulence models do not require modification. The RSM and the k-ε series models need wall functions that include Standard Wall Function (SWF), Non-equilibrium Wall Function (NWF), Enhanced Wall Treatment (EWT), and so on. The latter approach is to modify the turbulence model so that the region affected by viscosity can be analyzed by a fine near-wall mesh. As to the comparative study of near-wall treatments, Besagni et al. [8] stated that EWT allows achieving generally better results than SWF and NWF, and nearly no differences between the SWF and EWT results. Han et al. [14] compared SWF and EWT for the prediction of a steam ejector. They found that the latter was superior to the former. Wu et al. [16,17] considered the effect of fluid–structure interaction. However, no agreement about near-wall treatments has been reached.

According to the classical viewpoint, flow fields with shock waves must adopt the density-based solver(D-B) owing to its excellent adaptability of the simulation for compressible and super-sonic flows [18]. However, the pressure-based solver (P-B) is for low-speed incompressible flow, historically [11]. Nevertheless, in the works of Li and Li [19], Yazdani et al. [20], Croquer et al. [12], Sriveerakul et al. [21], Besagni et al. [7], and Bourhan et al. [22], the pressure-based solver was proposed to solve supersonic problems in ejectors. Therefore, a comparative study of the two solvers is a great lack that at present needs to be performed on steam ejectors.

The ideal-gas model is adopted in related comparative research on steam ejectors. Nevertheless, the flow information and performance of steam ejectors are both affected by non-equilibrium condensation [23]. Moreover, most scholars simplify steam ejectors into a 2-D axisymmetric model in simulations to save calculation costs, ignoring the 3-D disturbance in actual flow fields. Mazzelli et al. [24] compared the prediction of ejector coefficients by 2-D and 3-D standard k-ε, Realizable k-ε, SST k-ω, and RSM models. Results showed that only the simulation results of the 3-D model are in the reliable range under the sub-critical operating state. Therefore, the comparative research of the turbulence models with different near-wall treatments and solvers should consider both condensation effects and 3-D effects, which the current research lacks.

Consequently, a study that contributes to a comprehensive selection of turbulence models, near-wall treatments, geometrical modeling (2-D and 3-D), solvers, and models (condensation and ideal-gas) in the RANS equations approach for steam ejectors is urgently needed. In this article, a condensation model with seven turbulence models (k-ε series, k-ω series, Transition SST, linear Reynolds Stress Model), three near-wall treatments (SWF, NWF, EWT), and two solvers (D-B and P-B) is utilized to simulate different sizes of 3-D steam ejectors under variable working conditions. The root causes of their distinctions in simulation results, applicable conditions, convergence, and computational cost are explained and compared. The complex phenomena involving shock waves, choking, and vapor condensation captured by different models are discussed. The connections between the ejector performance and local flow phenomena are analyzed. The originality of the study is that both condensation and 3-D asymmetric effects on the simulation results are considered. These research achievements provide rational foundations for simulation research of steam ejectors.

2. Numerical Simulation Method

2.1. Mathematical Model

Below are three conservation equations of wet steam at steady state:

$$\frac{\partial}{\partial x_j}(\rho v_j) = 0 \tag{1}$$

$$\frac{\partial}{\partial x_j}(\rho v_j v_i) = \frac{\partial \tau_{ij}}{\partial x_j} - \frac{\partial P}{\partial x_i} \tag{2}$$

$$\frac{\partial}{\partial x_j}[v_j(\rho E + P)] = \frac{\partial}{\partial x_j}\left(\lambda_{eff}\frac{\partial T}{\partial x_j}\right) + \frac{\partial}{\partial x_j}(v_i \tau_{ij}) \tag{3}$$

where E is the total energy contained in the fluid per unit mass (equal to the sum of the specific internal energy and kinetic energy):

$$E = h - \frac{p}{\rho} + \frac{1}{2}v_j v_j \tag{4}$$

The real state of vapor is expressed by [25]:

$$P = \rho_{\rm v} RT\left(1 + B\rho_{\rm v} + C{\rho_{\rm v}}^2\right) \tag{5}$$

where B and C are the virial coefficients which are detailed in the reference [25].

The vapor properties, equations of saturated vapor and saturated liquid lines, and liquid properties are detailed in the references [26]. The wet steam density and the properties $\varphi_{\rm m}$ of wet steam are:

$$\rho = \rho_{\rm v}/(1-\beta) \tag{6}$$

$$\varphi_m = \varphi_l \beta + (1-\beta)\psi_{\rm v} \tag{7}$$

where β is the liquid mass fraction which is equal to the ratio of the mass of the condensate to the total mass.

Mach number is an important dimensionless number for determining sonic velocity flow. It is equal to the ratio of fluid speed to sonic velocity:

$$Ma = v/a \tag{8}$$

where a is the sonic velocity of wet steam [27]:

$$a = \left[\rho\left(\frac{\beta}{\rho_{\mathrm{v}}{a_{\mathrm{v}}}^2} + \frac{1-\beta}{\rho_{\mathrm{l}}{a_{\mathrm{l}}}^2}\right)\right]^{-1/2} \tag{9}$$

The governing equations for vapor–liquid two-phase mixtures are established based on the Euler coordinate system [28]:

$$\nabla \cdot \left(\rho \vec{v} \beta\right) = \Gamma \tag{10}$$

$$\nabla \cdot \left(\rho \vec{v} \eta\right) = \rho J \tag{11}$$

where Γ is the liquid mass generation rate that is composed of a two-part mass increase: the first term of the right end equation shows the mass of the liquid phase generated by spontaneous nucleation, and the second term shows the mass of the liquid phase generated by the droplet growth after nucleation [28]; J is nucleation rate [29]; and η is the number of droplets per unit volume:

$$\Gamma = \dot{m}_{\mathrm{l}} = -\dot{m}_v = \frac{4}{3}\pi\rho_{\mathrm{l}} J {r_*}^3 + 4\pi\rho_{\mathrm{l}}\eta\bar{r}^2\frac{\partial\bar{r}}{\partial t} \tag{12}$$

$$J = \frac{q_c}{(1+\theta)}\left(\frac{\rho_g^2}{\rho_{\mathrm{l}}}\right)\sqrt{\frac{2\sigma}{M_m^3\pi}}e^{-\left(\frac{4\pi r_*^2\sigma}{3k_B T_g}\right)} \tag{13}$$

$$\eta = \frac{\beta}{(1-\beta)V_{\mathrm{d}}(\rho_{\mathrm{l}}/\rho_{\mathrm{g}})} \tag{14}$$

$$V_d = \frac{4}{3}\pi\bar{r}^3 \tag{15}$$

where V_{d} is a mean droplet volume, and the average droplet radius is from V_{d} and η:

$$\bar{r} = \sqrt[3]{\frac{3\beta}{4\pi\eta(1-\beta)(\rho_{\mathrm{l}}-\rho_{\mathrm{g}})}} \tag{16}$$

The droplet's growth rate equation [29] is:

$$\frac{\partial\bar{r}}{\partial t} = \frac{P}{h_{\mathrm{lv}}\rho_{\mathrm{l}}\sqrt{2\pi RT}}\frac{\gamma+1}{2\gamma}\mathrm{Cl}\left(1-\frac{r_*}{\bar{r}}\right)\Delta T \tag{17}$$

where h_{lv} is the specific latent enthalpy and the ΔT is the degree of subcooling which is equal to:

$$\Delta T = T_{\mathrm{sat}} - T \tag{18}$$

2.2. Solution Scheme and Boundary Conditions

The finite-volume method was used to treat N-S equations at a steady state with the 7 turbulent models respectively based on Fluent. The implicit solver was based on density coupling. The variable gradient interpolation method was the Green–Gauss method based on node. The second-order upwind scheme was utilized. It is convergent when it meets the following 3 criteria:

(1) All the residual terms' convergence absolute criteria are <10^{-6} forever;
(2) The difference in the mass flow rate between the inlet and the outlet is <1×10^{-7};
(3) The calculated mass flow rate of two inlet sections and one outlet section is stable.

Figure 1 describes the ejector sizes and Table 1 illustrates the working conditions, using data from Chen and Sun's experiment [30]. The variables $\boldsymbol{P_m}$ and $\boldsymbol{T_m}$ represent the temperature and pressure of the motive fluid, while $\boldsymbol{P_s}$ and $\boldsymbol{T_s}$ denote the temperature and pressure of the suction fluid. Additionally, $\boldsymbol{P_d}$ signifies the pressure of the discharge fluid. The case labels are consistent with those in the reference. The simulation parameters are consistent with the experimental parameters to compare their results. The computation parameters in Table 1 are given as the boundary conditions for pressure inlet and pressure outlet, given that outlet pressures are constant. The turbulence boundary conditions are set as the turbulence intensity and the turbulent viscosity ratios. For the motive steam and the suction steam, their turbulent viscosity ratios are 500 and 100, respectively, and their turbulence intensity are 5.0% and 2.0%, respectively. There is no slip and no penetration in the adiabatic smooth solid walls, hypothetically.

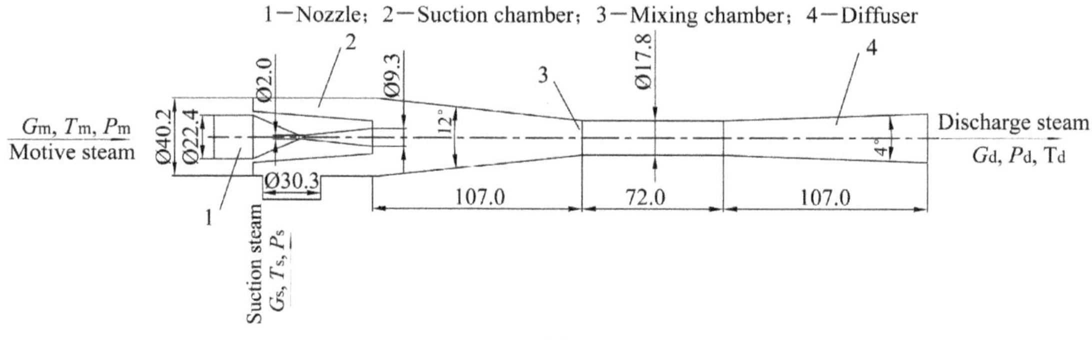

(**a**)

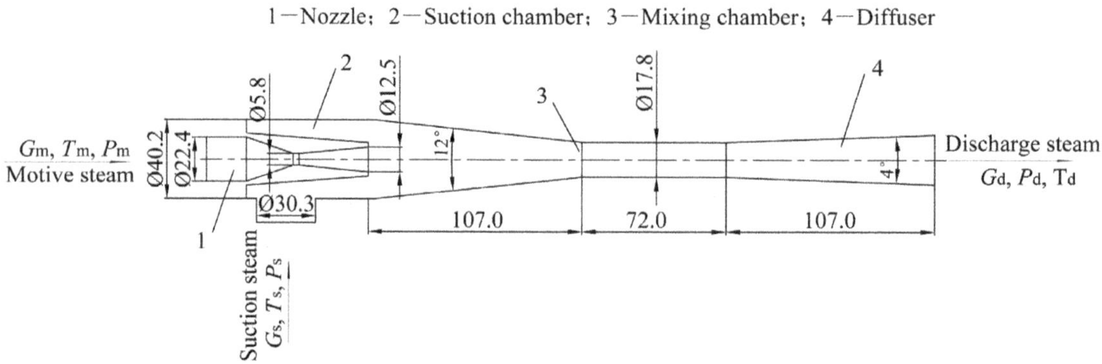

(**b**)

Figure 1. Two sizes of steam ejectors. (**a**) Size of E_1. (**b**) Size of E_2.

Table 1. Computation parameters.

Size	Case	P_m [kPa]	T_m [°C]	P_s [kPa]	T_s [°C]	P_d [kPa]
E_1	1	116.0	105.8	1.306	10.8	1.35~2.70
	2	153.0	114.2	1.306	10.8	1.35~2.90
	3	198.0	121.8	1.306	10.8	1.35~3.90
E_2	4	11.906	51.3	1.306	10.8	1.35~1.90
	5	15.465	56.6	1.306	10.8	1.35~2.40
	6	19.571	61.6	1.306	10.8	1.35~2.90
	7	19.571	61.6	1.813	15.8	1.85~2.90
	8	12.5	52.2	1.35	11.4	1.4
	ag	11.906	51.3	1.813	15.8	1.85~2.00
	bf	15.465	56.2	1.306	10.8	1.35~2.40
	cf	19.571	61.6	1.306	10.8	1.35~2.90
	df	24.837	66.8	1.306	10.8	1.35~3.20
	be	15.465	56.2	1.0	13.2	1.1~2.2
	ce	19.571	61.6	1.0	13.2	1.1~2.6
	de	24.837	66.8	1.0	13.2	1.1~3.3

2.3. Mesh Creation

The 3-D meshes of E_1 and E_2 with refinement both around boundary layers and the shock waves were created. Furthermore, each mesh was divided into two categories according to the y+ value that indicates the dimensionless distance of a first-layer node from the wall. The first mesh type with y+ ≈ 1 was used for EWT, and the second mesh type with $y+_{min} > 30$ was used for SWF and NWF. The first type of meshes of E_2 are shown in Figure 2a and their y+ values along the ejector wall are shown in Figure 2b. Except for the near-wall area, the two meshes have the same structures. The second type of mesh was obtained by locally adjusting mesh density around the near-wall area of the first type.

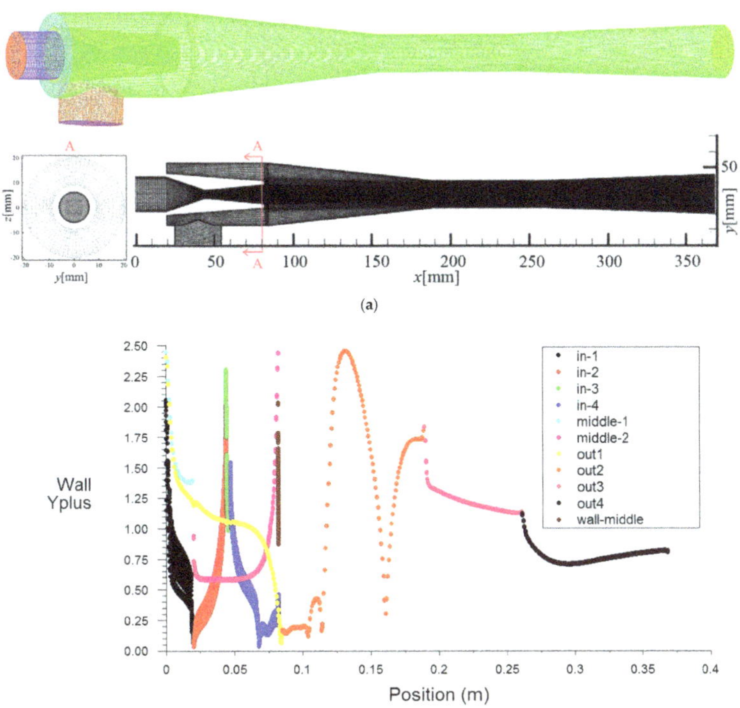

Figure 2. Information of steam ejector meshes. (**a**) Steam ejector meshes. (**b**) y+ profile along ejector wall.

Table 2 summarizes the number of cells of the two-type grids. Figures 3 and 4 illustrate that the ejection coefficients obtained by medium-level grids are similar to those of the fine-level grids when all other things are equal. Moreover, the time cost and the computation cost of the medium-level grids are lower than those of the fine-level grids. Accordingly, the medium-level grids were adopted in this paper.

Table 2. Number of cells.

	Type1, y+ ≈ 1			Type2, Minimum y+ = 30		
Mesh Level	**Coarse**	**Medium**	**Fine**	**Coarse**	**Medium**	**Fine**
Number [million]	0.8	2.0	2.9	0.2	0.4	0.9

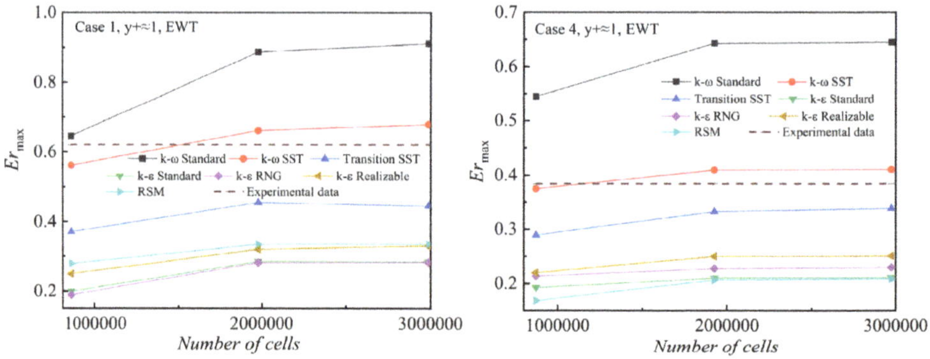

(**a**) (**b**)

Figure 3. First type mesh independence study of different models. (**a**) Grid independence study with Case 1. (**b**) Grid independence study with Case 4.

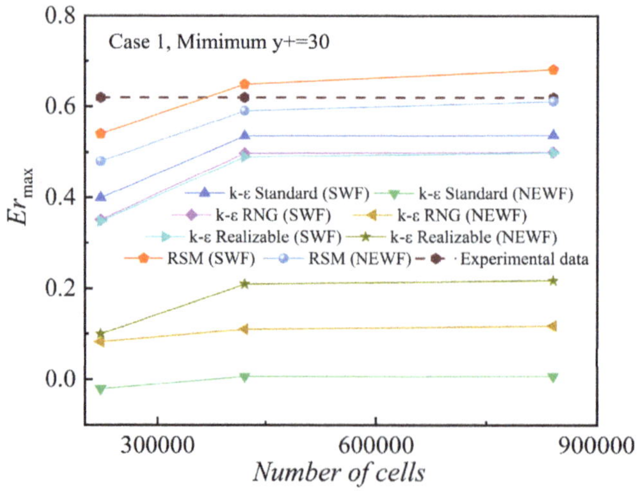

(**a**)

Figure 4. *Cont.*

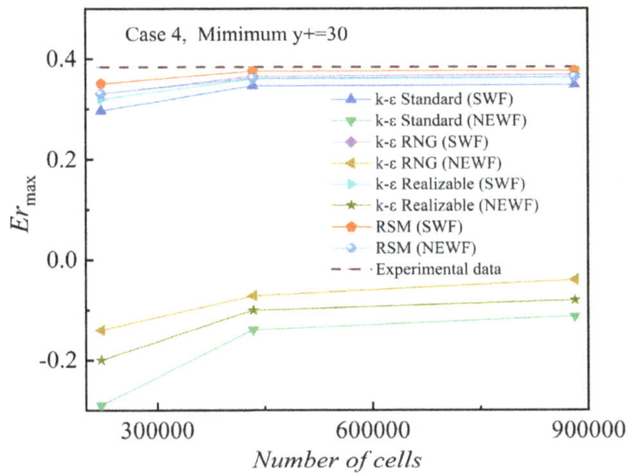

(b)

Figure 4. Second type mesh independence study of different models. (**a**) Grid independence study with Case 1. (**b**) Grid independence study with Case 4.

3. Results and Discussion

3.1. Validation, Comparison and Analysis of Different Near-Wall Treatments

Turbulent flow is divided into inner and outer layers according to a distance along the near wall. The flow in the outer layer is completely turbulent, and the flow in the inner layer can be affected by the wall obviously. According to the y+ value, the inner layer is divided into three sub-layers: a viscous sublayer (y+ < 5), a buffer layer ($5 \leq$ y+ ≤ 30), and a log-law region 30 < y+ < f(Re) [6].

The *k-ε* series and linear RSM require selecting near-wall treatments. The near-wall treatments commonly used for ejectors are SWF, NWF, and EWT. SWF and NWF cannot truly resolve the viscous sublayer and buffer layer, but utilize semi-empirical formulas to bridge the wall and the fully-turbulent region. Therefore, their first node of the near-wall grid must be arranged in the log-law region, namely $y+_{min} > 30$. The SWF can reasonably predict the bounded flow of most walls with high Reynolds numbers. The NWF can partially consider the influence of pressure gradient, which further expands the applicability. The EWT is suggested when the flow has special conditions, such as severe pressure gradient, boundary layer separation, and so on. This method can analyze the viscous sublayer if the near-wall region has a fine mesh [6]. Therefore, the first node of the near-wall mesh is arranged within the viscous sublayer, namely y+ $\approx$ 1.

The mathematical structures of the *k-ω* series and Transition SST already emphasize the near-wall flow. Therefore, there is no need to select any near-wall treatments. Their wall boundary conditions for the *k* equation are treated likewise as the *k* equation treated when EWT is applied in *k-ε* models [6]. When the grid in the near-wall region is very dense (y+ $\approx$ 1), these three turbulence models coupled with the low-Re boundary conditions can deal with the complex flow.

Figure 3a,b compare the Er_{max} of experimental data and seven different turbulence models combined with EWT. Results obtained from the *k-ω* SST are reasonable with an average relative error ($\bar{e}$) of 7%.

Figure 4a,b compare the Er_{max} of experimental data and seven turbulence models combined with SWF or NWF. The linear RSM combined with SWF or NWF can reasonably predict the Er_{max} with $\bar{e}$ of 1.5% and −5.5%, respectively, whereas the convergence of NEWF is worse than that of SWF, and the steps of iterations that reach the convergent condition are 1.8 times that of SWF.

The Er_{max} predicted by the EWT or NEWF combined with the k-ε series deviates greatly from the experimental data. Even the Er_{max} predicted is less than 0, which indicates the reverse flow as shown in Figure 4. The SWF combined with the k-ε series are just suitable for Case 4 with the lowest pressure, and their relative errors are −10%, −5%, and −6% respectively.

Therefore, the near-wall treatment method should be selected according to the specific turbulence model and working conditions. Moreover, it is also affected by the structure of the supersonic steam ejector according to the comparison results of E1 and E2.

3.2. Validation, Comparison and Analysis of Different Turbulence Models

Each Er_{max} predicted by k-ω SST is within an acceptable range (relative error of −5%~11%) for each case as shown in Figure 5. The Er_{max} predicted by the linear RSM has only one large deviation (−19%) compared to experimental data. The average relative error ($\overline{e}$) of Er_{max} obtained from the k-ω SST and linear RSM which are 4.0% and −5.2%, respectively, are within the acceptable range, as shown in Table 3. Their maximum relative errors are 11% and −19%, respectively. Therefore, the two models may have the potential to provide accurate predictions for steam ejectors.

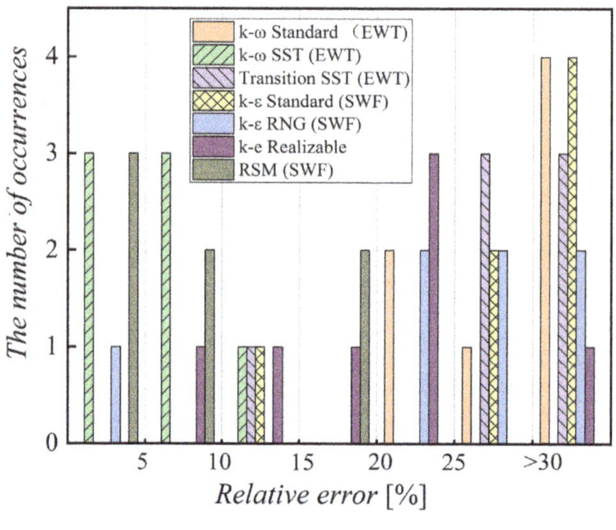

Figure 5. Histogram of the prediction error.

Table 3. Relative errors of maximum *Er*.

Case \ Model	$e(Er_{max})$						
	k-ω Standard (EWT)	k-ω SST (EWT)	Transition SST (EWT)	k-ε Standard (SWF)	k-ε RNG (SWF)	k-ε Realizable (SWF)	Linear RSM (SWF)
1	43%	7%	−26%	−24%	−20%	−21%	5%
2	47%	11%	−38%	−38%	−36%	−28%	11%
3	28%	2%	−31%	−40%	−37%	−30%	−19%
4	42%	9%	−13%	−10%	−5%	−6%	−2%
5	21%	6%	−49%	−62%	−52%	−38%	−19%
6	22%	−5%	−27%	−76%	−74%	−55%	−10%
7	38%	−2%	−30%	−27%	−32%	−22%	−2%
$\overline{e}(Er_{max})$	34.5%	4.0%	−30.6%	−39.5%	−36.5%	−28.6%	−5.2%

Compared to the experiments, Er_{max} predicted by *k-ω* Standard are greater for all cases with an $\bar{e}$ of 34.5%. The Er_{max} of the *k-ε* Standard, *k-ε* RNG, *k-ε* Realizable, and Transition SST have an $\bar{e}$ of −39.5%, −36.5%, −28.6%, and −30.6%, respectively, compared to experimental values. However, the minimum relative errors, which are −10%, −5%, −6%, and −13%, respectively, obtained by the three models under Case 4 are within the acceptable range. Case 4 has the lowest primary inlet pressure for E_2 of all cases. Furthermore, Case 1 has the lowest primary inlet pressure for E_1 and also has the lowest relative errors of all E_1 cases. Therefore, the simulation accuracy of the three turbulence models has a strong dependence on the motive pressure. It appears that the Er_{max} can be reasonably predicted by the three models only under relatively low-pressure conditions with the same physical dimensions. Moreover, pressure dependence has a different sensitivity to different structures of ejectors.

The difference in Er_{max} predicted by different models depends mainly on their mass flow rates of suction steam (G_s) as shown in Table 4. Under the same case, the mass flow rates of motive steam (G_m) predicted by different models are close. The maximum difference is 8% and the minimum difference is 2%. However, G_s predicted by different models are significantly different. The maximum difference is 129% and the minimum difference is 55%. The *k-ω* Standard predicts the maximum G_s under each case, which causes the overestimation of Er_{max}.

Table 4. The average difference of the mass flow rate predicted by seven turbulence models.

Case \ $\overline{\Delta G}$	$\overline{\Delta G_m}$	$\overline{\Delta G_s}$
1	8%	67%
2	2%	93%
3	5%	88%
4	4%	55%
5	7%	96%
6	2%	129%
7	7%	80%
average	5%	87%

As to the comparison of P_d^* by different turbulence models in Figure 6, they are less than the experimental data, which accords with the majority of results [31–36]. Since the P_d^* for Case 2 and Case 7 is too close to clearly illustrate in Figure 6, Case 2 in Figure 6 is omitted. The $\bar{e}$ of *k-ω* Standard and *k-ω* SST are within an acceptable range, which is −8.0% and −15%, respectively. Although the relative error of P_d^* predicted by *k-ω* Standard is the lowest, the $\bar{e}$ of Er_{max} predicted by *k-ω* Standard is as high as 34.5%. Therefore, as to the ejector performance, *k-ω* SST agrees with the experimental results.

Differences in P_d^* are explained through the local flow characteristics of the second choking in Figures 7 and 8. The black outline in Figure 8 is the sonic line whose Mach number is 1. The sonic line at choking mode is closest to the wall [9]. The appearance of the first shock wave in the diffuser reduces the fluid Mach number, causing the sonic line to move away from the wall, and the second choking phenomenon ends. The distance between the first shock wave position and the ejector outlet is positive with the second choking length (L_c). When L_c increases, the differences between P_d and P_d^* are getting greater [37]. Accordingly, when the working condition is the same, P_d^* predicted by different turbulence models increases with the distance between the first shock wave position and the ejector outlet by comparing Figures 6 and 7. The order of the distances from large to small is the *k-ω* Standard, *k-ω* SST, *k-ε* Realizable, *k-ε* RNG, *k-ε* Standard, linear RSM, and Transition SST, as shown in Figure 7. The same order of P_d^* is shown in Figure 6.

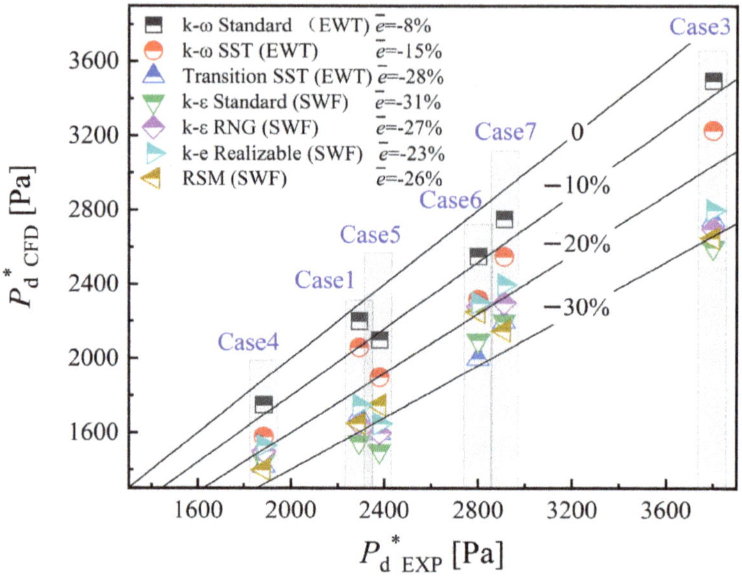

Figure 6. Relative errors of P_d^*.

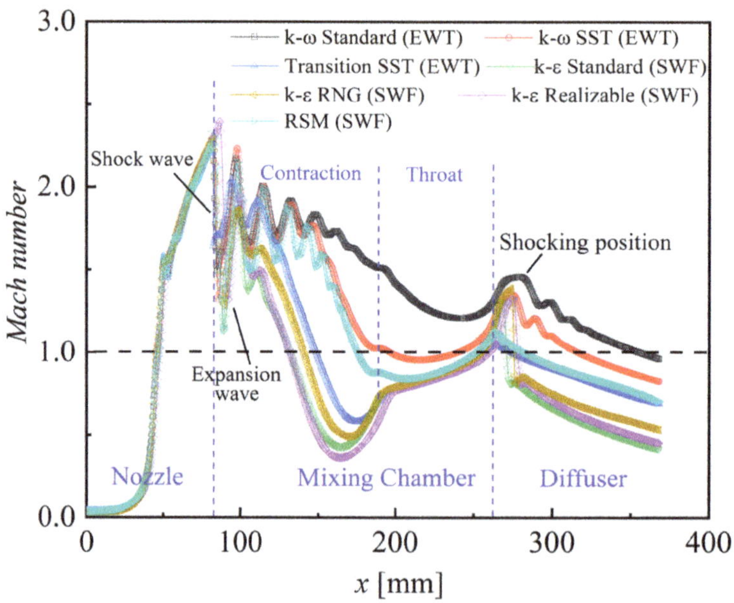

Figure 7. Mach number profiles of the axis under Case 4 with P_d = 1335 Pa.

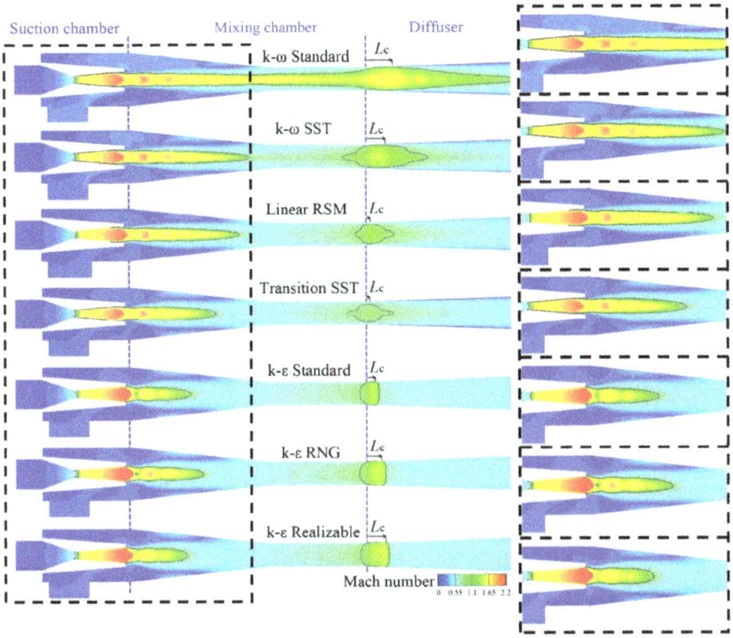

Figure 8. Mach number contours under Case 4 with P_d = 1335 Pa.

An abrupt decrease of *Ma* represents that an aerodynamic shock wave occurs, as shown in Figures 7 and 8. The aerodynamic shock wave whose shock surface is perpendicular to the flow direction and whose flow direction is unchanged after passing through the shock wave is a normal shock wave. Otherwise, it is an oblique shock. Velocity after a normal shock wave must be subsonic. However, velocity after an oblique shock wave is not certain. As can be seen in Figures 7 and 8, the *k-ε* series models get a normal shock, while the others obtain a series of oblique shock waves that accord with the experiments [30].

In steam ejectors, vapor experiences a transonic expansion process of rapid depressurization and cooling. When the vapor crosses a saturation line, it does not immediately condense due to the free energy barrier. The vapor expands continuously, resulting in the vapor temperature continuing to decrease, and the vapor diverges from the saturation equilibrium state. When the unbalanced state develops to a certain extent, the non-equilibrium condensation phenomenon occurs. Results of the linear RSM, *k-ω* Standard, *k-ω* SST, and Transition SST agree with Chen and Sun's observation [30], as can be seen in Figure 9.

The computational effort mainly depends on the turbulence model utilized when the discretization scheme, grid density, numerical solution method, and boundary conditions are fixed [8]. According to the convergence criteria reported above, taking the 3-D asymmetric mesh of E_2 with 2 million elements under Case 8 as an example, the steps of iterations with the same initialization condition for the seven models are used for the computational effort and convergence capability analysis.

As to the comparative study of the computational effort, the *k-ε* Standard with the least iteration number is taken as a reference. Their relative iteration number is summarized in Table 5. The calculations showed that among the seven turbulence models, the linear RSM model has the highest relative computational cost and is the most difficult to converge. This is due to the fact that it takes into account the anisotropy of turbulent viscosity, which requires an additional seven transport equations in three-dimensional flow, greatly increasing the computational cost and time cost, and weakening the convergence of iterative calculations in some cases, even failing to meet the convergence conditions. The relative computational cost of the transition SST model is only slightly less than that of the linear RSM model, as it has four transport equations. The remaining turbulence models are all two-equation turbulence models, so their relative computational cost is smaller. An item that reflects the time-averaged strain rate of the mainstream is added to the ε equation in the *k*-ε RNG. A content involving rotation and curvature is added to the *k*-ε Realizable's turbulent viscosity equation [6]. Therefore, the iteration number of Standard is the least of the *k*-ε series.

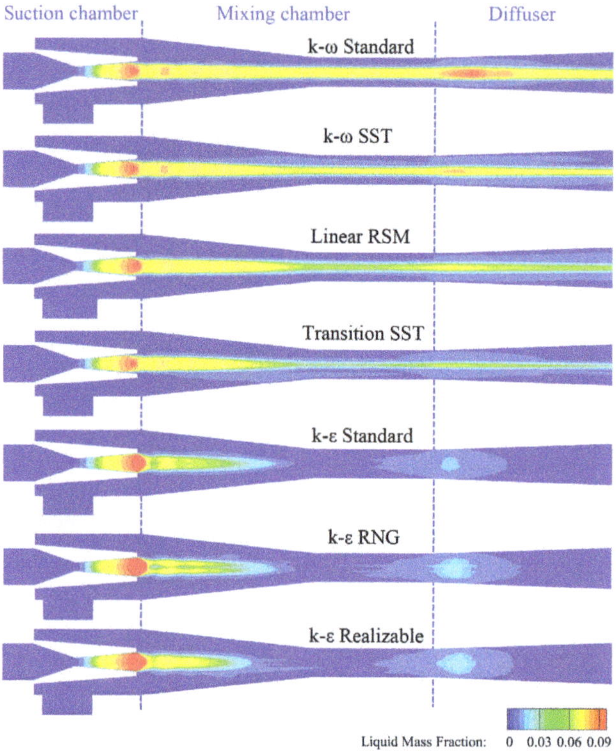

Figure 9. β contours under Case 4 with P_d = 1335 Pa.

Table 5. Relative computational effort required for solving Case 8 with different turbulence models.

k-ω Standard (EWT)	*k*-ω SST (EWT)	Transition SST (EWT)	*k*-ε Standard (EWT)	*k*-ε RNG (EWT)	*k*-ε Realizable (EWT)	RSM (EWT)
1.3	1.6	3.6	1	1.5	1.4	4.2

As to the convergence capability of the seven turbulence models, the reduction in mass or energy residuals is the slowest. The linear RSM is the most difficult model to converge, owing to its high degree of non-linearity. *k*-ω SST is the easiest to reach convergence, which is aligned with the conclusions of Besagni et al. [8], although the models compared are different.

In summary, considering condensation and 3-D effects, the *k*-ω SST model with low Reynolds number boundary conditions has the most accurate prediction results globally and locally, the best convergence, a low computational cost, and can be applied to the widest range of working conditions.

3.3. Comparison and Analysis of Density-Based Solver and Pressure-Based Solver

In this chapter, the non-equilibrium condensation model was combined with the k-ω SST turbulence model to perform a comparison of the density-based solver (D-B) and the P-B globally and locally under the same working conditions (Case 8).

P_d^* and Er_{max} obtained by the two solvers are nearly the same. Compared to D-B, Er, G_s, and G_m obtained by P-B are all slightly greater, with maximum differences of 1.2%, 1.7%, and 0.1%, respectively.

The flow structures that are reflected by contours of Mach number and static pressure simulated by both solvers are almost identical, as shown in Figures 10 and 11. While their liquid mass fraction contours are different around the wall region in the mixing chamber and the diffuser, they are also a little different around the axis, with a maximum relative error of 2.6%, as shown in Figure 12.

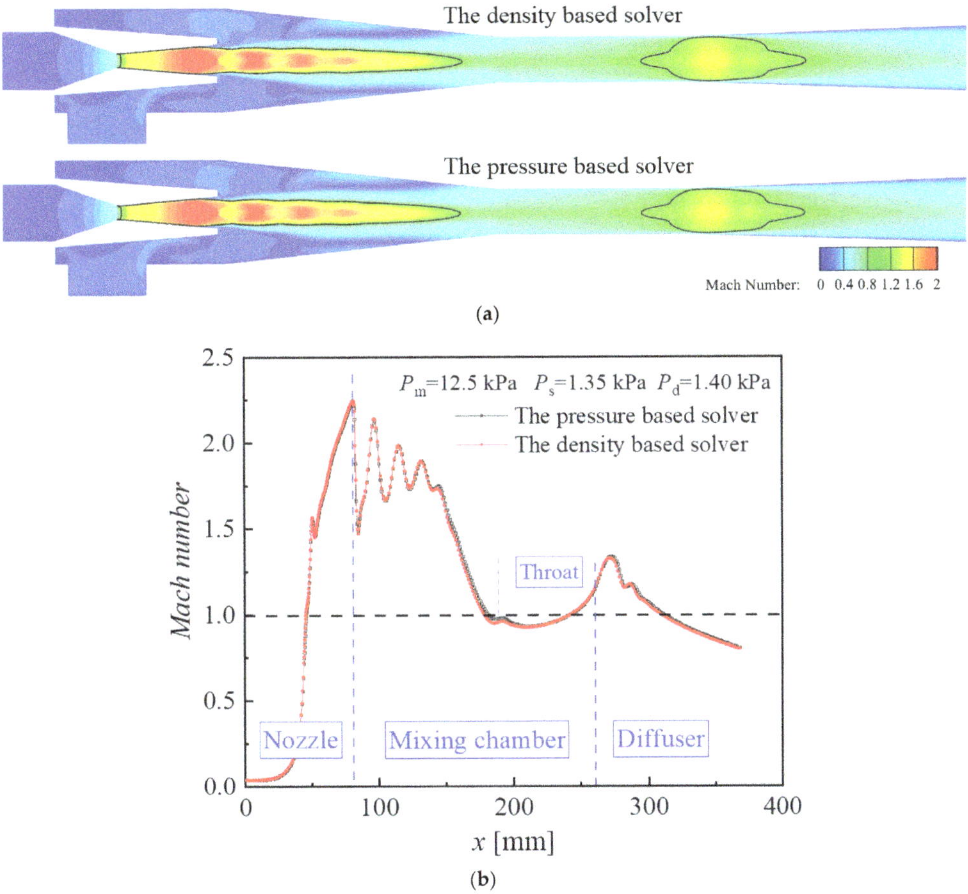

Figure 10. Mach number distributions with different solvers. (**a**) Mach number contours with different solvers. (**b**) Mach number profiles of the axis.

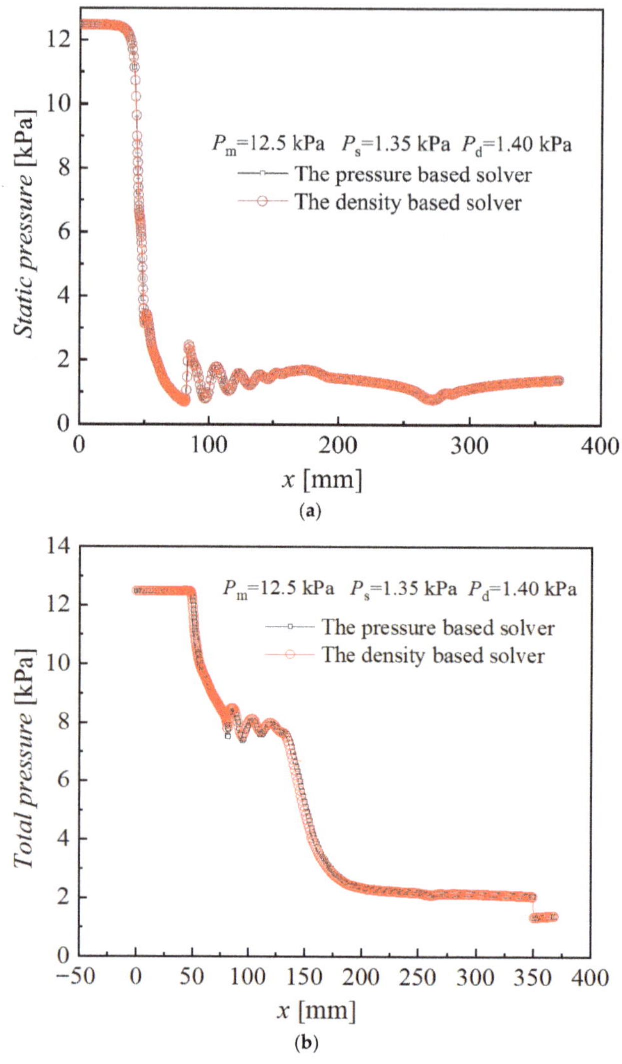

Figure 11. Pressure profiles of the axis with different solvers. (**a**) Static pressure profiles of the axis with different solvers. (**b**) Total pressure profiles of the axis with different solvers.

According to the convergence criteria reported above, the steps of iterations that reach the convergent condition with the same initialization condition for the two solvers are used for the computational effort and convergence capability analysis. The computational effort of P-B is one-sixth that of D-B. As to the convergence capability, the Courant Number needs to be reduced when D-B is adopted, which greatly increases the computation time and the number of iterations. Moreover, D-B is more sensitive to boundary conditions than P-B. In some other working conditions, D-B cannot achieve convergence while the P-B can. Therefore, the computational effort and convergence capability of P-B are much better than that of D-B, especially after coupling the non-equilibrium condensation model, which results in more computation and harder convergence than those of the ideal-gas model.

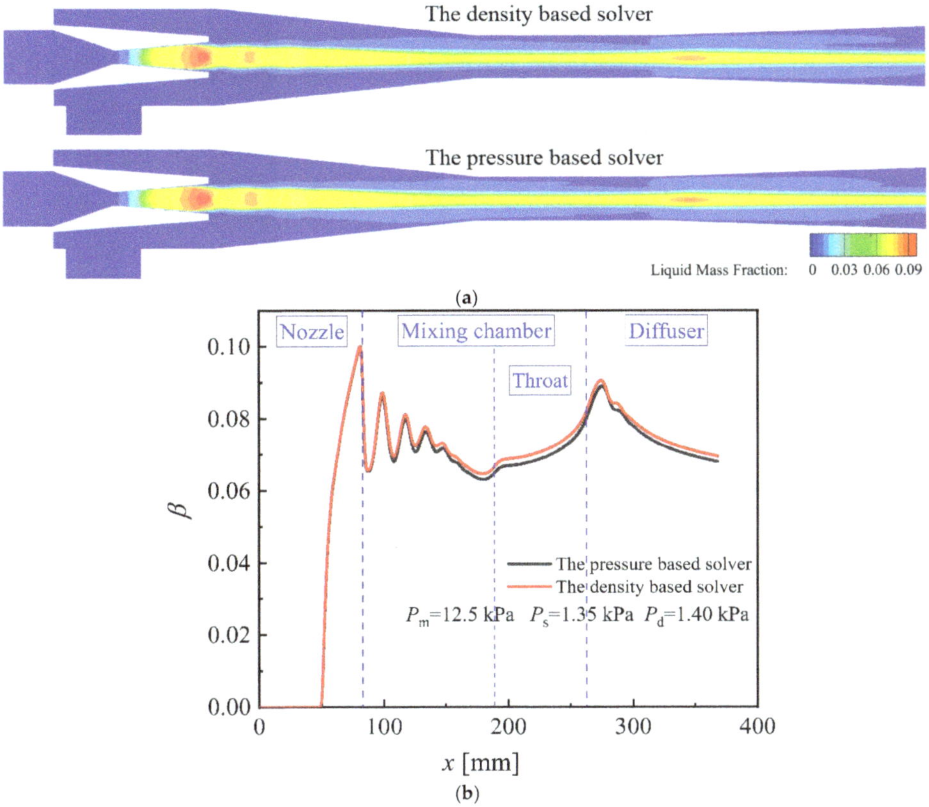

Figure 12. Liquid mass fraction distributions with different solvers. (**a**) Liquid mass fraction contours with different solvers. (**b**) Liquid mass fraction profiles of the axis.

In summary, in research and engineering not concerned with condensation phenomena, P-B can be used instead of D-B in simulating high compression with shock wave flow because of its excellent convergence properties and computation saving. However, since D-B can better reflect the physical nature of supersonic fluids, for the sake of rigor, it is still proposed to use D-B in scientific research considering condensation effects and three-dimensional effects.

3.4. Validation, Comparison, and Analysis of 3-D and 2-D Models, Condensation Model, and Ideal-Gas Model

Figure 13a illustrates the validation of ejector performance (Er_{max}) of the 3-D asymmetric model and 2-D model through validation with Chen and Sun's experimental data [30]. The average relative error ($\bar{e}$) of Er_{max} obtained by the 3-D and 2-D models with the experimental values are −1.9% and 12.8%, respectively, and the maximum relative errors are −8.8% and 25.6%, respectively. This not only verifies the reliability of the CFD model developed in this paper but also shows that, compared with the 2-D axisymmetric model, the 3-D model can reduce the $\bar{e}$ of the Er between the CFD results and the experimental value by 78.4% and the maximum relative error by 50.0%.

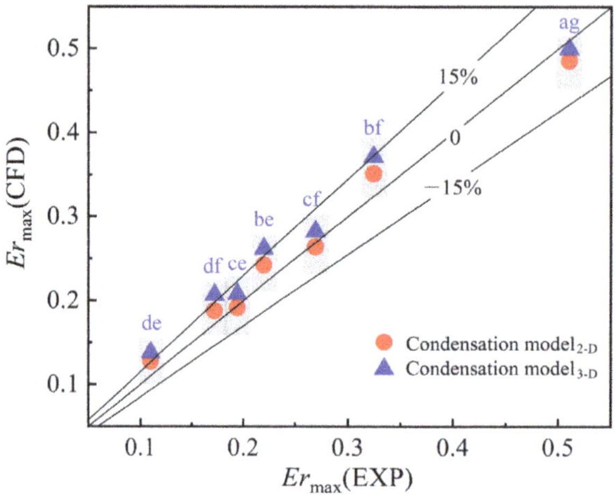

(**a**)

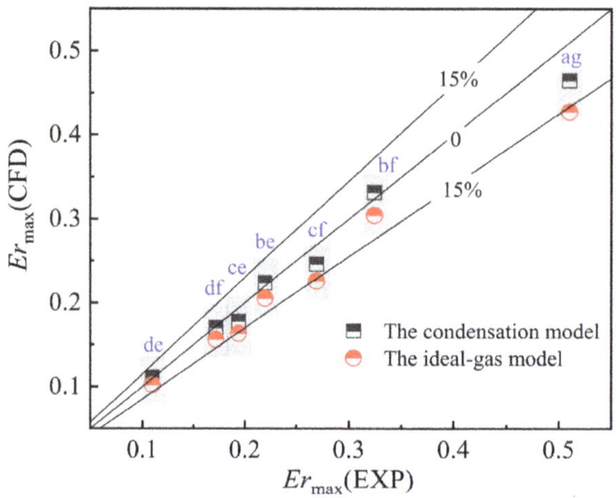

(**b**)

Figure 13. Er_{max} of different models and experiments comparison. (**a**) Comparison of the 2-D and 3-D models. (**b**) Comparison of the condensation model and the ideal-gas model.

Even though the computational effort and convergence capability of the ideal-gas model are much better than the condensation model, it is proposed to adopt it based on the following two points. As to the ejector performance, compared with the experimental values, the average relative error ($\overline{e}$) of the Er obtained by the non-equilibrium condensation model is 4.9%, which has a good agreement. In addition, the simulation results of the non-equilibrium condensation model are more consistent with the experimental values than those of the ideal-gas model. Compared with the ideal-gas model, the condensation model can reduce the $\overline{e}$ of the Er between the CFD results and the experimental value by 72.0% and the maximum relative error by 45.6%, as illustrated in Figure 13b. As to flow phenomena, the condensation model can obtain the condensation shock wave that cannot be obtained by the ideal-gas model, as shown in Figure 14 whose experimental values from the Nozzle A by Moore et al. [38]. When the condensation shock wave appears, the fluid pressure suddenly drops. $\overline{e}$ of P/P_{in} obtained by the 3-D model and 2-D model is 1.1% and 1.9%, respectively. The condensation shock wave occurrence position of the 3-D model accords with experiments quite well. In contrast, the position obtained by the 2-D model is 20.02 mm (2.67% of the nozzle length) ahead of the experimental results. The 3-D model results are closer to experiments compared to the 2-D model.

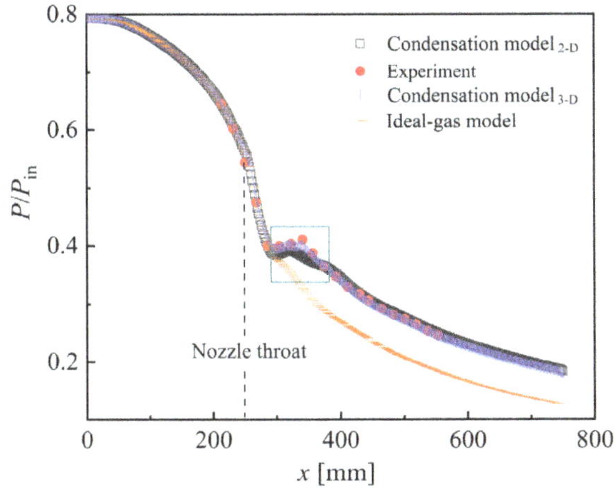

Figure 14. P/P_{in} of different models and experiments comparison.

In conclusion, a 3-D asymmetric condensation model with a density-based solver is recommended for simulating steam ejectors accurately.

4. Conclusions

This paper's originality is that both condensation and 3-D effects on the simulation results have been considered in the comprehensive comparative study of turbulence models, near-wall treatments, and solvers in the RANS equations approach for steam ejectors. Future research can perform a comparative study of the RANS equations and Large Eddy Simulation method based on this paper. The condensation and 3-D effects should be considered. The following conclusions have emerged:

(1) Considering non-equilibrium condensation and 3-D effects, the k-ω SST model with the low-Re boundary conditions has the most accurate prediction results globally and locally compared with the experiments: (a) The average relative error of Er_{max} and P_d^* are 4% and −15%, (b) the shock wave type and the condensation position are well predicted. Moreover, it has the widest adaptability to working conditions and the best convergence of the seven turbulence models studied in this paper.

(2) The difference in Er_{max} predicted by different models depends mainly on their mass flow rates of suction steam (G_s). The minimum difference in G_s predicted by different models is greater than 55%, while the maximum difference in G_m is less than 8%.

(3) The number and types of shock waves, and the location of condensation phenomenon of different turbulence models, are various. The k-ε series models predict a normal shock in the diffuser while the other models predict a different number of oblique shock waves.

(4) All other things being equal, the 3-D model can reduce the average relative error of Er by 78.4% compared to the 2-D model and the condensation model can reduce the average relative error of Er by 72.0% compared to the ideal-gas model.

(5) A 3-D asymmetric condensation model with a density-based solver is recommended for simulating steam ejectors accurately.

Author Contributions: Conceptualization, Y.L. and S.S.; Methodology, Y.L., H.H., D.D., S.S. and D.Z.; Software, S.S.; Validation, D.D. and S.L.; Formal analysis, Y.L., H.H., D.D., D.Z. and S.L.; Investigation, H.H., D.D. and D.Z.; Resources, Y.L. and S.S.; Data curation, H.H.; Writing—original draft, Y.L.; Writing—review & editing, H.H., D.D. and D.Z.; Visualization, D.Z. and S.L.; Supervision, Y.L.; Project administration, Y.L. and S.S.; Funding acquisition, Y.L. All authors have read and agreed to the published version of the manuscript.

Funding: The project was supported by "the Fundamental Research Funds for the Provincial Universities of Liaoning" (No.LJ212410150010).

Data Availability Statement: The raw data supporting the conclusions of this article will be made available by the authors on request.

Conflicts of Interest: Authors Yiqiao Li and Dan Zhou were employed by the company Bingshan Refrigeration & Heat Transfer Technologies Co., Ltd. The remaining authors declare that the research was conducted in the absence of any commercial or financial relationships that could be construed as a potential conflict of interest.

Nomenclature

B	[m^3/kg]	Virial coefficients
C	[m^6/kg^2]	Virial coefficients
C_P	[J/(kg·K)]	Isobaric heat capacity
E	[J]	Total energy
E_r	[-]	Entrainment ratio
e	[%]	Relative error
F	[N/m^3]	Source term
G	[kg/s]	Mass flow rate
h	[J/kg]	Specific enthalpy
h_{lv}	[J/kg]	Latent heat of condensation
J	[1/s]	Nucleation rate
K	[W/(m^2·K)]	Heat transfer coefficient
k	[m^2/s^2]	Turbulent kinetic energy
k_B	[-]	Boltzmann constant
Ma	[-]	Mach number
P	[Pa]	Pressure
R	[-]	Gas-law constant
r	[m]	Droplet radius
S	[-]	Super-saturation ratio
s	[J/(kg·mol·K)]	Specific entropy
S_T	[kg·K/ (m^3·s)]	Viscous dissipative term
T	[K]	Temperatures
Greek letters		
β	[-]	Liquid mass fraction
Γ	[kg/s]	Liquid mass generation rate
ρ	[kg/m^3]	Density
γ	[-]	Specific heat capacities ratio
μ	[Pa/s]	Dynamic viscosity
σ	[N/m]	Liquid surface tension
η	[1/m^3]	Droplet number density
θ	[-]	Non-isothermal correction factor
ν	[m^2/s]	Kinematic viscosity
ε	[m^2/s^3]	Turbulent dissipation rate
τ	[N/m^2]	Stress tensor
v	[m/s]	Velocity

Subscripts	
sat	Saturation
m	Motive steam
s	Suction steam
d	Discharge steam
l	Liquid
v	Vapor
max	Maximum
*	Critical
-	Average
eff	Effective
i,j	Space components
Abbreviations	
D-B	Density-based solver
P-B	Pressure-based solver
RSM	Reynolds Stress Model
SWF	Standard Wall Function
NWF	Non-equilibrium Wall Function
EWT	Enhanced Wall Treatment
N-S	Navier-Stokes
RANS	Reynolds-Averaged Navier–Stokes

References

1. Li, Y.; Shen, S.; Niu, C.; Mu, X.; Zhang, L. The effect of variable motive pressures on the performance and shock waves in a supersonic steam ejector with non-equilibrium condensing. *Int. J. Therm. Sci.* **2023**, *185*, 108034. [CrossRef]
2. Li, Y.; Shen, S.; Yang, Y. Three-dimensional characteristics of aerodynamic shockwave and condensation shockwave in steam ejectors. *Desalination* **2024**, *581*, 117606. [CrossRef]
3. Wang, X.; Dong, J.; Zhang, G.; Fu, Q.; Li, H.; Han, Y.; Tu, J. The primary pseudo-shock pattern of steam ejector and its influence on pumping efficiency based on CFD approach. *Energy* **2019**, *167*, 224–234. [CrossRef]
4. Han, Y.; Wang, X.; Yuen, A.C.Y.; Li, A.; Guo, L.; Yeoh, G.H.; Tu, J. Characterization of choking flow behaviors inside steam ejectors based on the ejector refrigeration system. *Int. J. Refrig.* **2020**, *113*, 296–307. [CrossRef]
5. Wen, C.; Karvounis, N.; Walther, J.H.; Ding, H.; Yang, Y. Non-equilibrium condensation of water vapour in supersonic flows with shock waves. *Int. J. Heat Mass Transf.* **2020**, *149*, 119109. [CrossRef]
6. Ansys, Inc. *Ansys FLUENT 18.0—Theory Guide*; Ansys Fluent: Canonsburg, PA, USA, 2019.
7. Besagni, G.; Cristiani, N.; Croci, L.; Guédon, G.R.; Inzoli, F. Computational fluid-dynamics modelling of supersonic ejectors: Screening of modelling approaches, comprehensive validation and assessment of ejector component efficiencies. *Appl. Therm. Eng.* **2020**, *186*, 116431. [CrossRef]
8. Besagni, G.; Inzoli, F. Computational fluid-dynamics modeling of supersonic ejectors: Screening of turbulence modeling approaches. *Appl. Therm. Eng.* **2017**, *117*, 122–144. [CrossRef]
9. Hemidi, A.; Henry, F.; Leclaire, S.; Seynhaeve, J.M.; Bartosiewicz, Y. CFD analysis of a supersonic air ejector. Part II: Relation between global operation and local flow features. *Appl. Therm. Eng.* **2009**, *29*, 2990–2998. [CrossRef]
10. Gagan, J.; Smierciew, K.; Butrymowicz, D.; Karwacki, J. Comparative study of turbulence models in application to gas ejectors. *Int. J. Therm. Sci.* **2014**, *78*, 9–15. [CrossRef]
11. Zhu, Y.; Jiang, P. Experimental and numerical investigation of the effect of shock wave characteristics on the ejector performance. *Int. J. Refrig.* **2014**, *40*, 31–42. [CrossRef]
12. Croquer, S.; Poncet, S.; Aidoun, Z. Turbulence modeling of a single-phase R134a supersonic ejector. Part 1: Numerical benchmark. *Int. J. Refrig.* **2016**, *61*, 140–152. [CrossRef]
13. Ruangtrakoon, N.; Thongtip, T.; Aphornratana, S.; Sriveerakul, T. CFD simulation on the effect of primary nozzle geometries for a steam ejector in refrigeration cycle. *Int. J. Therm. Sci.* **2013**, *63*, 133–145. [CrossRef]
14. Han, Y.; Wang, X.; Sun, H.; Zhang, G.; Guo, L.; Tu, J. CFD simulation on the boundary layer separation in the steam ejector and its influence on the pumping performance. *Energy* **2019**, *167*, 469–483. [CrossRef]
15. Varga, S.; Soares, J.; Lima, R.; Oliveira, A.C. On the selection of a turbulence model for the simulation of steam ejectors using CFD. *Int. J. Low-Carbon Technol.* **2017**, *12*, 233–243. [CrossRef]
16. Wu, J.; Xu, P.; Li, L.; Li, Z.; Qi, H.; Wang, C.; Zhang, Y.; Xie, Y.; Tan, D. Multiphase dynamic interfaces and abrasive transport dynamics for abrasive flow machining in shear thickening transition states. *Powder Technol.* **2024**, *446*, 120150. [CrossRef]
17. Tan, Y.; Ni, Y.; Xu, W.; Xie, Y.; Li, L.; Tan, D. Key technologies and development trends of the soft abrasive flow finishing method. *J. Zhejiang Univ.-Sci. A* **2023**, *24*, 1043–1064. [CrossRef]
18. Aidoun, Z.; Ameur, K.; Falsafioon, M.; Badache, M. Current Advances in Ejector Modeling, Experimentation and Applications for Refrigeration and Heat Pumps. Part 1: Single-Phase Ejectors. *Inventions* **2019**, *4*, 15. [CrossRef]

19. Li, C.; Li, Y.Z. Investigation of entrainment behavior and characteristics of gas-liquid ejectors based on CFD simulation. *Chem. Eng. Sci.* **2011**, *66*, 405–416. [CrossRef]
20. Yazdani, M.; Alahyari, A.A.; Radcliff, T.D. Numerical modeling of two-phase supersonic ejectors for work-recovery applications. *Int. J. Heat Mass Transf.* **2012**, *55*, 5744–5753. [CrossRef]
21. Sriveerakul, T.; Aphornratana, S.; Chunnanond, K. Performance prediction of steam ejector using computational fluid dynamics: Part 1. Validation of the CFD results. *Int. J. Therm. Sci.* **2007**, *46*, 812–822. [CrossRef]
22. Tashtoush, B.M.; Moh'd A, A.N.; Khasawneh, M.A. A comprehensive review of ejector design, performance, and applications. *Appl. Energy* **2019**, *240*, 138–172. [CrossRef]
23. Cai, L.; He, M.; Huang, K.Z.; Xiong, W. Computational fluid dynamics simulation of the supersonic steam ejector using different condensation model. *Therm. Sci.* **2019**, *23*, 236. [CrossRef]
24. Mazzelli, F.; Little, A.B.; Garimella, S.; Bartosiewicz, Y. Computational and experimental analysis of supersonic air ejector: Turbulence modeling and assessment of 3D effects. *Int. J. Heat Fluid Flow* **2015**, *56*, 305–316. [CrossRef]
25. Young, J.B. An equation of state for steam for turbomachinery and other flow calculations. *J. Eng. Gas Turbines Power* **1988**, *110*, 1–7. [CrossRef]
26. Young, J.B. Spontaneous condensation of steam in supersonic nozzles. *PCH Physicochem. Hydrodyn.* **1982**, *3*, 57–82.
27. Kong, N.; Qi, Z. Influence of speed of sound in two-phase region on 1-D ejector performance modelling. *Appl. Therm. Eng.* **2018**, *139*, 352–355. [CrossRef]
28. Daiguji, H.; Ishazaki, K.; Ikohagi, T. A high-resolution numerical method for transonic non-equilibrium condensation flows through a steam turbine cascade. *Proc. 6th Int. Symp. Comput. Fluid Dyn.* **1995**, *1*, 479–484.
29. Senoo, S.; Shikano, Y. Two-dimensional analysis for non-equilibrium homogeneously condensing flows through steam turbine cascade. *JSME Int. J. Ser. B Fluids Therm. Eng.* **2002**, *45*, 865–871. [CrossRef]
30. Chen, Y.; Sun, C. Experimental study of the performance characteristics of a steam-ejector refrigeration system. *Exp. Therm. Fluid Sci.* **1997**, *15*, 384–394. [CrossRef]
31. Zhang, G.; Zhang, X.; Wang, D.; Jin, Z.; Qin, X. Performance evaluation and operation optimization of the steam ejector based on modified model. *Appl. Therm. Eng.* **2019**, *163*, 114388. [CrossRef]
32. Zhang, G.; Dykas, S.; Li, P.; Li, H.; Wang, J. Accurate condensing steam flow modeling in the ejector of the solar-driven refrigeration system. *Energy* **2020**, *212*, 118690. [CrossRef]
33. Wang, X.; Dong, J.; Li, A.; Lei, H.; Tu, J. Numerical study of primary steam superheating effects on steam ejector flow and its pumping performance. *Energy* **2014**, *78*, 205–211. [CrossRef]
34. Ariafar, K.; Buttsworth, D.; Al-Doori, G.; Malpress, R. Effect of mixing on the performance of wet steam ejectors. *Energy* **2015**, *93*, 2030–2041. [CrossRef]
35. Mazzelli, F.; Giacomelli, F.; Milazzo, A. CFD modeling of condensing steam ejectors: Comparison with an experimental test-case. *Int. J. Therm. Sci.* **2018**, *127*, 7–18. [CrossRef]
36. Zhang, G.; Wang, X.; Pourranjbar, D.; Dykas, S.; Li, H.; Chen, J. The comprehensive analysis of the relationship between the latent heat, entrainment ratio, and ejector performance under different superheating degree conditions considering the non-equilibrium condensation. *Appl. Therm. Eng.* **2022**, *200*, 117701. [CrossRef]
37. Li, Y.; Niu, C.; Shen, S.; Mu, X.; Zhang, L. Double choking characteristics of three-dimensional steam ejector with non-equilibrium condensing. *Appl. Therm. Eng.* **2022**, *211*, 118446. [CrossRef]
38. Moore, M.J.; Walters, P.T.; Crane, R.I.; Davidson, B.J. Predicting the fog-drop size in wet-steam turbines. In Proceedings of the IMechE Conference on Heat and Fluid Flow in Steam and Gas Turbine Plant, Coventry, UK, 3–5 April 1973; pp. C37–C73.

energies

MDPI

Article

The Effect of Facade Orientation on the Electrical Performance of a BIPV System: A Case Study in João Pessoa, Brazil

Michelli Mayara de Medeiros Gomes [1], Renan Douglas Lopes da Silva Cavalcante [2], Oswaldo Hideo Ando Junior [3], Claudio Del Pero [4], João Alves de Lima [1] and Taynara Geysa Silva do Lago [1,2,*]

[1] Multiphysics Energy Simulation Laboratory (LABSIMER), Department of Renewable Energy Engineering, Center for Alternative and Renewable Energies (CEAR), Federal University of Paraíba (UFPB), João Pessoa 58051-900, Brazil; michelli.gomes@estudante.cear.ufpb.br (M.M.d.M.G.); jalima@cear.ufpb.br (J.A.d.L.)

[2] Graduate Program in Renewable Energy (PPGER), Center for Alternative and Renewable Energies (CEAR), Federal University of Paraíba (UFPB), João Pessoa 58051-900, Brazil; renan.cavalcante@estudante.cear.ufpb.br

[3] Research Group on Energy & Energy Sustainability (GPEnSE), Academic Unit of Cabo de Santo Agostinho (UACSA), Federal Rural University of Pernambuco (UFRPE), Cabo de Santo Agostinho 54518-430, Brazil; oswaldo.ando@ufrpe.br

[4] Department of Architecture, Built Environment and Construction Engineering, Politecnico di Milano (POLIMI), 20133 Milan, Italy; claudio.delpero@polimi.it

* Correspondence: taynara@cear.ufpb.br

Abstract: Photovoltaic solar energy has emerged as a vital solution in energy transition, especially in buildings with building-integrated photovoltaic (BIPV) systems. This study evaluated a BIPV system in João Pessoa, Brazil, using PVsyst 7.4 software for simulations in 2023 and experimental tests in 2024 to compare real and simulated energy values injected into the grid. The experiments were conducted in April, June, July, and August, covering seasonal variations. The east and west orientations showed the best performance, with annual production of 65.5 and 81.6 kWh, respectively, due to the balanced solar irradiance. Conversely, the north and south orientations generated less electricity, with 51.3 and 42.1 kWh, respectively, due to lower direct solar exposure. The average injected energy was 0.2445 to 0.2530 kWh/day in April, 0.2851 to 0.2470 kWh/day in June, 0.2816 to 0.2400 kWh/day in July, and 0.1556 to 0.1700 kWh/day in August. Performance ratios were 44.3% for the north, 44.4% for the south, 47.4% for the east, and 56.8% for the west, highlighting the impact of orientation. It is concluded that the east and west facades are more suitable for BIPV systems in João Pessoa, emphasizing the importance of local factors in optimizing urban energy efficiency.

Keywords: BIPV; energy efficiency; photovoltaic energy

1. Introduction

Currently, photovoltaic technologies have gained global prominence because of increasing awareness of environmental issues and the demand for sustainable energy solutions. These technologies demonstrate significant competitiveness compared to other renewable energy sources, particularly due to their unique ability to directly convert solar energy into electricity through photovoltaic effects. These characteristics position solar photovoltaic energy as one of the primary alternatives for diversifying the global energy matrix and substantially contribute to sustainable energy transition [1,2].

However, the implementation of PV systems in urban environments presents significant challenges, primarily due to the limited availability of space for installing solar panels.

In densely built areas, rooftops and coverings are often the only available locations, thus limiting the potential for energy generation. To overcome these restrictions, it is essential to explore and maximize the use of other surfaces, such as building facades, by integrating photovoltaic systems into architectural design [3].

The integration of photovoltaic modules into facades (building-integrated photovoltaic—BIPV) is an innovative approach that transforms architectural elements into solar energy generators, utilizing vertical surfaces that are traditionally not used for this purpose. These modules can be applied in transparent or semi-transparent forms on glass, such as windows, or integrated into cladding materials, such as tiles and metal panels. In addition to adding a modern esthetic to buildings, local energy generation from these vertical surfaces can significantly reduce long-term energy costs, thereby promoting sustainability and energy self-sufficiency in buildings [4].

The field of photovoltaic integration in buildings is constantly evolving because of advances in photovoltaic materials, energy storage systems, and integration technologies. These developments can make BIPV systems more accessible, efficient, and applicable to a wide range of architectural projects, from small constructions to large urban developments [5].

The performance of PV systems on windows and facades is affected by factors such as geographic location and climatic conditions. Parameters such as solar irradiance, ambient temperature, and dirt accumulation on the surface of the modules play crucial roles in the efficiency and longevity of photovoltaic conversion. Therefore, a detailed analysis of these factors is essential for the planning and optimization of integrated photovoltaic systems in buildings, especially in urban contexts where environmental variables can vary significantly [6–9].

Although the analysis of these systems can be performed manually, the use of simulation software allows for more detailed and precise evaluation. PVsyst, for example, is a widely used tool for analyzing multiple performance parameters of photovoltaic systems. The proposed software provides detailed estimates of electricity generation considering local characteristics and climate patterns throughout the year [10].

Recent studies have demonstrated the effectiveness of the PVsyst software in the detailed analysis of building-integrated photovoltaic systems (BIPV). Essah et al. [11] conducted PVsyst simulations to assess the performance, yield, and aging characteristics of a polycrystalline silicon cell PV system integrated into the facade of the Engineering Building at the University of Reading. Shakouri et al. [12] investigated the effects of a double-skin photovoltaic facade on energy generation, cooling, and thermal load reduction. They used PVsyst to evaluate energy generation and DesignBuilder to calculate the demand, focusing on incorporating PV modules as a retrofit strategy.

Jeong et al. [13] compared the output power of a colored BIPV system simulated by PVsyst using meteorological data from the Korea Meteorological Administration (KMA) and measured irradiance data. Hamzah and Go [14] analyzed the performance of colored BIPV systems in a building located in a tropical region of Malaysia, combining 3D modeling in Building Information Modeling (BIM) software with detailed solar analysis in PVsyst to obtain precise results. Yang et al. [15] modeled complex BIPV systems installed on roofs, facades, and coverings of buildings in Zhuhai, China, using different simulation tools, including PVsyst, to investigate the capabilities and limitations of these technologies.

Demir and Aktacir [16] investigated strategies to maximize the efficiency of building-integrated photovoltaic (BIPV) systems under specific meteorological conditions in Şanlıurfa, Turkey. The study focused on comparing the performance of monofacial and bifacial panels installed on different building surfaces using experimental methods and simulations with the PVsyst software. Yu et al. [17] optimized the operation of BIPV systems and evaluated their environmental and economic benefits. The study used PVsyst to simulate the energy performance of building-integrated photovoltaic systems and PV-BILD to assess the environmental and economic impacts throughout the lifecycle of BIPV systems.

Zhang et al. [18] optimized building-integrated photovoltaic (BIPV) systems in a high-rise building in Shenzhen, China, to improve its energy efficiency. The PVsyst software was used to simulate the energy performance of the PV system in different scenarios, considering the location, solar irradiance, and shading. The analysis involved optimizing the panel layout and assessing its technical and economic feasibility.

In this context, this paper aims to analyze the electrical performance of a BIPV system consisting of a monocrystalline silicon photovoltaic window installed on a facade in the city of João Pessoa, Paraíba, Brazil, using PVsyst software. Additionally, the analysis involves the design and construction of an experimental setup, focusing on evaluating efficiency under specific climatic conditions, considering factors such as solar irradiance, ambient temperature, and other critical parameters for optimizing energy performance.

This type of BIPV system is often referred to as a PV window. The adoption of semi-transparent photovoltaic (PV) windows in buildings has emerged as a relevant trend in sustainable design because they simultaneously combine thermal insulation, daylighting, and energy generation. However, this application requires careful analysis to determine the optimal window orientation.

Jamala et al. [19] performed computational modeling and simulation using Rhinoceros and Grasshopper for a prototype building with a single semi-transparent monocrystalline PV window, considering the climate of Gowa, Indonesia (5.2° S). They investigated the uncertainty of weighting factors in optimizing the PV window orientation by considering several annual daylight and energy performance metrics. The results indicated that the south orientation was the most advantageous in 73% of the cases, followed by the east (24%) and north (3%). When each group of metrics was given a single weight, the south orientation remained the best in 62% of the cases.

Guo et al. [20] conducted experimental tests using naturally ventilated double-pane PV windows and validated the model in EnergyPlus. Using the validated model, the energy performance of the windows was analyzed in five typical climates in China. The south-facing windows showed the lowest electricity consumption in the Harbin, Beijing, Shanghai, and Lhasa climates. However, east-facing windows exhibited the lowest electricity consumption in the Shenzhen climate.

This work's main contribution is the identification of the optimal orientation of the monocrystalline PV window that provides the best annual electrical performance for a BIPV system. This study, the first of its kind in this city, offers important insights and establishes a reliable criterion for optimizing the design of PV windows in the southern hemisphere.

2. Location Characterization

The present study was conducted at the Center for Alternative and Renewable Energies (CEAR) at the Federal University of Paraíba (UFPB), located at the João Pessoa campus, in the capital of the State of Paraíba, Brazil, at approximate coordinates of 7.12° south and 34.86° west, near the Equator. This geographic location ensures high solar radiation incidence throughout the year, thereby favoring the efficiency of photovoltaic systems. The choice of João Pessoa for the installation of the building-integrated photovoltaic system was made not only because of the favorable climatic conditions but also because of the scarcity of regional studies on photovoltaic systems applied to vertical facades (90° orientation), contributing to filling gaps in the scientific literature and promoting the development of this technology under the typical climatic conditions of Northeast Brazil.

The Paraíba state is located in the Northeast Region of Brazil and borders Rio Grande do Norte to the north, Pernambuco to the south, Ceará to the west, and the Atlantic Ocean to the east. It comprises 223 municipalities, according to the Brazilian Institute of Geography and Statistics (IBGE) [21] and exhibits semi-arid and tropical humid climates. It is situated in bioclimatic zone 8 according to the NBR 15220-3 standard [22]. This classification defines climatic regions in Brazil based on average monthly maximum and minimum temperatures and relative humidity.

Bioclimatic zone 8 covers 53.7% of Brazil's territory, as shown in Figure 1, and is concentrated in the northernmost areas, including humid equatorial climate zones such as the Amazon. The main characteristics of this region include high year-round temperatures, high relative humidity, and minimal daily temperature variation. These regions are located between the Tropic of Capricorn and the Equator [23].

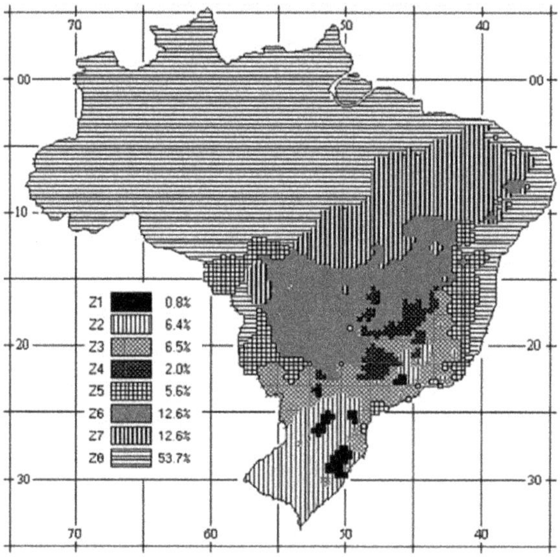

Figure 1. Brazilian bioclimatic zoning [22].

The rainfall regime in Paraíba exhibits two distinct rainy seasons that are influenced by regional atmospheric systems. In the Sertão and Alto Sertão regions, which cover the west and north of the Borborema Plateau, rainfall occurs between February and May. The rainy season occurs from April to July on the coast and in adjacent areas to the east. This seasonal variation is modulated by the trade wind circulation, the Intertropical Convergence Zone (ZCIT), and other regional atmospheric systems and is influenced by the geographic diversity across the Paraíba state [24].

Paraíba has high solar radiation levels, indicating its significant potential for renewable energy use. In João Pessoa, the annual average solar radiation is 5.53 kWh/(m^{2}day), according to data from the Reference Center for Solar and Wind Energy Sérgio de S. Brito (CRESESB) [25], which positioned the city as a strategic location for the application of photovoltaic technologies. This scenario stimulates scientific research and promotes the integration of PV systems, contributing to the diversification of the energy matrix and strengthening sustainability in urban areas.

The levels of Horizontal Global Irradiance (GHI) in the territory of Paraíba, with a focus on João Pessoa, are presented in Figure 2, which illustrates the spatial distribution of solar radiation across the state. The graph was created using QGIS 3.34.12 software [26], a Geographic Information System (GIS) tool widely used for processing and visualizing georeferenced data. QGIS allows for the integration and analysis of multiple information layers to create thematic maps and precise geospatial models.

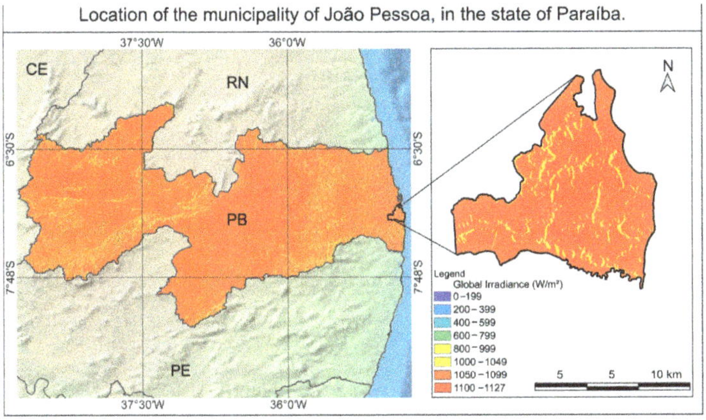

Figure 2. State of Paraíba—Horizontal Global Irradiance.

For the creation of Figure 2, data from the National Institute for Space Research (INPE) [27] and the Brazilian Institute of Geography and Statistics (IBGE) [28] were used, including information on altitude, slope, relief, and global radiation indexes across the state. This integrated approach enabled the identification of areas with higher solar energy potential, with an emphasis on the urban regions of João Pessoa.

3. Materials and Methods

3.1. BIPV System Characterization

The present study focuses on the electrical performance analysis of a BIPV system composed of a 132 Wp photovoltaic module with monocrystalline silicon (Si-mono) cells and 25% efficiency under standard test conditions (STC). The module, encapsulated between two 5 mm layers of translucent glass, was installed vertically at a 90° angle on the facade of a container, functioning as a double-glazed window.

The installation was designed with a 50 mm gap between the module and the inner glass, creating a stagnant air gap. The system, shown in Figure 3, is connected to the Paraíba electrical grid via a 130 Wp microinverter with maximum power point tracking (MPPT) and is currently oriented toward the north. Information about the key system components is presented in Table 1.

Figure 3. BIPV system.

Table 1. Photovoltaic module and microinverter specifications.

Photovoltaic Module		Microinverter	
Manufacturer	Solid PV (São Paulo, Brazil)	Manufacturer	SunPV (Shenzhen, China)
Dimensions (mm)	990 × 990	Dimensions (mm)	203 × 76
Thickness (±0.5) (mm)	11	Thickness (mm)	35
Pmax. (Wp)	132	Pnom. (W)	130
Voc (V)	23.1	V_{mppt}(min) (V)	13
Isc (A)	7.1	V_{mppt}(max) (V)	25
Vmax. (V)	19.8	Vmax. (V)	30
Imax. (A)	6.7	Efficiency max. (%)	92.50
Call Efficiency (%)	25	Efficiency CEC (%)	90.50

This configuration was chosen to evaluate the performance of the system integrated into buildings considering local environmental variables such as solar irradiance and ambient temperature through two complementary methods: simulation and experimental. In the simulation method, simulations were performed to estimate energy production throughout 2023, considering the orientations north, south, east, and west, aiming to identify the most efficient orientation for 90° photovoltaic modules on vertical facades in João Pessoa, as well as to assess the influence of climatic variations on electrical performance.

The experimental method involved collecting energy production data on specific days in 2024 to improve the understanding of the real operational conditions of the system. This combined approach will allow for a detailed analysis of the electrical performance of the BIPV system, considering both simulations based on annual climatic models and field-observed results.

The methodological flow of this study is schematically represented in the flowchart shown in Figure 4.

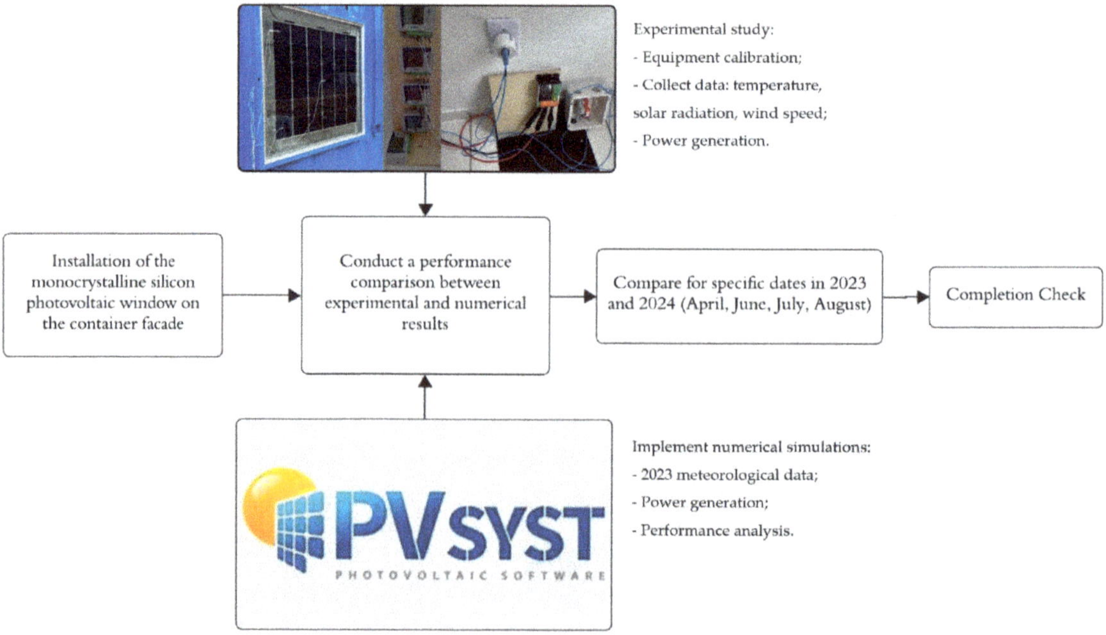

Figure 4. Performance analysis method for BIPV system.

3.2. Experimental Method

In the experimental method, data on temperature, wind speed, and solar radiation on the surface of the photovoltaic module were collected to ensure greater accuracy in the results. Temperatures were measured using 13 type K thermocouples strategically positioned and connected to Novus dataloggers. Three thermocouples were installed on the external surface of the module, three on the internal surface, four on the internal and external surfaces of the window glass, and three others to monitor the air temperatures inside and outside the container, as well as in the cavity air gap.

The thermocouples were calibrated using a high-resolution thermometer with an accuracy of 0.1 °C. The thermometer was stabilized at ambient temperature, and the thermocouples were immersed in water until they reached thermal equilibrium with the thermometer. The water temperature was gradually increased using a magnetic stirrer with a heating function (Figure 5) until reaching values close to 100 °C.

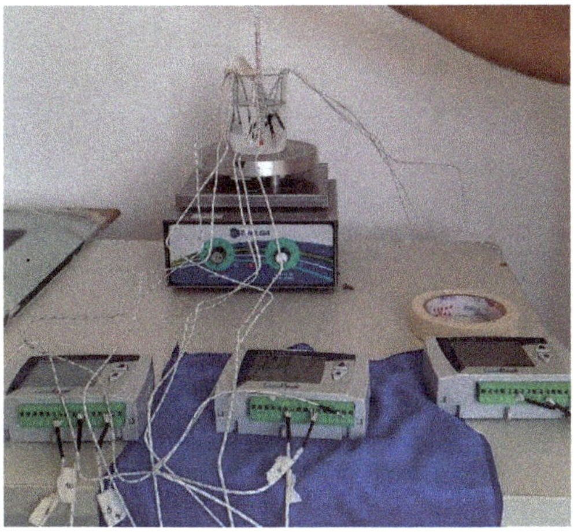

Figure 5. Thermocouple calibration in thermal stirrer.

Wind speeds were measured at 1.5 m above the ground using a digital anemometer (MDA-20), while global direct (Hor.Global) and diffuse (Hor.Diffuse) horizontal solar radiation was recorded by a digital radiometer positioned in front of the module, ensuring measurement accuracy. Finally, the amount of energy injected into the power distribution grid was recorded using a smart plug, which collected data on the produced power, current, and voltage.

Figure 6 shows the arrangement of the thermocouples and the instruments. Figure 7 shows a smart plug for recording energy injected into the grid. Table 2 shows the instrument's precision in detail.

Table 2. Technical details of instruments used.

Instrument	Manufacturer	Technical Data	Precision
Temperature sensor	Type K thermocouple (São Paulo, Brazil)	−50 °C to 204 °C	±2.5 °C
Digital radiometer	Instrutherm (São Paulo, Brazil)	Measuring range: 2000 W/m^2 Spectral Response: 400 nm to 1000 nm	±10 W/m^2
Digital anemometer MDA-20	Minipa (Belo Horizonte, Brazil)	0.8 m/s to 40 m/s	±2%

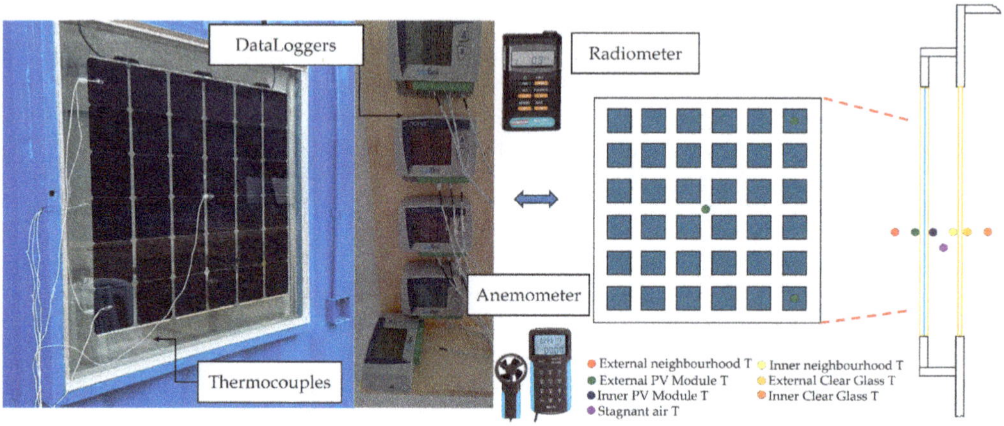

Figure 6. Schematic diagram of thermocouple arrangements and measurement instruments.

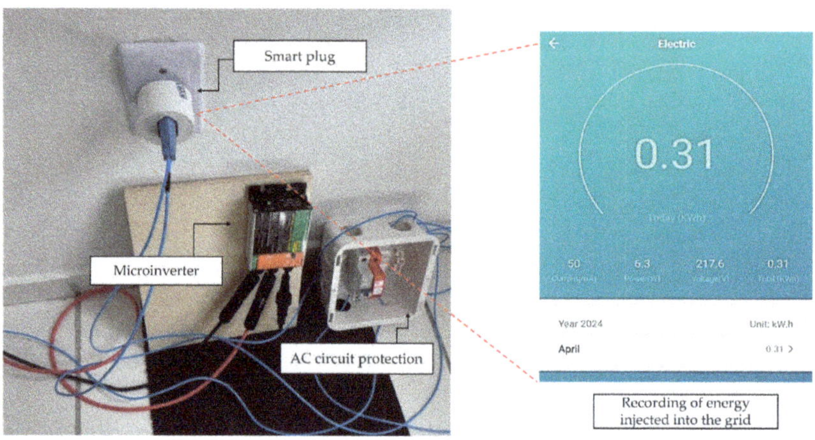

Figure 7. Smart plug for recording energy injected into grid.

Data collection took place in 2024 on 23–25 April; 21, 26, and 28 June; 17–19 July; and 23, 26, and 27 August, between 8:00 AM and 4:00 PM, with measurements taken every 20 min, which can be consulted in Table 3.

Table 3. Weather data for João Pessoa.

	Experimental (2024)				Simulated (2023)			
Period	**Hor. Global (kWh/(m^{2}day))**	**Hor. Global (W/m^2)**	**Ambient Temperature (°C)**	**Wind Speed (m/s)**	**Hor. Global (kWh/(m^{2}day))**	**Hor. Global (W/m^2)**	**Ambient Temperature (°C)**	**Wind Speed (m/s)**
23 April	10.37	432.19	33.9	1.50	14.72	613.30	31.0	2.63
24 April	11.75	489.60	33.7	1.76	12.91	537.90	30.2	2.67
25 April	10.94	445.80	34.0	1.49	12.98	540.80	30.0	2.51
21 June	12.81	533.80	30.1	2.02	11.51	479.60	27.6	2.96
26 June	10.99	457.90	30.1	2.19	12.18	508.50	28.1	3.32
28 June	10.43	434.60	30.3	2.02	11.80	491.70	28.2	3.16
17 July	12.03	501.30	30.3	2.36	13.92	580.00	27.8	4.17
18 July	10.73	447.10	29.1	2.27	12.09	503.80	27.4	4.08
19 July	9.04	376.70	29.1	2.26	12.20	508.30	27.2	3.84
23 August	7.04	293.30	29.0	2.14	11.02	459.20	27.3	4.12
24 August	8.99	374.60	29.9	2.68	14.24	593.30	27.6	4.20
25 August	6.89	287.10	30.4	2.18	12.36	515.00	27.5	3.38

3.3. Simulation Method

For the simulation method, an electrical simulation was performed using PVsyst software to analyze the performance of a PV system under different orientations (north, south, east, and west), considering the amount of energy produced and injected into the grid under these conditions. Pvsyst was selected because of its wide recognition in the solar energy industry and its established application in academic studies, as discussed in the first section of this article. Although it is commercial software, it was used during the 30-day free trial period, enabling analysis without additional costs.

Pvsyst allows for the consideration of essential variables such as geographic location, module tilt, the technical specifications of the system components, and local climatic variations, making the simulation highly representative of real operating conditions. Figure 8 shows a diagram of the simulation steps performed.

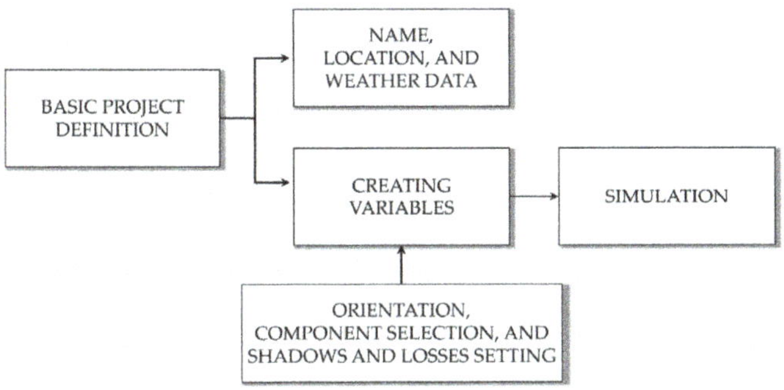

Figure 8. Pvsyst simulation diagram.

The location was defined using the software's interactive map, and its interface is presented in Figure 9. Based on the selected location, Pvsyst automatically imports the meteorological data required for the simulation into its monthly meteorological database. These data were sourced from reliable internal databases, including NASA, Meteonorm, and SAM, as well as external databases such as those provided by INPE and field measurements.

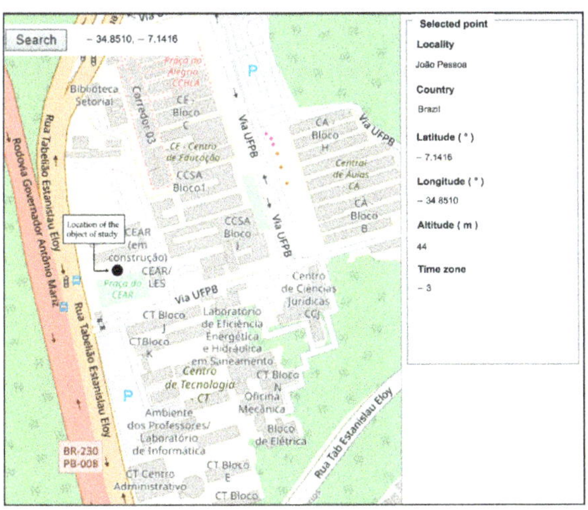

Figure 9. Pvsyst interactive map interface.

For this analysis, meteorological data collected in 2023 from the Solarimetric Station of the Federal University of Paraíba were used. These data include measurements of direct (Hor.Global) and diffuse horizontal irradiation (Hor.Diffuse) (kWh/m^2), ambient temperature (°C), and wind speed (m/s), which are essential for the software simulation. Data were collected at 10 min intervals throughout the year, ensuring adequate temporal resolution for analyzing climatic variations and their impact on the photovoltaic system's performance, and are presented in Table 4.

Table 4. Weather data for João Pessoa, year 2023.

Period	Hor. Global (kWh/(m^{2}day))	Hor. Diffuse (kWh/(m^{2}day))	Ambient Temperature (°C)	Wind Speed (m/s)
January 2023	6.13	2.80	27.3	2.2
February 2023	5.73	2.86	26.8	2.4
March 2023	5.44	2.82	27.3	2.1
April 2023	5.59	2.37	27.4	1.8
May 2023	5.24	2.23	27.1	2.0
June 2023	4.29	1.96	24.7	2.1
July 2023	4.88	2.19	25.4	2.4
August 2023	5.35	2.57	25.5	2.3
September 2023	5.79	2,85	25.9	2.6
October 2023	6.56	2.52	26.8	2.6
November 2023	6.54	2.61	27.3	2.5
December 2023	6.08	2.77	27.6	2.3
Year 2023	5.64	2.54	26.6	2.3

After inputting the meteorological data, Pvsyst enables the configuration of module orientations, specification of system components (such as inverters and photovoltaic modules), and calculation of system losses during operation. These losses are determined based on factors such as shading, inverter efficiency, thermal losses, and electrical losses, thereby allowing for a precise estimation of energy production and system performance under various operating conditions.

The performance evaluation of a photovoltaic system considers several factors, such as pollution, ohmic losses, variations in the angle of incidence, and temperature, which affect the energy conversion efficiency. These factors contribute to the system's annual losses and are essential for accurately estimating solar yield.

The calculation begins with the global irradiance on the horizontal plane, which is obtained by summing the monthly local irradiance values throughout the year. When directed toward the photovoltaic module plane, losses such as shading and soiling are considered, which reduce the conversion efficiency. In this study, the soiling loss over a year was set at 3%, as determined using the Pvsyst software.

Pvsyst also considers a 3D model of the system, accounting for the system's geometry, the sun's varying position throughout the year, and the potential shading caused by surrounding obstacles, which allows for precise tracking of the solar movement and its impact on the photovoltaic system's performance.

Losses related to the angle of incidence reflect how the inclination of the photovoltaic module affects its absorption capacity, with this relationship being inversely proportional. In other words, when sunlight strikes the module at a steep angle, the absorption efficiency decreases. According to Abella et al. [29], the Pvsyst software calculates these losses using the ASHRAE (American Society of Heating, Refrigerating, and Air-Conditioning Engineers) [30] parameterization, as shown in Equation (1).

$$FIAM = 1 - b_0 \cdot \left(\frac{1}{\cos i - 1} \right) \quad (1)$$

where b_0 = 0.05 for crystalline modules, and I corresponds to the angle of incidence.

Furthermore, the photovoltaic process involves losses due to fluctuations in solar irradiance and increased temperature, which reduce the open-circuit voltage and decrease the energy conversion efficiency. Additional losses occur in the wiring due to the electrical resistance of the conductors and connectors, causing heat dissipation and further reducing the efficiency.

The system efficiency is also influenced by the module temperature, as higher temperatures reduce the open-circuit voltage and energy conversion. The heat transfer coefficient, U, which depends on the wind conditions, was set at 15 $W/m^2 \cdot K$ for modules with an insulated rear side, which is common in BIPV systems. Thermal losses were evaluated using Faiman's [31] heat transfer model, as shown in Equation (2).

$$T_{cell} = T_{amb} + \frac{1}{U}(\alpha \cdot G_{inc} \cdot (1 - \eta_0)) \quad (2)$$

where G_{inc} is the irradiance of the panel [W/m^2]; U is the heat transfer coefficient [$W/(m^2K)$]; α is the absorptivity of the photovoltaic panel, adopted as 0.90; η_0 is the efficiency of the photovoltaic panel; and ant is the ambient temperature [°C].

The ohmic losses were calculated based on the resistances of the direct current (DC) and alternating current (AC) cables. Under standard test conditions (STC), the software estimated an ohmic loss of 1.5%. When the conditions differ from the STC, the resistance is adjusted according to the length and cross-section of the cables, resulting in resistive losses of approximately 60% of the STC value. These losses vary quadratically with the current in the circuit, as shown in Equation (3).

$$P_{ohm} = RI^2 \quad (3)$$

Losses in the microinverters were calculated based on their efficiency in converting direct current (DC) to alternating current (AC) using the efficiency curve of the devices, represented by Equation (4). The nighttime consumption was estimated at 700 mW, which is necessary to keep the system operational overnight. This energy loss is particularly relevant for systems with microinverters, as they continue to consume energy even without solar generation.

$$\eta_{inv} = \frac{P_{out}}{P_{in}} \quad (4)$$

where P_{out} is the output power [W], and P_{in} is the input power [W].

With all the losses identified, the system's performance index is calculated using the performance ratio (PR), as shown in Equation (5), which measures the efficiency of the photovoltaic system relative to its theoretical capacity. PR reflects the impact of factors such as temperature, shading, module technology, and tilt angle. It is crucial for assessing the system's quality, as it indicates the total losses during the conversion from direct current (DC) to alternating current (AC). PR allows for evaluating the amount of energy injected into the grid (E_Grid) and making comparisons between photovoltaic systems, regardless of location or installed capacity, and assesses how close the system is to its ideal performance, considering module temperature and the efficiency of incident solar radiation [32].

$$PR = \frac{\gamma_f}{\gamma_r} = \frac{\gamma_f}{G_{inc} \cdot P_{nom}} \quad (5)$$

where γ_f is the available energy for use [kWp/kWp/day], γ_r is the energy produced in the STC power, G_{inc} is the incident irradiance [$kWp/m^2/day$], and P_{nom} is the STC installed power [kWp].

The equivalent units are explained by the fact that with the STC, 1 kWh/kWp of electricity should be generated for every kWh/m^2 of irradiation.

Several challenges were identified during the study. One notable issue was the high variability of solar radiation in tropical regions, such as João Pessoa, which can be attributed to the frequent presence of clouds and rapid climatic changes. This variability significantly affected the consistency of data collection over time because it directly impacted both direct and diffuse solar radiation, consequently influencing the electrical performance of the system. The fixed 90° inclination of the PV module further constrained the system's energy efficiency, limiting its performance and precluding direct comparisons with other inclination angles commonly studied in BIPV research.

Additionally, shading conditions posed a significant challenge, as shadows cast by architectural elements, vegetation, and nearby structures caused fluctuations in the solar radiation reaching the module. These variations introduced complexities in accurately assessing the system's electrical performance under real-life conditions. Finally, ensuring instrumentation accuracy, including solar radiation, current, and voltage sensors, was difficult due to their exposure to adverse environmental conditions, further highlighting the need for robust methodologies and more precise sensors in similar studies.

4. Results and Discussion

To assess the electrical performance of the designed system, it is essential to consider the losses calculated by Pvsyst, which are detailed in Table 5.

Table 5. Loss accounted for in Pvsyst.

	North	South	East	West
Horiz. Irradiation total (kWh/m^2)	2057	2057	2057	2057
Loss due to nearby shadows (%)	18.1	10.9	23.5	6.9
Global IAM loss (%)	9.6	11.3	5.0	4.3
Loss due to dirt (%)	3.0	3.0	3.0	3.0
Loss due to low irradiance level (%)	3.3	4.0	2.4	2.1
Loss due to temperature (%)	6.1	5.3	9.0	9.8
Ohmic loss	0.2	0.1	0.3	0.3
Loss of microinverter efficiency (%)	28.3	32.4	23.80	21.60
Energy injected into the grid (kWh/year)	51.3	42.1	65.5	81.6

Considering the losses described in Table 5, it is possible to evaluate the performance ratio (PR). Figure 10 shows some variations in the performance of photovoltaic systems throughout the year. The average yield recorded was 44.3% for the north façade, 44.4% for the south, 47.4% for the east, and 56.8% for the west.

In the study by Lauffer et al. [33], a simulation of energy generation in vertical residential photovoltaic façades was presented for the cities of Recife and Porto Alegre, also located in Brazil, with different types of photovoltaic modules and climatic conditions, taking into account the city of Recife, as it has climatic aspects much closer to the city of João Pessoa; the research data were used as a comparative effect for the applicability of the model in João Pessoa.

The system installed on the north façade (Figure 10a) achieved its best performance between April and July, during the winter solstice in the southern hemisphere, when the sun is positioned to the north, directly over the Tropic of Capricorn, benefiting the north façade of João Pessoa. During this period, a low irradiance loss of 3.3% and temperature loss of approximately 6.6% were observed.

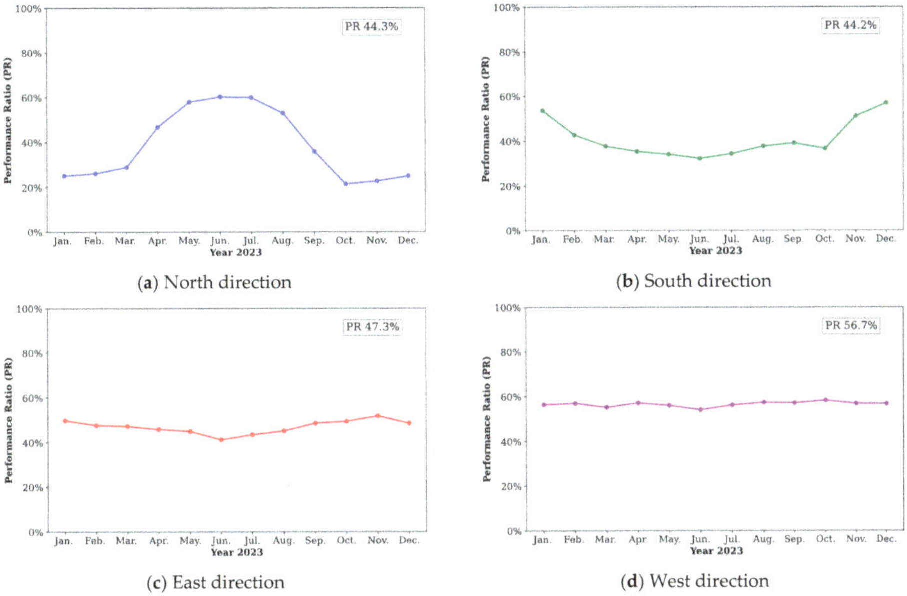

(**a**) North direction (**b**) South direction (**c**) East direction (**d**) West direction

Figure 10. Performance index for north orientation.

For south orientation (Figure 10b), the best performance occurred between November and February, during the summer solstice, when the sun shines directly over the Tropic of Cancer. However, the lower solar altitude created longer shadows, reducing the incident radiation and the system's efficiency. The south facade exhibited the highest irradiance loss among the four directions analyzed (4.0%) but the lowest temperature loss (5.3%).

The east and west facades, as shown in Figure 10c,d, demonstrated more consistent performance throughout the year, receiving solar radiation primarily during specific times, such as mornings and afternoons. In these orientations, the irradiance losses were 2.4% and 2.1%, respectively. However, these façades also experienced higher temperature losses (9.0% on the east and 9.8% on the west) due to greater solar exposure, which led to increased module heating and a reduction in photovoltaic conversion efficiency.

For the city of João Pessoa, it was initially estimated that the north-facing façade would exhibit the highest performance ratio (PR) throughout the year. However, the east and west façades demonstrated superior performances. This behavior, as observed in Lauffer et al. [33], can be explained by the higher and more consistent solar radiation incidence during the afternoon period, particularly between 2 p.m. and 4 p.m. This result is directly related to the solar trajectory, which, due to steeper angles during this time interval, enhances energy generation in these orientations.

The PR value showed subtle variations, which is expected in tropical regions, where the losses are primarily associated with the BIPV system and local climatic conditions. The performance ratio (PR) is directly related to the amount of energy injected into the grid (E_Grid) [34], with facades having higher PR values generating more energy, as shown by the values presented in Table 6.

Table 6. The energy injected into the grid according to the performance of the system.

	North		South		East		West	
Period	**PR**	**E_Grid (kWh/day)**	**PR**	**E_Grid (kWh/day)**	**PR**	**E_Grid (kWh/day)**	**PR**	**E_Grid (kWh/day)**
January 2023	25.40%	1.626	54.00%	7.025	50.00%	6.404	56.20%	7.122
February 2023	63.00%	1.475	43.00%	3.589	47.80%	4.907	56.70%	6.421
March 2023	29.20%	2.217	38.00%	2.369	47.40%	5.484	55.00%	6.430
April 2023	47.10%	5.498	35.50%	1.949	46.00%	5.127	56.90%	7.026
May 2023	58.10%	8.847	34.40%	1.859	45.00%	4.942	55.80%	6.808
June 2023	60.30%	8.066	32.40%	1.472	41.40%	3.558	53.80%	5.053
July 2023	60.00%	8.932	34.50%	1.812	43.60%	4.414	56.00%	6.225
September 2023	36.10%	3.289	39.40%	2.382	48.70%	5.718	56.90%	6.882
October 2023	21.60%	1.363	36.90%	3.021	49.50%	6.792	58.00%	8.454
November 2023	22.90%	1.392	51.50%	6.389	51.90%	7.213	56.60%	7.367
December 2023	25.20%	1.607	57.20%	8.098	48.60%	5.967	56.50%	7.533
Year 2023	44.30%	51.254	44.40%	42.173	47.50%	65.493	56.80%	82.268

Figure 11 shows the relationship between the energy injected into the grid as a function of the global irradiance on the plane of the system for the north, south, east, and west orientations, respectively.

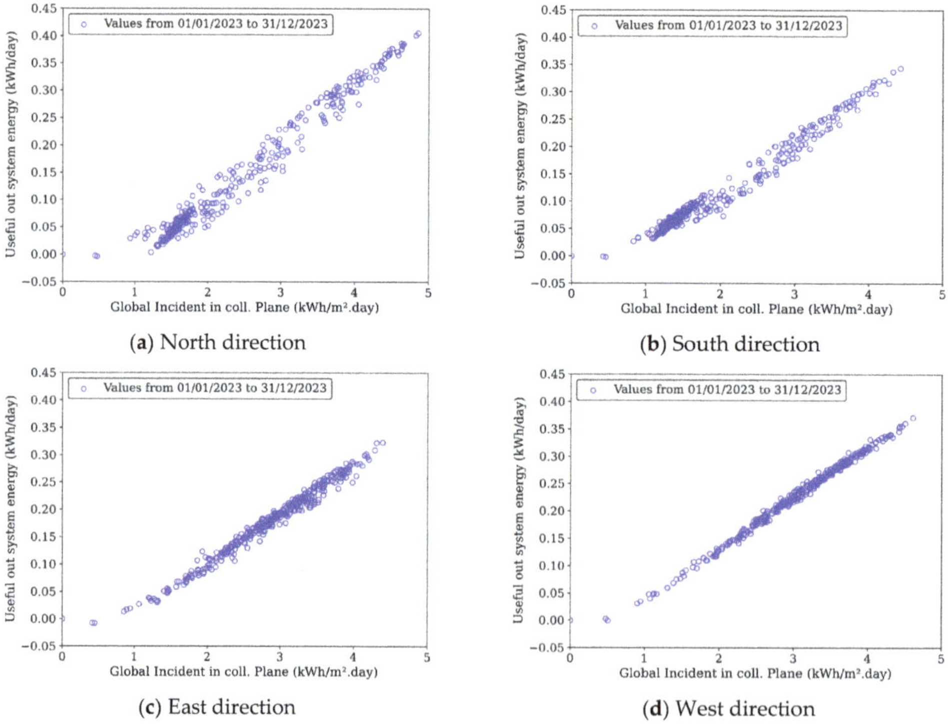

(**a**) North direction (**b**) South direction (**c**) East direction (**d**) West direction

Figure 11. Daily input/output diagram.

For orientations with higher performance indexes, the injected energy curves appear more linear and upward-trending, reflecting a proportional increase in the radiation received and greater efficiency in photovoltaic conversion. A detailed analysis of these graphs shows that the west (Figure 11d) and east (Figure 11c) orientations produce more energy, as their curves are more linear and with more agglutination of points for greater radiation, reinforcing the importance of the correct orientation and tilt of photovoltaic modules to maximize the system's energy efficiency.

An experimental study was conducted to provide a solid basis for validating the simulations. In the case of a photovoltaic window, real environmental conditions (such as temperature, solar radiation, and incidence angles) can be complex and vary significantly. The experimentation allows the simulations to be compared with the observed data, ensuring that the model is accurate and applicable to local reality.

When comparing the grid-injected energy values obtained through the simulation method with the experimental method for the days 23–25 April; 21, 26, and 28 June; 17, 18, and 19 July; and 23, 26, and 27 August, a substantial correlation between the results is observed, as shown in Figure 12. It is important to note that the comparisons are made for the north orientation, which is the actual orientation of the study object.

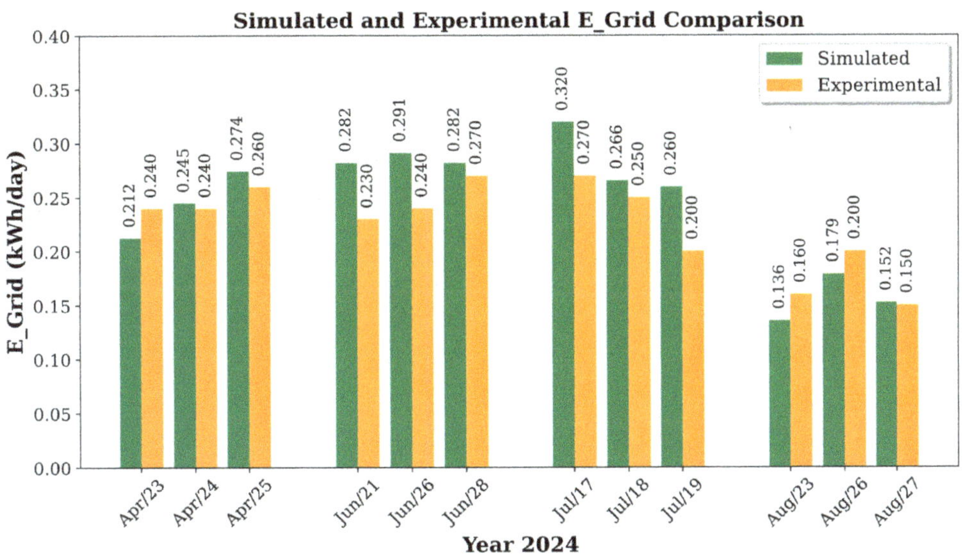

Figure 12. Comparison of energy injected into grid between experimental and simulation methods.

For the analyzed days in April, there was practically no discrepancy between the simulated and experimental values. The experimental values were 0.2400, 0.2400, and 0.2600 kWh/day, resulting in an average of 0.2530 kWh/day. The simulated values obtained through the simulation method were 0.2912, 0.2820, and 0.2851 kWh/day, with an average of 0.2445 kWh/day. This slight difference can be attributed to the typical climatic conditions of April in João Pessoa, which, despite the beginning of winter, offer good availability of direct solar radiation and low climatic variability, favoring greater agreement between the calculation methods.

In contrast, during June, a more significant difference was observed between the energy injected into the grid obtained from the simulation and experimental methods. On 21, 26, and 28 June, the simulated values were 0.2820, 0.2912, and 0.2820 kWh/day, resulting in an average of 0.2851 kWh/day, while the experimental values were 0.2300, 0.2400, and 0.2700 kWh/day, with an average of 0.2467 kWh/day.

In the days analyzed in July 2024, the discrepancy between the simulated and experimental results became more pronounced. The experimental energy injected into the grid was 0.2700, 0.2500, and 0.2000 kWh/day, resulting in an average of 0.2400 kWh/day. In contrast, the simulated values obtained through the simulation method were 0.3197, 0.2656, and 0.2596 kWh/day, with an average of 0.2816 kWh/day.

This June–July discrepancy can be explained by the seasonal climatic conditions of João Pessoa during winter, which is characterized by increased cloud cover and reduced direct solar irradiance. These conditions negatively affect the module's performance, which relies heavily on direct radiation to achieve high efficiency levels. Although robust, the PVsyst simulation model may underestimate these variations because it uses average climatic data as the basis for its estimates.

However, in August, a greater alignment between the simulated and experimental values was observed. On 23, 26, and 27 August, the simulated values were 0.1357, 0.1788, and 0.1525 kWh/day, resulting in an average of 0.1556 kWh/day. In comparison, the experimental values were 0.1600, 0.2000, and 0.1500 kWh/day, with an average of 0.1700 kWh/day. This realignment can be attributed to the stabilization of the climatic conditions typical of late winter, when cloudiness begins to decrease and solar radiation levels become more consistent.

João Pessoa is located in Northeastern Brazil. It has a tropical climate with a high solar incidence. However, during the winter period (April to July), rainfall and cloudiness are common. These factors reduce the efficiency of the photovoltaic system because of the decrease in direct incident radiation and the increase in thermal losses caused by higher module temperatures under low ventilation conditions.

Overall, the results demonstrate the consistency of the simulation model in estimating the energy injected into the grid, especially during periods of greater climatic stability. However, the analysis also highlights the importance of considering regional and seasonal specificity when adjusting the simulations and improving the accuracy of the results.

The robust experimental data collection also serves as a basis for future research. This validated data set can serve as a starting point for other researchers who wish to expand or adapt the study to other regions or contexts.

5. Conclusions

This study fills a gap in the literature by providing detailed experimental model data on the electrical performance of monocrystalline silicon BIPV systems in a tropical climate, such as João Pessoa, Brazil. The location has specific characteristics, such as high solar radiation throughout the year, mainly diffuse, and rapid climate changes. However, by investigating the photovoltaic module on a vertical façade with a fixed angle of 90 degrees, this article contributes to a better assessment of the impact of this system on energy efficiency, as it has challenges related to interference from nearby structures and less direct solar incidence, which are key points for systems integrated into buildings due to the predominance of vertical facades.

The orientation of the photovoltaic module significantly influences the amount of solar radiation and energy generation. In 2023, the south-facing facade recorded the highest irradiance losses, which were attributed to the lower solar altitude during the southern hemisphere's winter, resulting in reduced solar radiation intensity. Conversely, the east and west facades experienced lower irradiance losses throughout the year, benefiting from a more balanced distribution of solar radiation during the day. However, these facades also faced higher thermal losses due to prolonged direct sunlight exposure, which increased module temperatures and reduced the photovoltaic conversion efficiency.

The performance ratio (PR) of the modules showed slight variations throughout the year, reflecting the interaction between solar radiation and local climatic conditions. The north-facing facade exhibited the best performance between April and August during the winter solstice, with irradiance and thermal losses of 3.3% and thermal losses of 6.6%. On the south-facing facade, optimal performance was observed between November and February during the summer solstice, with irradiance losses of 4.0% and thermal losses of 5.3%. The east and west facades maintained more consistent performance, with lower irradiance losses (2.4% and 2.1%, respectively) but higher thermal losses (9.0% and 9.8%) due to increased solar exposure.

An analysis of the energy injected into the grid in 2023 revealed that the east and west facades displayed more stable and consistent performance. Although the north-facing facade generated higher amounts of energy in certain months, its values were more variable, reducing the predictability of energy generation in BIPV systems with modules installed at a 90° tilt.

A comparison of the simulated values with the experimental data from 2024 revealed minor discrepancies due to differences in the data collection methods. The experimental data were collected directly from the photovoltaic module, while the meteorological data were obtained from a solarimetric station located at different heights and distances from the study object. Despite these limitations, the results confirmed the accuracy of the PVsyst model under stable climatic conditions but also highlighted the challenges of simulating scenarios with seasonal variability and adverse atmospheric conditions.

These findings underscore the impact of seasonal climatic conditions on the performance of photovoltaic systems and emphasize the need to improve simulation models to better account for extreme atmospheric variations. Furthermore, they highlight the importance of appropriate module orientation and tilt to optimize energy production throughout the year.

It was concluded that in the tropical climate of João Pessoa, east- and west-facing facades are more advantageous for installing monocrystalline silicon photovoltaic modules with a 90° tilt, as these orientations provide a more balanced capture of solar radiation throughout the day.

This approach provides real data collected under practical conditions, increasing its applicability. The article's contribution goes beyond a one-off analysis, as it includes important information for urban planners, architects, and engineers interested in applying sustainable technologies in tropical regions.

Although the data presented are significant, several questions remain unanswered, highlighting the need for further research into the impact of orientation on the electrical performance of building-integrated photovoltaic (BIPV) façades in predominantly tropical climates, such as that of João Pessoa, Brazil. Such research should include investigations into varying the inclination angle of photovoltaic windows beyond the traditional 90° configuration.

Additionally, future studies could delve deeper into seasonal variations in solar radiation, considering both direct and diffuse components, as well as the shading effects caused by architectural elements. The exploration of advanced photovoltaic technologies, such as hybrid modules that integrate energy generation with thermal insulation properties, offers promising potential for enhancing the efficiency of such systems. In this context, combining sophisticated computational models with empirical experiments in diverse urban scenarios can significantly improve our understanding of the performance of BIPV systems across different configurations. This approach strengthens their applicability in regions with varied climatic conditions, paving the way for optimized and context-specific solutions.

Author Contributions: Conceptualization: M.M.d.M.G., R.D.L.d.S.C., T.G.S.d.L. and O.H.A.J.; Methodology: M.M.d.M.G., R.D.L.d.S.C., T.G.S.d.L. and C.D.P.; Validation: M.M.d.M.G., R.D.L.d.S.C. and T.G.S.d.L.; Investigation and Simulation: M.M.d.M.G., R.D.L.d.S.C. and T.G.S.d.L.; Writing—Original Draft Preparation: M.M.d.M.G., R.D.L.d.S.C., T.G.S.d.L., C.D.P., J.A.d.L. and O.H.A.J.; Review and Editing: M.M.d.M.G., R.D.L.d.S.C., T.G.S.d.L., C.D.P., J.A.d.L. and O.H.A.J.; Project Administration: T.G.S.d.L.; Funding Acquisition: T.G.S.d.L. and O.H.A.J. All authors have read and agreed to the published version of the manuscript.

Funding: This research was supported by the Federal University of Paraíba, and by the FAPESQ agency (Fundação de Apoio à Pesquisa do Estado da Paraíba) throughout the project (Term No. 066/2023, Announcement No. 19-2022). This project was also supported by the FACEPE agency (Fundação de Amparo a Pesquisa de Pernambuco) throughout the project (APQ-0616-9.25/21 and APQ-0642-9.25/22. O.H.A.J. was funded by the Brazilian National Council for Scientific and Technological Development (CNPq), grant numbers 407531/2018-1, 303293/2020-9, 405385/2022-6, 405350/2022-8, and 406662/2022-3.

Data Availability Statement: The original contributions presented in this study are included in the article. Further inquiries can be directed to the corresponding author.

Acknowledgments: The authors would like to thank the Foundation for the Promotion of Research of the State of Paraíba—FAPESQ Brazil (Term No. 066/2023, Announcement No. 19-2022).

Conflicts of Interest: The authors declare no conflicts of interest.

References

1. Scognamiglio, A.; Garde, F. Photovoltaics' architectural and landscape design options for Net Zero Energy Buildings, towards Net Zero Energy Communities: Spatial features and outdoor thermal comfort related considerations. *Prog. Photovolt. Res. Appl.* **2014**, *24*, 477–495. [CrossRef]
2. Frankel, M. *National Renewable Energy Laboratory: Solar Has the Most Potential of Any Renewable Energy Source*; ThinkProgress: Washington, DC, USA, 2012.
3. Chivelet, N.M.; Solla, I.F. *Técnicas de Vedação Fotovoltaica na Arquitetura. Tradução: Alexandre Salvaterra*; Bookman: Porto Alegre, Brazil, 2010; ISBN 978-85-7780646-1.
4. Ching, F.D.K.; Shapiro, I.M. *Edificações Sustentáveis Ilustráveis. Tradução: Alexandre Salvaterra*; Bookman: Porto Alegre, Brazil, 2017; ISBN 978-85-8260-429-8.
5. Karteris, M.; Slini, T.; Papadopoulos, A.M. Urban solar energy potential in Greece: A statistical calculation model of suitable built roof areas for photovoltaics. *Energ. Build.* **2013**, *62*, 459–468. [CrossRef]
6. Eltawil, M.A.; Zhao, Z. Grid-connected photovoltaic power systems: Technical and potential problems—A review. *Renew Sustain. Energy Rev.* **2010**, *14*, 112–129. [CrossRef]
7. Al-Sabounchi, A.M.; Yalyali, S.A.; Al-Thani, H.A. Design and performance evaluation of a photovoltaic grid-connected system in hot weather conditions. *Renew Energy* **2013**, *53*, 71–78. [CrossRef]
8. Kim, J.-Y.; Jeon, G.-Y.; Hong, W.-H. The performance and economical analysis of grid- connected photovoltaic systems in Daegu. *Korea Appl. Energy* **2009**, *86*, 265–272. [CrossRef]
9. Hachicha, A.A.; Al-Sawafta, I.; Said, Z. Impact of dust on the performance of solar photovoltaic (PV) systems under United Arab Emirates weather conditions. *Renew Energy* **2019**, *141*, 287–297. [CrossRef]
10. PVsyst Software. Available online: https://www.pvsyst.com/about-us/ (accessed on 23 October 2023).
11. Essah, E.A.; Arguelles, A.R.; Neil Glover, N. Assessing the performance of a building integrated BP c-Si PV system. *Renew. Energy* **2015**, *73*, 36–45. [CrossRef]
12. Shakouri, M.; Ghadamian, H.; Noorpoor, A. Quasi-dynamic energy performance analysis of building integrated photovoltaic thermal double skin façade for middle eastern climate case. *Appl. Therm. Eng.* **2020**, *179*, 115724. [CrossRef]
13. Jeong, S.; Jang, J.; Park, I.; Jo, S.K.; Lim, J.; Lee, S.; Lee, H.; Oh, W. Performance analysis of coloured BIPV systems depending on reliability of irradiance data. *Microelectron. Reliab.* **2022**, *138*, 114704. [CrossRef]
14. Hamzah, A.H.; Go, Y.I. Design and assessment of building integrated PV (BIPV) system towards net zero energy building for tropical climate. *e-Prime—Adv. Electr. Eng. Electron. Energy* **2023**, *3*, 100105. [CrossRef]

15. Yang, R.J.; Zhao, Y.; Jayakumari, S.D.S.; Schneider, A.; Rajan, P.; Leloux, J.; Alamy, P.; Raharjo, G.P.; Rende, F.; Samarasinghalage, T.; et al. Digitalising BIPV energy simulation: A cross tool investigation. *Energy Build.* **2024**, *318*, 114484. [CrossRef]
16. Demir, Y.C.; Aktacir, M.A. Experimental Investigation of the BIPV System under Şanlıurfa Meteorological Conditions. *Appl. Sci.* **2023**, *13*, 11286. [CrossRef]
17. Yu, Z.; Lu, Z.; Xin, W.; Pu, H.; Jiang, J.; Ye, B. Environmental and economic evaluation of urban building-integrated photovoltaic and electric vehicle system. *J. Build. Eng.* **2024**, *95*, 110205. [CrossRef]
18. Zhang, Y.; Herr, C.; Guo, Y. Optimization and Design of Building-Integrated Photovoltaic Systems for a High-Rise Building in Shenzhen. 2024. Available online: https://link.springer.com/chapter/10.1007/978-981-97-0621-1_31 (accessed on 5 February 2025).
19. Jamala, N.; Mangkuto, R.A.; Paramita, B. Uncertainty analysis of annual daylight metrics and energy yield weighting factors in optimising photovoltaic window orientation. *Sol. Energy* **2023**, *262*, 111901. [CrossRef]
20. Guo, W.; Kong, L.; Chow, T.; Li, C.; Zhu, Q.; Qiu, Z.; Li, L.; Wang, Y.; Riffat, S.B. Energy performance of photovoltaic (PV) windows under typical climates of China in terms of transmittance and orientation. *Energy* **2020**, *213*, 118794. [CrossRef]
21. Brazilian Institute of Geography and Statistics (IBGE). Regional Division of the State of Paraíba. Available online: https://cidades.ibge.gov.br/brasil/pb/panorama (accessed on 15 November 2024).
22. *Brazilian Standard (NBR) 15220-3*; Thermal Performance in Buildings—Part 3: Brazilian Bioclimatic Zones and Building Guidelines for Low-Cost Houses. Brazilian Standards: Rio de Janeiro, Brazil; São Paulo, Brazil, 2025.
23. Solarimetric Atlas of Paraíba. *Center for Research, Development, and Innovation*; CGPDI: Sao Paulo, Brazil, 2023; ISBN 978-65-980791-0-9.
24. Roriz, M. Climate Classification of Brazil—Version 2. 2013. Available online: https://labeee.ufsc.br/projetos/proposta-de-revisao-do-zoneamento-bioclimatico-brasileiro (accessed on 15 November 2024).
25. de Sérgio, S.; Brito Reference Center for Solar and Wind Energy (CRESESB). Solar Potential—SunData v3.0. Available online: https://cresesb.cepel.br/index.php?section=sundata (accessed on 15 November 2024).
26. QGIS Overwiew Software. Available online: https://qgis.org/project/overview/ (accessed on 15 November 2024).
27. National Institute for Space Research (INPE). Irradiation Data for the State of Paraíba. Available online: https://labren.ccst.inpe.br/atlas_2017_PB.html# (accessed on 15 November 2024).
28. Brazilian Institute of Geography and Statistics (IBGE). Organization of the Territory: Territorial Grids. Available online: https://www.ibge.gov.br/geociencias/organizacao-do-territorio/malhas-territoriais/15774-malhas.html (accessed on 15 November 2024).
29. Abella, M.A.; Lorenzo, E.; Chenlo, F. Effective Irradiance Estimation for PV Applications. In Proceedings of the 3rd Word Conference on Photovoltaic Energy Conversion, Osaka, Japan, 11–18 May 2003; Volume 2, pp. 2085–2089.
30. *ASHRAE Standard 93-77*; Methods of Testing to Determine the Thermal Performance of Solar Collectors. American Society of Heating, Refrigeration and Air Conditioning Engineers: New York, NY, USA, 1977.
31. Faiman, D. Assessing the outdoor operating temperature of photovoltaic modules. *Prog. Photovolt.* **2008**, *16*, 307–315. [CrossRef]
32. Irshad, K.; Algarni, S.; Jamil, B.; Ahmad, M.T.; Khan, M.A. Effect of gender difference on sleeping comfort and building energy utilization: Field study on test chamber with thermoelectric air-cooling system. *Build Environ.* **2019**, *152*, 214–227. [CrossRef]
33. Lauffer, H.A.; Tomaszewski, G.A.; Haeberle, F.; Dias, J.B.; Wander, P.R. Simulação e Análise de Diferentes Tecnologias de Módulos Fotovoltaicos Aplicados em uma Edificação Residencial Vertical. In Proceedings of the Brazilian Solar Energy Conference."—CBENS. 2020. Available online: https://anaiscbens.emnuvens.com.br/cbens/article/view/934 (accessed on 15 November 2024).
34. Rachchh, R.; Kumar, M.; Tripathi, B. Solar photovoltaic system design optimization by shading analysis to maximize energy generation from limited urban area. *Energy Convers. Manag.* **2016**, *115*, 244–252. [CrossRef]

energies

MDPI

Article

Optimization Research on a Novel Community Integrated Energy System Based on Solar Energy Utilization and Energy Storage

Xunwen Zhao [1], Hailin Mu [1], Nan Li [1,*], Xue Kong [2] and Xunpeng Shi [3]

1 Key Laboratory of Ocean Energy Utilization and Energy Conservation of Ministry of Education, Dalian University of Technology, Dalian 116024, China
2 School of New Energy, Ningbo University of Technology, Ningbo 315336, China
3 Australia-China Relations Institute, University of Technology Sydney, Sydney, NSW 2007, Australia
* Correspondence: nanli_dlut@163.com

Abstract: Integrated energy systems (IESs) are essential for enabling the energy transition in communities and reducing CO_2 emissions. This paper proposes a novel IES that combines photovoltaic (PV) and solar thermal energy with coordinated electrical and thermal energy storage to meet the energy demands of residential communities. The system also incorporates hydrogen production for fuel cell vehicles. A dual-objective optimization model was developed, minimizing both economic costs and CO_2 emissions. The system's performance was evaluated using data from a case study in Dalian, which showed that the IES successfully reduced the annual total cost and CO_2 emissions compared to conventional systems. The key findings showed that PV electrolysis for hydrogen production provides both economic and environmental advantages. The system's integration of solar thermal energy offers higher economic efficiency, while PV energy supplies enhance coordination. Additionally, carbon trading prices effectively reduce emissions, but excessively high prices do not always lead to better emission outcomes. This study introduces a comprehensive, multi-energy approach for optimizing the energy supply, contributing novel insights to the field of sustainable energy systems.

Keywords: integrated energy system; solar thermal energy; photovoltaic energy; hydrogen production; energy storage; dual-objective optimization

1. Introduction

The transition to renewable energy sources is critical for addressing energy shortages and reducing environmental pollution, especially in urbanization and industrialization. Integrated Energy Systems (IESs) have emerged as a promising solution to facilitate this transition by optimizing the use of local energy resources by combining renewable energy technologies with energy storage and management systems [1–4]. IESs provide a comprehensive approach to meeting the diverse energy demands of communities while reducing dependence on fossil fuels and lowering carbon emissions. However, implementing IESs faces several challenges, including high initial costs, integration with existing infrastructure, and policy and regulatory barriers [5–9].

Recent studies have focused on optimizing IESs to enhance energy efficiency and reduce carbon emissions. For example, Liu et al. [10] proposed an optimal scheduling model to minimize operational costs, while Zhao et al. [11] improved the dynamic energy efficiency of IESs by optimizing system capacity configurations. Babaei et al. [12] demonstrated the

https://doi.org/10.3390/en18051151

feasibility and optimization of hybrid energy systems by focusing on sustainable electricity and heat. Additionally, Wang et al. [13] developed scheduling methods based on the N-1 security criterion to ensure the reliability of system operations. Shu et al. [14] and Yang et al. [15] also discussed multi-objective optimization frameworks, addressing operational efficiency and environmental concerns such as carbon emissions and intensity. Despite these advancements, many studies have primarily focused on optimizing specific components of IESs, such as management of energy loads, optimization of equipment configurations, or integration with renewable energy sources [16–20]. Limited work has been performed on integrating multiple energy types within a unified system that can meet both electrical and thermal demands.

A key research gap exists in integrating photovoltaic (PV) and solar thermal energy with energy storage systems, and hydrogen production for fuel cell vehicles [21,22]. While recent studies have explored hydrogen production from renewable energy sources through electrolysis [23,24], its role in supporting community-based IESs has not been fully addressed. Recent breakthroughs in hydrogen production have shown the potential of hydrogen as an intermediate carrier in IES [25–30], but integrating this production seamlessly into community-level energy systems remains underexplored. Furthermore, coupling energy storage systems with multiple energy sources to optimize the electrical and thermal energy supply, particularly across different seasons, remains an underexplored area of research. Zhen et al. [31] proposed a framework for integrating photovoltaic–wind–hydrogen systems with urban community energy storage, highlighting the potential of combining renewable energy sources but failing to explore the full integration of energy types in community energy applications. Other research, such as that by Liu et al. [32], Tubussum et al. [33], and Bazdar et al. [34], focused on optimizing hybrid systems for buildings or bridging gaps between renewable and thermal energy systems but lacked a comprehensive community-scale integration approach.

The explosive growth of installed renewable energy capacity has emphasized the importance of addressing intermittency issues, making energy storage a fundamental solution [35–40]. Many studies have highlighted the critical role of energy storage in optimizing renewable energy systems. Wang et al. [41] studied the stable multi-frequency operation of IESs and proposed models to handle high-frequency fluctuations in renewable energy generation. These systems help manage grid fluctuations and improve system resilience, but they do not fully explore the operational relationship between electrical and thermal storage across different seasons, as pointed out by Hao et al. [42]. Energy storage systems, particularly when integrated with PV and thermal energy, have proven essential in balancing supply and demand effectively, especially in regions with high penetration of intermittent renewable energy [43–46].

One key gap in the existing research is the limited exploration of real-time hydrogen production in community-based IESs, especially regarding fuel cell applications. While several studies have investigated hydrogen production through renewable electricity, the role of hydrogen in integrated community energy systems to support diverse energy demands—such as cooling, heating, power, and hydrogen—has not been comprehensively addressed. In particular, the coupling of photovoltaic, solar thermal, and energy storage systems, which could optimize the electrical and thermal energy supply across different seasons, has received little attention.

This study aimed to address these research gaps by proposing a novel IES that integrates photovoltaic and solar thermal energy with coordinated electrical and thermal energy storage. The system was designed to meet various community energy demands, including cooling, heating, power, and hydrogen, while minimizing economic costs and carbon emissions. The key contributions of this study are as follows:

1. We propose an innovative integrated energy system combining multiple renewable energy sources with coordinated energy storage, addressing electrical and thermal energy needs while incorporating hydrogen production for fuel cell vehicles.
2. We introduce a dual-objective optimization model that simultaneously minimizes economic costs and carbon emissions, providing a more balanced and sustainable solution than traditional models focused on only one of these factors.

The structure of this paper is organized as follows. Section 2 introduces the proposed integrated energy system model, optimization objectives, and evaluation indicators. Section 3 presents specific case studies detailing each component's technical and economic parameters. Section 4 analyzes the performance of the IES in terms of cost, CO_2 emissions, and energy consumption based on the optimization results. Section 5 summarizes the findings and offers prospects for future research.

2. Methodology

2.1. System Model

The energy sources include solar energy, natural gas, and grid electricity. The energy supply devices in the IES include PV panels, solar collectors (SC), power generation units (PGU), heat recovery units (HRU), and gas boilers (GB). The storage devices comprise batteries and thermal storage tanks (TST). The utilization devices include electrolyzers, electric boilers (EB), absorption chillers (AC), electric chillers (EC), and heat exchangers (HE). Additional supporting facilities include converters, hydrogen refueling stations, and hydrogen storage tanks. The energy flow diagram of the integrated energy system is shown in Figure 1.

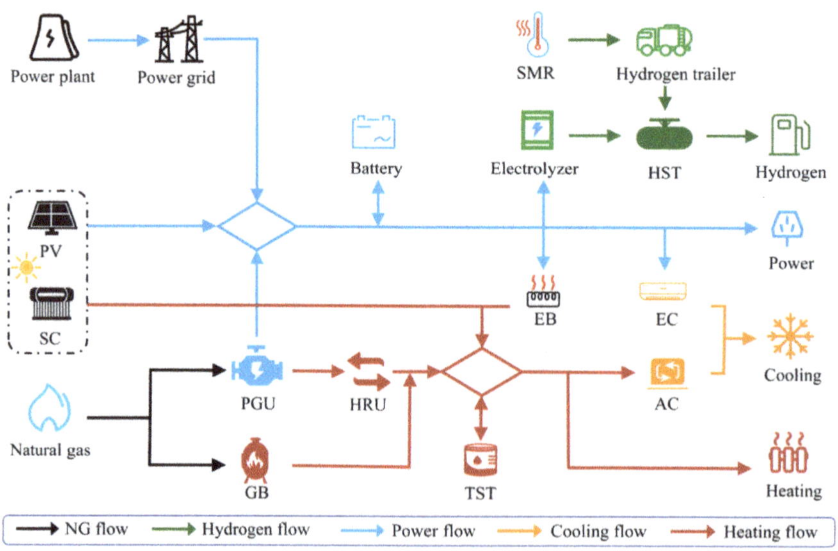

Figure 1. Energy flow diagram of the integrated energy system.

The electricity generated by PV panels and PGU supplies the electrical devices in the IES. When the electricity generated is insufficient, stored energy from the batteries is released. If battery storage is inadequate, the system relies on power from the grid. Conversely, when surplus electricity is available, the excess recharges the batteries. Similarly, the thermal supply is managed by the SC, HRU, EB, and GB through HE to meet the thermal load. When thermal output is insufficient, heat is released from the TST, while surplus heat replenishes the thermal storage when adequate. Both EC and AC meet the cooling load. Electrolysis is employed to produce hydrogen to satisfy the hydrogen storage requirements of the refueling stations. When hydrogen production is inadequate, external hydrogen trailers supplement the supply.

The IES can be divided into a cooling system, a heating system, a power system, and a hydrogen system according to the four types of energy demand. The systems coordinate and function through the balance of energy supply and demand. Each system's equipment selection is crucial for a proper energy supply. Each component is described next in a systematic order.

2.1.1. Power System

The electricity sources include PV panels, PGU, and the grid, supplying power to residential demand, EB, EC, and electrolyzers. When the generated power exceeds the consumption, the surplus electricity is directed to the batteries; conversely, when consumption surpasses generation, the batteries provide power to the system. The electricity supply–demand balance is expressed as follows:

$$E_{PV} + E_{PGU} + E_{GP} + a \cdot E_{out} = E_{LD} + E_{EB} + E_{EC} + E_{EL} + (1-a) \cdot E_{in} \tag{1}$$

where E_{PV} and E_{PGU} represent the electricity generated by PV panels and PGU, respectively; E_{GP} denotes the amount of electricity purchased from the grid; E_{LD} is the residential electricity load; E_{EB}, E_{EC}, and E_{EL} refer to the electricity consumption of EB, EC, and electrolyzers, respectively; E_{out} and E_{in} indicate the amount of electricity released from and charged into the batteries, respectively; and a is a Boolean variable.

(1) PV

The unit power output of photovoltaic panels is mainly influenced by the solar radiation intensity and ambient temperature [47]. The rated power of the PV panels denoted as Cap_{PV}, is determined by the installation area of the photovoltaic panels A_{PV} and the area coefficient η_{PV}.

$$E_{PV} = Cap_{PV} \times f_{PV} \times \frac{G_{amb}}{G_{ref}}\left[1 + K_t\left(T_{amb} + 0.0256 \times G - T_{ref}\right)\right] \cdot \eta_{CON} \tag{2}$$

$$Cap_{PV} = A_{PV} \cdot \eta_{PV} \tag{3}$$

where f_{PV} is the degradation coefficient; G_{amb} and T_{amb} are the actual solar irradiance and temperature, respectively; G_{ref} and T_{ref} are the irradiance and ambient temperature under standard conditions; K_t is the power temperature coefficient; and η_{CON} and η_{PV} are the efficiencies of the converter and PV panel, respectively.

(2) PGU

A gas turbine, known for its quick startup and responsive performance, was chosen as the PGU to ensure a stable and coordinated power supply. Given the significant residual heat in the high-temperature exhaust gases, a heat recovery unit was employed to capture the waste energy [48].

$$E_{PGU} = \eta_{PGU} \cdot W_{PGU} \tag{4}$$

$$Q_{HRU} = W_{PGU} \cdot (1 - \eta_{PGU}) \cdot \eta_{HRU} \tag{5}$$

where η_{PGU} and η_{HRU} represent the power generation efficiency of the gas turbine and the thermal recovery efficiency of the HRU, respectively.

(3) Battery

Energy storage batteries can only either charge or discharge at any given moment, necessitating the introduction of a Boolean variable a. When $a = 0$, the storage battery absorbs surplus energy from the system; when $a = 1$, energy flows out to supply the system. The energy in the battery at time $t + 1$ equals the energy stored at time t plus the energy entering the battery at time $t + 1$ minus the energy flowing out at that time. The battery's maximum storage capacity corresponds to its rated capacity, while the maximum depth of discharge is defined as $N\%$ of the battery capacity. The remaining capacity determines the maximum energy input at time $t + 1$. In contrast, the maximum energy output is the difference between the energy at time t and the allowable minimum energy of the battery.

$$E_{B,t+1} = E_{B,t} - a \cdot E_{out,t+1}/\eta_B + (1-a) \cdot E_{in,t+1} \cdot \eta_B \tag{6}$$

$$(1 - N\%)Cap_B = \min(E_{B,t}) \tag{7}$$

$$\max(E_{B,t}) = Cap_B \tag{8}$$

$$\max(E_{in,t+1} \cdot \eta_B) = Cap_B - E_{B,t} \tag{9}$$

$$\max(E_{out,t+1}/\eta_B) = E_{B,t} - (1 - N\%)Cap_B \tag{10}$$

where $E_{B,\,t}$ and $E_{B,\,t+1}$ are the energy levels in the battery at times t and $t + 1$, respectively; Cap_B represents the capacity of the energy storage battery; and η_B is the battery's efficiency during charging or discharging.

2.1.2. Heating System

The heat sources include solar collectors, gas boilers, electric boilers, waste heat recovery from gas turbines, and thermal storage tanks. Thermal demand is met through heat exchangers, absorption chillers, and thermal storage tanks. The balance between heat supply and demand within the system can be expressed as:

$$Q_{SC} + Q_{GB} + Q_{EB} + Q_{HRU} + bQ_{out} = Q_{HE} + Q_{AC} + (1-b) \cdot Q_{in} \tag{11}$$

where Q_{SC} is the heat output from the SC; Q_{GB} and Q_{EB} are the heat outputs from the GB and EB, respectively; Q_{out} and Q_{in} are the heat released and charged into the thermal storage tank, respectively; Q_{HE} represents the heat obtained by the heat exchanger; and b is a Boolean variable.

(1) SC

Solar irradiance affects the solar collector's heat output, depending on the collector's area and efficiency.

$$Q_{SC} = A_{SC} \cdot G_{amb} \cdot \eta_{SC} \tag{12}$$

where A_{SC} is the area of the SC and η_{SC} is the collector's efficiency.

The total solar irradiation area is the sum of the PV panel installation and SC areas.

$$A = A_{SC} + A_{PV} \tag{13}$$

(2) Boiler

The gas boiler combusts natural gas to provide a stable heat supply, ensuring a reliable thermal supply within the integrated energy system. The electric boiler enables the power supply system to rapidly supplement the heating system through energy conversion, thus increasing the contribution of electric heating as a heat source.

$$Q_{GB} = W_{GB} \cdot \eta_{GB} \tag{14}$$

$$Q_{EB} = E_{EB} \cdot \eta_{EB} \tag{15}$$

where W_{GB} represents the natural gas consumption of the GB and η_{GB} and η_{EB} denote the GB's and EB's heat exchange efficiencies, respectively.

(3) HE

Heat is transferred from the heat exchanger to residential buildings to meet thermal demand.

$$Q_{LD} = Q_{HE} \cdot \eta_{HE} \tag{16}$$

where Q_{LD} is the thermal load of the residential buildings and η_{HE} is the efficiency of the heat exchanger.

(4) TST

The principles governing the thermal storage tank are analogous to those of the energy storage battery.

$$Q_{TST,t+1} = Q_{TST,t} - b \cdot Q_{out,t}/\eta_{TST} + (1-b) \cdot Q_{in,t} \cdot \eta_{TST} \tag{17}$$

$$(1 - M\%)Cap_{TST} = \min(Q_{TST,t}) \tag{18}$$

$$\max(Q_{TST,t}) = Cap_{TST} \tag{19}$$

$$\max(Q_{in,t+1} \cdot \eta_{TST}) = Cap_{TST} - Q_{TST,t} \tag{20}$$

$$\max(Q_{out,t+1}/\eta_{TST}) = Q_{TST,t} - (1 - M\%)Cap_{TST} \tag{21}$$

where $Q_{TST,t}$ and $Q_{TST,t+1}$ are the thermal energy levels in the storage tank at times t and t + 1, respectively; Cap_{TST} is the capacity of the thermal storage tank; η_{TST} is the efficiency of charging or discharging thermal energy; and $M\%$ represents the maximum depth of thermal discharge for the storage tank.

2.1.3. Hydrogen System

The on-site hydrogen production and refueling station comprises an electrolyzer, hydrogen storage tanks, refueling compressors, and auxiliaries. To ensure the hydrogen supply and meet the daily operation of the hydrogen refueling station, hydrogen produced by steam methane reforming (SMR) is transported to the station in trailers. The amount of hydrogen in the hydrogen storage tank at time t + 1 is calculated as the amount at time t plus the hydrogen produced and procured at time t + 1 minus the amount of hydrogen demand. The station has a daily hydrogen refueling capacity of L kg and a maximum refueling depth of $Y\%$ of the hydrogen storage tank's capacity.

$$H_{HST,t+1} = H_{HST,t} - H_{LD,t+1} + H_{EL,t+1} + H_{PH,t+1} \tag{22}$$

$$\sum_{t}^{24}(H_{LD,t}) = L \quad (23)$$

where H_{EL} is the hydrogen quantity produced by the electrolyzer; $H_{HST,t}$ and $H_{HST,t+1}$ are the amounts of hydrogen in the storage tank at times t and $t+1$, respectively; and H_{LD} and H_{PH} are the hydrogen load and the amount of purchased hydrogen, respectively.

For electrolytic hydrogen production, the converter changes alternating current to direct current, which is electrolyzed in the electrolyzer to produce hydrogen. The compressor subsequently fills the hydrogen storage tanks with the produced hydrogen. A proton exchange membrane (PEM) electrolyzer was selected. This device is known for its effective absorption of fluctuating electrical energy, small footprint, and environmental friendliness, making it suitable for residential neighborhoods. The hydrogen production amount is the product of the electricity consumption for hydrogen production, the converter efficiency, and the electrolyzer's hydrogen production efficiency. The hydrogen production by the electrolyzer is calculated as follows.

$$H_{EL} = E_{EL} \cdot \eta_{CON} \cdot \eta_{EL} \quad (24)$$

where η_{CON} and η_{EL} are the efficiencies of the converter and electrolyzer, respectively.

2.1.4. Cooling System

The cooling system operates on the power and heating systems, comprising both EC and AC.

$$CG_{EC} + CG_{AC} = CG_{LD} \quad (25)$$

$$CG_{EC} = E_{EC} \cdot \eta_{EC} \quad (26)$$

$$CG_{AC} = Q_{AC} \cdot \eta_{AC} \quad (27)$$

where CG_{LD} is the total cooling load of the system; CG_{EC} and CG_{AC} are the cooling outputs of the EC and the AC, respectively; and η_{EC} and η_{AC} represent the coefficients of performance for the EC and the AC, respectively.

2.1.5. Assumptions in the Model

The optimization model relies on the following assumptions:

1. The wind speed, solar radiation intensity, temperature, and load are assumed to be deterministic.
2. The system operates in steady-state conditions throughout the year.
3. The system's equipment is assumed to operate safely and reliably.
4. The equipment capacities are assumed to be continuously distributed.
5. There is no power loss in the system.

2.1.6. Optimization Objectives

A dual-objective optimization model was employed in this study to simultaneously minimize both annual total cost (ATC) and carbon dioxide emission (CDE), addressing both economic and environmental concerns. The optimization process prioritizes ATC as the primary objective while ensuring that CDE remains minimized. CDE is treated as a secondary objective to reduce carbon emissions while keeping costs within an acceptable range. A priority-based approach was used in the optimization, allowing for the primary objective (ATC) to be optimized first, followed by the minimization of the secondary objective (CDE). The theoretical framework for the optimization process is illustrated in Figure 2.

The ATC encompasses various components, including equipment costs C_{Cap}, operating and maintenance costs C_{OM}, natural gas costs C_{NG}, the grid power costs C_{GP}, costs for purchased hydrogen C_{PH}, and carbon trading costs C_{CT}.

The equipment costs were calculated using the capital recovery method, while the operating and maintenance costs were estimated at 5% of the annual equipment costs.

The emission factor method was used to calculate the CDE. These emissions encompass the combustion of natural gas and those from purchased electricity and purchased hydrogen.

$$ATC = C_{Cap} + C_{OM} + C_{NG} + C_{GP} + C_{PH} + C_{CT} \tag{28}$$

$$CDE = \sum_{t}^{T}\left(\frac{E_{GP,t}}{\eta_{GP}}\right) \times EF_{GP} + \sum_{t}^{T}(W_{GB,t} + W_{PGU,t}) \times EF_{NG} + \sum_{t}^{T}\frac{H_{PH,t}}{\eta_{SMR}} \times EF_{PH} \tag{29}$$

$$\max(w_i) = Cap_i \tag{30}$$

$$C_{Cap} = \frac{r \cdot (1+r)^n}{(1+r)^n - 1} \cdot \sum Cap_i \cdot p_i \tag{31}$$

$$C_{NG} = p_{NG} \cdot \sum_{t}^{T}(W_{GB,t} + W_{PGU,t}) \tag{32}$$

$$C_{GP} = p_{GP} \cdot \sum_{t}^{T} E_{GP,t} \tag{33}$$

$$C_{PH} = p_{PH} \cdot \sum_{t}^{T} H_{PH,t} \tag{34}$$

$$C_{CT} = p_{CT} \cdot CDE \tag{35}$$

where EF_{GP}, EF_{NG}, and EF_{PH} represent the emission factors for grid electricity, natural gas, and hydrogen produced via SMR, respectively; η_{GP} and η_{SMR} denote the efficiencies of grid power generation and SMR of hydrogen; w_i denotes the energy or material output of device i; Cap_i signifies the rated power or capacity of device i; r represents the bank interest rate; n indicates the lifespan of the equipment; and p_i refers to the unit price of device i. The unit market prices for natural gas, grid electricity, purchased hydrogen, and carbon trading are represented by p_{NG}, p_{GP}, p_{PH}, and p_{CT}, respectively.

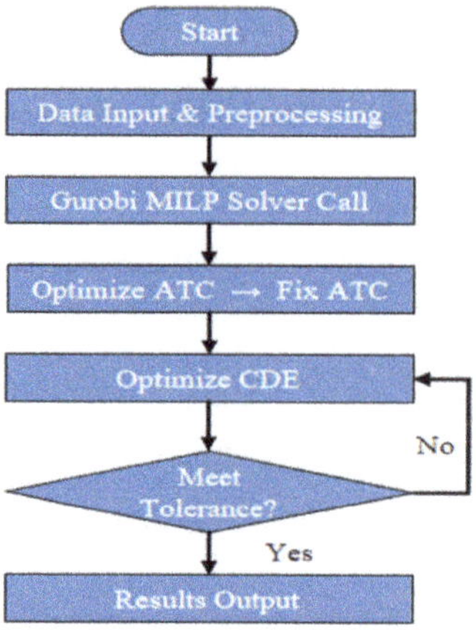

Figure 2. Theoretical framework for dual-objective optimization process.

2.1.7. Evaluation Indicators

ATC, CDE, and energy consumption were selected to evaluate the system benefits. The energy consumption, F, includes the energy consumption from PV, solar thermal, natural gas, purchased grid power, and hydrogen.

$$F = \sum_{t}^{T}(E_{PV,t} + Q_{SC,t}) + \sum_{t}^{T}(W_{GB,t} + W_{PGU,t}) + \sum_{t}^{T}\left(\frac{E_{GP,t}}{\eta_{GP}}\right) + \sum_{t}^{T}\frac{H_{PH,t}}{\eta_{SMR}} \tag{36}$$

3. Case Study

In the case study, the city of Dalian, located at approximately 38.91° N latitude and 121.61° E longitude, was selected as the subject of the study. Dalian is a coastal city on the Liaodong Peninsula. It experiences a humid continental climate with warm, wet summers and cold, dry winters. Dalian also has significant renewable energy resources, including substantial wind energy potential along its coastline and favorable solar energy conditions, making it a suitable location for the proposed integrated energy system.

This study focused on a residential community equipped with an onsite hydrogen production and refueling station within a future scenario. The thermal and electrical loads correspond to the community's residential needs, while the hydrogen load pertains to a refueling station with a daily hydrogen production capacity of 500 kg. Three typical days were selected to facilitate the optimization calculation and analysis of hourly conditions in each season: 156 days in winter, 119 days in the transitional season, and 90 days in summer. Heat and electrical loads were derived from simulated residential building demand, while the hydrogen load was converted from hourly refueling data from a small gas station. The residential community's heating, cooling, power, and hydrogen loads are shown in Figure 3, illustrating the hourly energy demand for heating, cooling, electricity, and hydrogen over typical days in winter, transition, and summer seasons. The hourly ambient temperature and radiation intensity for Dalian city on typical days of the three seasons are shown in Figure 4.

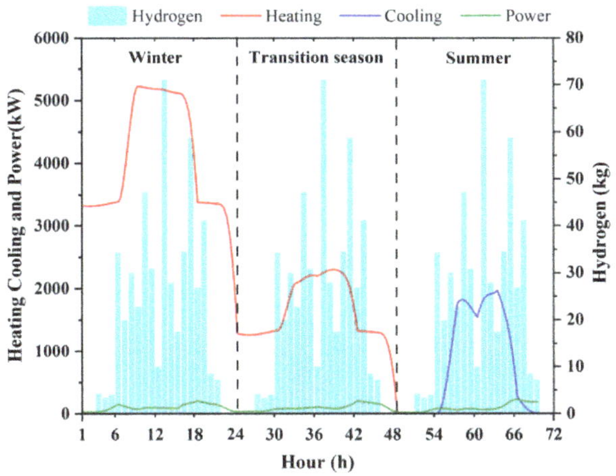

Figure 3. Heating, cooling, power, and hydrogen loads.

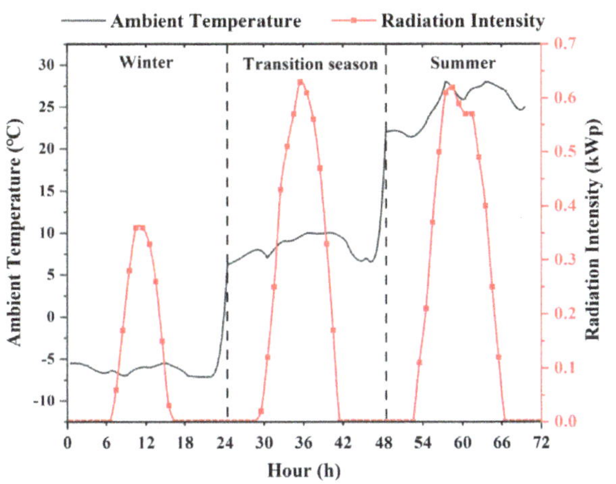

Figure 4. Hourly ambient temperature and radiation intensity.

The direct specification of the initial storage levels for batteries, hydrogen tanks, and thermal storage tanks can significantly influence the optimization results. Therefore, these initial values are defined as variables, and their costs are incorporated into the overall cost. The values of these variables are determined through the optimization process. This research disregards influences within the same season, assuming that the storage levels at the end of each typical day are equal to the initial values. The maximum energy release depth for the battery, hydrogen storage tank, and thermal storage tank is 90%. Gas turbines operate relatively inefficiently at low loads, so their start–stop operation factor is 20%. Each device's operational efficiencies and unit prices are detailed in Table 1. The area coefficient for PV panels is taken as seven. The emission factor for grid electricity is referenced from coal-fired power generation, and the electricity price is based on a time-of-use tiered pricing model. The carbon trading price is set at 90 CNY/t. The energy parameters are shown in Table 2. The electricity purchase price from the grid follows the peak and valley electricity prices specified by Liaoning Province, with the hourly prices shown in Figure 5.

Table 1. Economic and technical parameters of equipment.

Main Equipment	Efficiency/COP	Price
Power generation unit	36%	4200 CNY/kW
Photovoltaic panel	/	2400 CNY/kW
Solar collector	50%	1500 CNY/kW
Electric chiller	4	950 CNY/kW
Absorption chiller	0.70	1500 CNY/kW
Heat exchanger	80%	200 CNY/kW
Electric boiler	92%	1200 CNY/kW
Gas boiler	92%	1200 CNY/kW
Thermal storage tank	95%	200 CNY/kW
Battery	98.5%	1600 CNY/kW
PEM electrolyzer and its auxiliaries	57.9 kWh/kg H_2	9000 CNY/kW
Converter	98.5%	600 CNY/kW
H_2 refueling station	/	2×10^7 CNY

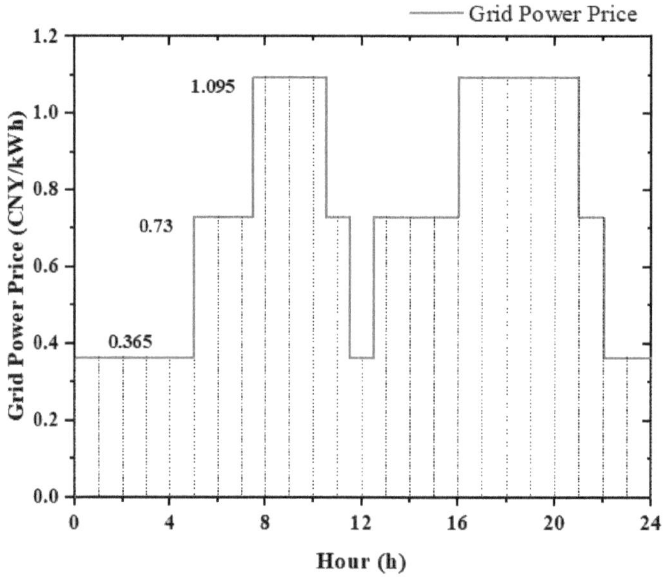

Figure 5. Hourly electricity price from the grid.

Table 2. Energy parameters.

Gas	Price	Carbon Dioxide Emission Factors
Grid power	/	1.2 kg/kWh
Natural gas	4.18 CNY/Nm^3	2.11 kg/Nm^3
Purchased hydrogen	36.4 CNY/kg	36.8 kg/kg

A Python 3.9.7 program was developed based on the system model, utilizing the Gurobi optimization solver to obtain solutions, followed by a multidimensional analysis of the results.

4. Results and Discussion

4.1. Comparison of Scenarios

Building upon the research in the above scenarios, three control systems—S1, S2, and S3—were established. S1 excludes solar power generation and thermal energy, S2 removes the energy storage devices (batteries and thermal storage tank), and S3 focuses on single-objective optimization with ATC as the only goal. S4 represents the entire proposed system. The performance of these systems in terms of economy, carbon emissions, and energy consumption is shown in Figure 6 and Table 3. The proposed system significantly outperformed the control systems regarding energy consumption and CO_2 emissions. At the same time, its cost was only slightly higher than that of the S3 system, which is solely focused on cost.

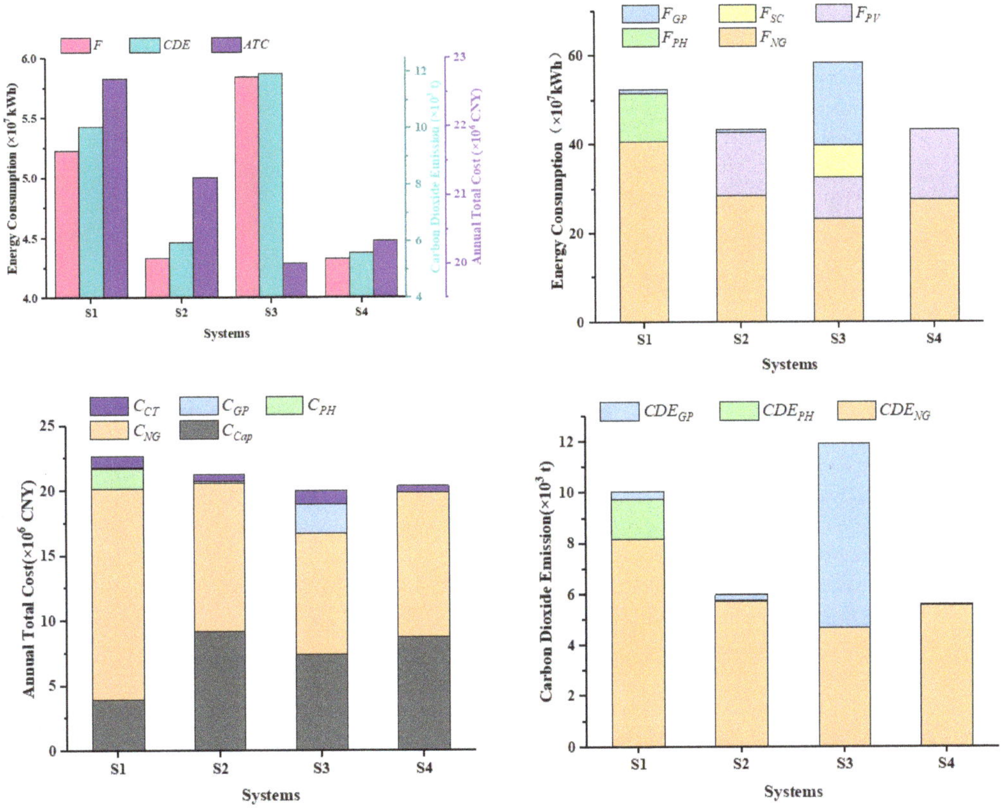

Figure 6. The performance of each system.

Table 3. Rated power of equipment configured for each system (kW).

System	PGU	GB	EB	EC	AC	EL	B	TST	PV	SC
S1	1450	4180	80	410	390	1280	20	4490	0	0
S2	2360	960	3280	230	2020	3780	0	0	12,460	0
S3	780	3200	1580	40	1860	2240	680	6530	8000	6320
S4	2290	610	4230	0	2020	2490	1230	7360	13,590	0

In all systems, a significant portion of the annual total costs is allocated to purchasing natural gas, which remains the primary energy source. S1, which lacks solar power, relies heavily on natural gas and external hydrogen, resulting in high external energy dependency. The S1 system also has the lowest-rated electrolyzer power, making it economically less viable for hydrogen production without solar energy. The S2 system lacks energy storage and has a higher-rated gas turbine and electrolyzer power, leading to higher grid electricity purchases. Energy storage in S4 alleviates this pressure by balancing energy use and supply across different times.

The S3 system utilizes solar collectors for thermal loads, demonstrating higher thermal efficiency than photovoltaic panels when CO_2 emissions are not considered. The S3 system also uses absorption chillers that can work with various heat sources, including gas boilers and electric boilers. In contrast, S4 prioritizes photovoltaic panels and electric boilers, optimizing the energy supply by integrating storage devices. Despite no solar collectors being used in S4, the system's ability to generate electricity and convert it to thermal energy via electric boilers makes it highly efficient and cost-effective.

The full utilization of photovoltaic and solar thermal energy can significantly reduce carbon emissions. The development of green energy should not only focus on the energy production sector but also emphasize the coordinated development of energy storage. Compared to cost-based single-objective optimization, dual-objective optimization is bound to increase costs. Therefore, the development of dual-objective optimization requires appropriate policy incentives.

4.2. Energy Production and Consumption

The proposed system demonstrates the coordinated supply of thermal, cooling, electrical, and hydrogen energy sources. The primary energy sources—electricity and thermal energy—were analyzed in terms of production and consumption. Figure 7 illustrates the hourly trends in power production and consumption by different devices across seasons. Similarly, Figure 8 presents the hourly trends in heating production and consumption by different devices across seasons. In both figures, the names in the legends correspond to the abbreviations used earlier in the equations.

Seasonal variations primarily influence the operational behavior of the IES in terms of solar radiation, ambient temperature, and energy demand. These factors affect the system's performance and operation at a physical level.

In winter, shorter daylight hours and lower solar angles reduce photovoltaic output. During this period, the gas turbine plays a more dominant role in covering the electrical load, while its waste heat is used to meet heating demands, reducing the need for additional fuel consumption in the gas boiler. In the transition season, as solar irradiance improves, the PV output increases, and electricity from the PV helps power the electrolyzers and electric boilers. The system's energy storage capacity is used more efficiently, helping to smooth out peak demands. In summer, longer daylight hours and higher solar angles maximize PV generation, significantly contributing to meeting electricity and cooling demands. Despite some reduction in efficiency due to high ambient temperatures, surplus solar power is used to meet cooling needs via electric or absorption chillers and to produce hydrogen.

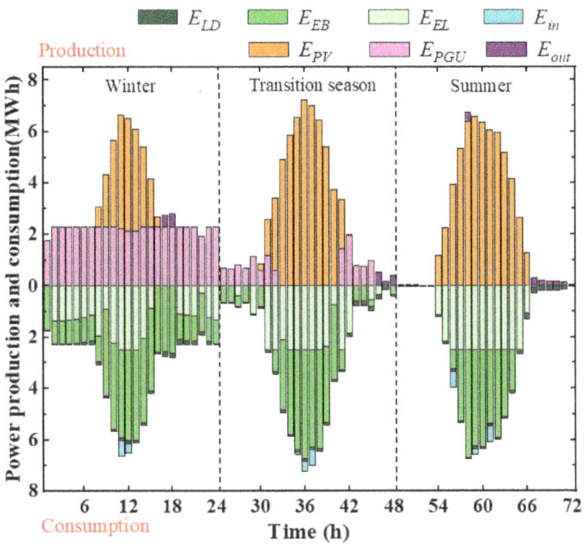

Figure 7. Hourly trends in power production and consumption by different devices across seasons.

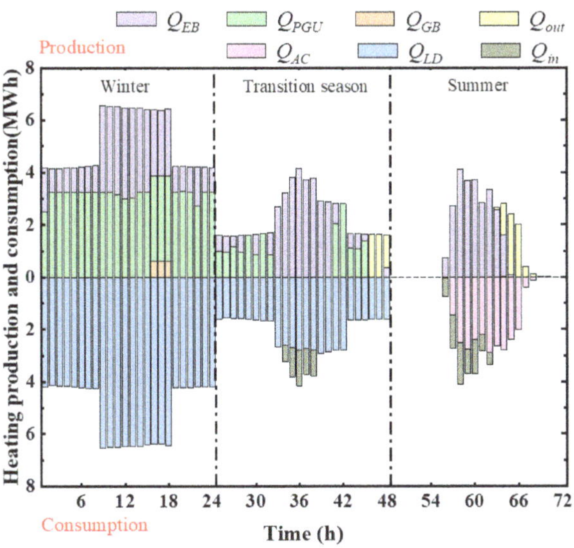

Figure 8. Hourly trends in heating production and consumption by different devices across seasons.

The system also manages the interplay between electrical and thermal loads by converting surplus electricity from PV generation into thermal energy through electric boilers or storing it in batteries. This flexibility allows the system to dispatch electricity when needed for cooling or heating. In winter, the gas turbine's waste heat becomes particularly valuable, providing thermal energy without additional fuel consumption. In summer, batteries and thermal storage handle the electrical and cooling loads more effectively, reducing dependence on the gas turbine.

Finally, hydrogen production via electrolysis is most efficient during high solar irradiance periods, allowing the system to minimize its carbon footprint using clean PV power. However, when solar resources are lower in winter, the system may rely more on the gas turbine's output to meet hydrogen demand, emphasizing the need for flexibility in the system's design.

The system optimally balances energy supply and demand by accounting for these seasonal factors and the interaction between electricity, thermal loads, and hydrogen production. This approach maximizes renewable energy utilization, reduces reliance on fossil fuels, and enhances overall system flexibility, ensuring reliable and cost-effective operation year-round.

4.3. Impact of Illumination Area

The available illumination area significantly impacts the energy output of both photovoltaic panels and collectors. Given the limited illumination area in residential communities, analyzing the effects of varying area constraints is essential. The system is optimized without area constraints, with the PV panels rated at 13,590 kW and 95,130 m^2. The energy system performance was calculated for restricted areas ranging from 0 to 90,000 m^2, and the results are shown in Figure 9. In the figure, UR represents the integrated energy system with an unlimited light area.

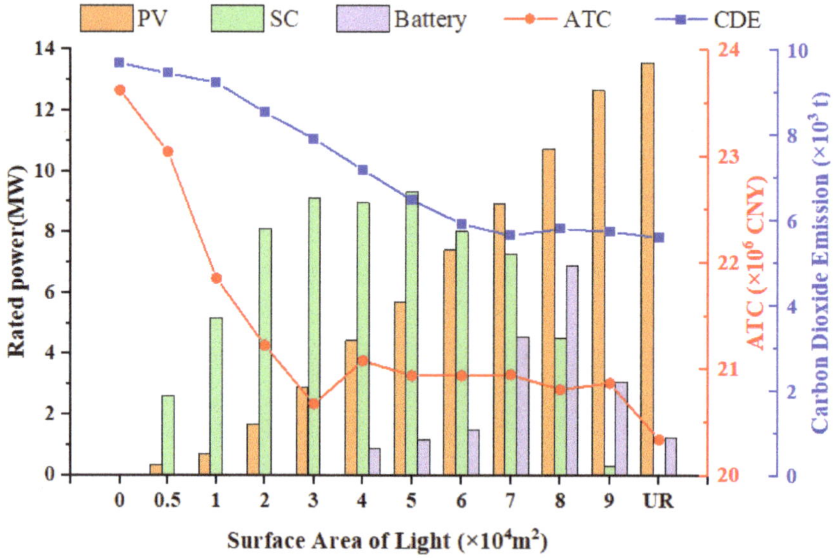

Figure 9. Critical data on the impact of illumination area.

The calculated actual light areas are all closely approximated to the restricted area. As the illumination area increases, the rated power of the PV panels increases steadily. Initially, the solar collector's rated power rises but then decreases, eventually reaching zero. Meanwhile, the battery capacity increases with the PV power until a certain threshold, after which the excess photovoltaic energy can be converted to heat through the electric boiler and stored in the thermal storage tank, reducing the need for battery capacity. As the illumination area expands, both the ATC and CO_2 emissions decrease, indicating the economic and emission reduction advantages of increased PV and solar collector use. Under different light areas, an appropriate combination of PV and SC can significantly reduce costs and carbon emissions. Operators should adapt to local conditions and adjust the PV and SC installed capacity according to energy demand.

4.4. Impact of Carbon Trading Prices

The carbon trading market is gradually maturing, but the price is still fluctuating, and it is of great significance to study the potential impact of the price on the IES. China's carbon trading price fluctuates at 90 CNY/t, so the interval of 0–200 was selected for this study. The primary data for that impact are shown in Figure 10. In the figure, Grid Power indicates the electricity purchased from the grid.

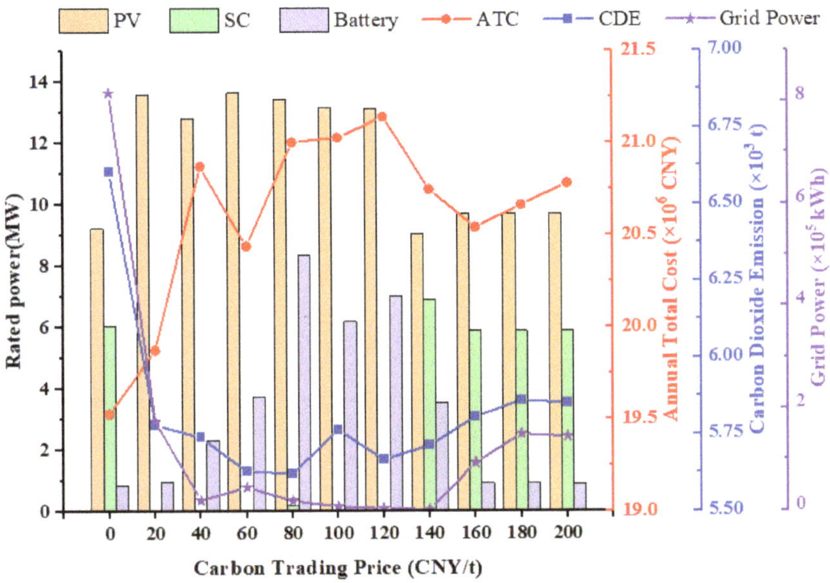

Figure 10. Critical data on the impact of carbon trading prices.

As carbon trading prices gradually increase, the rated power of the battery initially rises before declining, while the rated power of the collector transitions from zero to a positive value. The rated power of the PV panels sharply increases to a fixed range before dropping to another fixed range, and electricity consumption from the grid first decreases and then increases. Introducing carbon trading prices increases the rated power of the photovoltaic panels significantly, highlighting their strong correlation with these prices. As carbon trading prices rise, the battery capacity gradually increases, resulting in a reduced need for purchasing electricity from the grid. When the carbon trading price is 60 CNY/t, the system's cost and carbon emissions are relatively low.

When the carbon trading price reaches a certain threshold, the system's optimization favors the more efficient solar collector for thermal supply. Concurrently, the rated power of the PV panels and batteries progressively diminishes. When the carbon trading price hits 160 CNY/t, the rated power of all devices stabilizes. At this point, further increases in carbon trading prices fail to enhance solar energy utilization and lead the system to opt for more electricity purchases from the grid, thereby reducing natural gas usage. This finding contradicts the traditional belief that higher carbon trading prices always lead to more significant CO_2 emissions reduction. Therefore, policymakers should be sufficiently aware of the dual impact of carbon trading prices when formulating regulations.

5. Conclusions

This study proposes a novel integrated energy system that combines photovoltaic and solar thermal energy generation with coordinated electrical and thermal energy storage, designed to meet the diverse energy demands of community residential buildings, including heating, cooling, power, and hydrogen production. The system's dual-objective optimization minimizes economic costs and CO_2 emissions, offering a sustainable and efficient solution. The study concludes with the following:

1. The proposed integrated system, combining PV, solar thermal, and energy storage, reduces reliance on fossil fuels and grid electricity, resulting in significant cost and CO_2 emission reductions.
2. Integrating green hydrogen production via photovoltaic energy presents a cost-effective and environmentally beneficial alternative to conventional grid power and gas turbines.
3. The effective coordination of electrical and thermal energy storage systems ensures optimal energy management, especially during peak demand periods, reducing dependency on external energy sources.
4. An appropriate carbon trading price positively influences emissions reduction; however, a higher price does not necessarily equate to more significant benefits for emissions reduction. When the carbon trading price is 60 CNY/t, the system's cost and carbon emissions are relatively low.

Future research will build upon the integrated energy systems discussed in this study, focusing on developing a comprehensive energy hub that incorporates hydrogen refueling, gas refueling, and charging capabilities. This energy hub would enable the conversion and storage of various energy resources, providing stable energy supplies to different sectors, including transportation, residential communities, schools, hospitals, and factories. Further work will explore optimizing the coordination among these energy sources to enhance system resilience and efficiency.

Author Contributions: Conceptualization, X.Z. and X.S.; Methodology, X.Z., N.L. and X.K.; Software, X.Z. and X.K.; Validation, X.Z.; Formal analysis, X.Z.; Resources, H.M.; Data curation, X.Z.; Writing—original draft, X.Z.; Writing—review & editing, X.Z. and X.S.; Visualization, X.Z.; Supervision, H.M. and N.L.; Project administration, H.M. and N.L.; Funding acquisition, H.M. All authors have read and agreed to the published version of the manuscript.

Funding: This study was supported by the National Natural Science Foundation of China: Study on Optimal Control and Comprehensive Evaluation Model of Multi-microgrid System Based on Uncertain Environment (No. 51976020).

Data Availability Statement: The original contributions presented in this study are included in the article. Further inquiries can be directed to the corresponding author.

Conflicts of Interest: The authors declare no conflict of interest.

Nomenclature

IES	Integrated energy system
PV	Photovoltaic
SC	Solar collector
PGU	Power generation unit
HRU	Heat recovery unit
B	Battery
GB	Gas boiler
EB	Electric boiler
EL	Electrolyzer
TST	Thermal storage tank
HST	Hydrogen storage tank
AC	Absorption chille

EC	Electric chiller
CON	Converter
HE	Heat exchanger
SMR	Steam methane reforming
PEM	Proton exchange membrane
NG	Natural gas
GP	Grid power
PH	Purchased hydrogen
LD	Energy load of the system
CT	Carbon trading
in\out	Energy flow into or out of the energy storage device
ATC	Annual total cost
CDE	Carbon dioxide emission
E_i	Electrical energy of device i
Q_i	Thermal energy of device i
H_i	Hydrogen energy of device i
CG_i	Cooling energy of device i
η_i	Efficiency of device i
Cap_i	Capacity of device i
W_i	Natural gas consumption of device i
$a\backslash b$	Boolean variable
$A\backslash A_{PV}\backslash A_{SC}$	Total area, PV panel area, and SC area
$T_{amb}\backslash T_{ref}$	Actual and standard ambient temperature
$G_{amb}\backslash G_{ref}$	Actual and standard solar irradiance
f_{PV}	Degradation coefficient
K_t	Power temperature coefficient
C_i	Cost of project i
C_E	Equipment cost
C_{OM}	Operating and maintenance cost
EF_i	Carbon emission factor of i
w_i	Energy or material output of device i
r	Bank interest rate
n	Lifespan of the equipment
p_i	Unit price of device i or item i
F	Energy consumption

References

1. Lu, Q.; Liu, M. A multi-criteria compromise ranking decision-making approach for analysis and evaluation of community-integrated energy service system. *Energy* **2024**, *306*, 132439. [CrossRef]
2. Lv, H.; Wu, Q.; Ren, H.; Li, Q.; Zhou, W. A two-stage decision-making approach for optimal design and operation of integrated energy systems considering multiple uncertainties and diverse resilience needs. *Energy* **2024**, *305*, 132375. [CrossRef]
3. Agrawal, A.; Bakshi, B.R.; Kodamana, H.; Ramteke, M. Renewables-Integrated Energy Systems Can Provide Electricity at Lower Cost with Less Environmental and Social Damage. *ACS Sustain. Chem. Eng.* **2022**, *10*, 13390–13401. [CrossRef]
4. Wu, Y.; Jin, P.; Li, Q. Research on regulation strategy of integrated energy system based on game theory and divide-and-conquer algorithm. *Energy* **2025**, *319*, 134860. [CrossRef]
5. Liu, X.; Li, N.; Mu, H.; Li, M.; Liu, X. Techno-energy-economic assessment of a high capacity offshore wind-pumped-storage hybrid power system for regional power system. *J. Energy Storage* **2021**, *41*, 102892. [CrossRef]
6. Guo, Z.; Gao, J.; Dong, Y.; Du, J. Integrated batch production planning and scheduling with machine-learning based production capacity region considering wind energy system. *J. Clean. Prod.* **2024**, *477*, 143804. [CrossRef]
7. Li, Z.; Zhao, Y.; Wu, P.; Zhang, H.; Dong, N.; Chang, Y.; Du, Y. Enhancing efficiency and economy of hydrogen-based integrated energy system: A green dispatch approach based on exergy analysis. *Energy* **2024**, *310*, 133315. [CrossRef]
8. Wang, J.; Cui, Y.; Liu, Z.; Zeng, L.; Yue, C.; Agbodjan, Y.S. Multi-energy complementary integrated energy system optimization with electric vehicle participation considering uncertainties. *Energy* **2024**, *309*, 133109. [CrossRef]

9. Xiong, Y.; Wang, Y.; Zhang, Y.; Chen, J.; Wu, M. Production, investment and financial plan for a "natural gas+" integrated energy enterprise: An assessment using system dynamics and multi-objective optimization model. *Energy Rep.* **2025**, *13*, 1859–1874. [CrossRef]
10. Liu, C.; Fan, Y.; Yu, W.; Liu, Y.; Li, C.; Chi, Y. Low carbon operation and evaluation methods for integrated energy system counting EVs with whole life cycle considering demand response. *Renew. Energy* **2024**, *236*, 121447. [CrossRef]
11. Zhao, H.; Yao, Y.; Peng, D.; Zhao, P. A preference adjustable capacity configuration optimization method for hydrogen-containing integrated energy system considering dynamic energy efficiency improvement and load fast tracking. *Renew. Energy* **2024**, *235*, 121314. [CrossRef]
12. Babaei, R.; Ting, D.S.; Carriveau, R. Feasibility and optimization of hybrid energy systems for sustainable electricity, heat, and fresh water production in a rural community. *Int. J. Green Energy* **2025**, 1–19. [CrossRef]
13. Wang, Z.; Hao, Y.; Zhao, G.; Wang, M.; Su, Z.; Zhang, J. Day-ahead optimal dispatch method of integrated electric-heat-cool-gas energy system based on N-1 safety criterion. *Energy Build.* **2024**, *323*, 114800. [CrossRef]
14. Shu, Y.; Dai, Y. The synergetic competition of cross-regional integrated energy systems with low-carbon technology heterogeneity under carbon emission trading mechanism to achieve carbon reduction target. *Renew. Energy* **2024**, *236*, 121394. [CrossRef]
15. Yang, J.; Zhao, B.; Ma, K.; Zhong, J.; Xu, W. A carbon integrated energy pricing strategy based on non-cooperative game for energy hub in seaport energy system. *Energy* **2024**, *309*, 133009. [CrossRef]
16. Gao, L.; Yang, S.; Chen, N.; Gao, J. Integrated Energy System Dispatch Considering Carbon Trading Mechanisms and Refined Demand Response for Electricity, Heat, and Gas. *Energies* **2024**, *17*, 4705. [CrossRef]
17. Yao, Z.; Ran, L.; Wang, Z.; Guo, X. Integrated management of electric vehicle sharing system operations and Internet of Vehicles energy scheduling. *Energy* **2024**, *309*, 132498. [CrossRef]
18. Zhang, Y.; Li, J.; Ji, X.; Yang, M.; Ye, P. Optimal scheduling of electricity-gas-heat integrated energy system with coordination of flexibility and reliability. *Sustain. Energy Technol.* **2024**, *71*, 103968. [CrossRef]
19. Faisal, S.; Gao, C. A Comprehensive Review of Integrated Energy Systems Considering Power-to-Gas Technology. *Energies* **2024**, *17*, 4551. [CrossRef]
20. Nandi, S.; Ghatak, S.R.; Sannigrahi, S.; Acharjee, P. Coordinated planning and operation of PV- hydrogen integrated distribution network incorporating daily-seasonal green hydrogen storage and EV charging station. *Int. J. Hydrogen Energy* **2024**, *90*, 134–158. [CrossRef]
21. Liu, B.; Yang, L.; Lv, T.; Zhu, L.; Ji, M.; Hu, W. Application of PVT Coupled Solar Heat Pump System in the Renovation of Existing Campus Buildings. *Energies* **2024**, *17*, 4922. [CrossRef]
22. Abd Elbar, A.R.; Baz, F.B.; Marzouk, S.A.; Eldesoukey, A.; Salem, A.M. Analyzing solar still performance with PV electricity storage: Energy, exergy, exergoeconomic, and enviroeconomic perspectives. *Desalination* **2025**, *597*, 118400. [CrossRef]
23. Zhu, X.; Chen, J.; Wang, L.; Wang, P.; Fang, Z.; Du, X.; Gao, F. Optimum power control and coordinate sizing for the stand-alone wind-energy storage integrated hydrogen production system. *Int. J. Hydrogen Energy* **2024**, *88*, 803–814. [CrossRef]
24. Huo, S.; Li, Q.; Pu, Y.; Xie, S.; Chen, W. Low carbon dispatch method for hydrogen-containing integrated energy system considering seasonal carbon trading and energy sharing mechanism. *Energy* **2024**, *308*, 132794. [CrossRef]
25. Zhou, J.; Jiang, X.; Li, S.; Zhu, J.; Zhou, Z.; Liang, G. Operation optimization for gas-electric integrated energy system considering hydrogen fluctuations and user acceptance. *Int. J. Hydrogen Energy* **2024**, *88*, 666–687. [CrossRef]
26. Han, L.; Wang, S.; Cheng, Y.; Chen, S.; Wang, X. Multi-timescale scheduling of an integrated electric-hydrogen energy system with multiple types of electrolysis cells operating in concert with fuel cells. *Energy* **2024**, *307*, 132707. [CrossRef]
27. Ye, S.; Wang, J.; Yin, Z.; Kang, J.; Ma, Z. Optimization of a solar-driven community integrated energy system based on dynamic hybrid hydrogen-electric energy storage strategy. *J. Energy Storage* **2024**, *101*, 113917. [CrossRef]
28. Rahman, J.; Jacob, R.A.; Zhang, J. Multi-timescale power system operations for electrolytic hydrogen generation in integrated nuclear-renewable energy systems. *Appl. Energy* **2025**, *377*, 124346. [CrossRef]
29. Liu, B.; Zheng, S.; Lai, M. Carbon emission reduction potential analysis of fuel cell vehicles in China: Based on GRA-BiLSTM prediction model. *Int. J. Hydrogen Energy* **2024**, *66*, 110–121. [CrossRef]
30. Abdullah, Z.; Keeley, A.R.; Coulibaly, T.Y.; Managi, S. The impact of fuel cell vehicles deployment on road transport greenhouse gas emissions through 2050: Evidence from 15 G20 countries. *J. Environ. Manag.* **2024**, *370*, 122660. [CrossRef]
31. Zhen, J.; Liu, X.; Wu, C.; Ji, L.; Huang, G. Operation optimization and performance evaluation of photovoltaic-wind-hydrogen-based integrated energy system under carbon trading mechanism and uncertainty for urban communities. *J. Clean. Prod.* **2024**, *476*, 143688. [CrossRef]
32. Liu, W.; Wang, J.; Wang, Y. Building solar integrated energy systems considering power and heat coordination: Optimization and evaluation. *Sol. Energy* **2024**, *279*, 112821. [CrossRef]
33. Tubussum, H.; Aslam, M.; Hashim, H.M.; Marimuthu, M.; Mahmoud, K.H.; Alsubaie, A.S.; Bhutto, A.W.; Bazmi, A.A.; Bokhari, A.; Nizami, A. Reorientation of energy policy and management: Integrated Energy Planning (IEP) framework implementation for Renewable and Thermal Energy Systems (RTES). *Process Saf. Environ.* **2024**, *191*, 780–798. [CrossRef]

34. Bazdar, E.; Nasiri, F.; Haghighat, F. Resilience-centered optimal sizing and scheduling of a building-integrated PV-based energy system with hybrid adiabatic-compressed air energy storage and battery systems. *Energy* **2024**, *308*, 132836. [CrossRef]
35. Kim, Y.; Oh, S.; Kim, D.; Hong, S.; Park, J. Development of a hybrid energy storage system for heat and electricity: Application to green hydrogen production process integrated with a municipal solid waste incinerator. *Energ. Convers. Manag.* **2024**, *321*, 119009. [CrossRef]
36. Chong, Z.; Yang, L.; Jiang, Y.; Zhou, W. Hybrid-timescale optimal dispatch strategy for electricity and heat integrated energy system considering integrated demand response. *Renew. Energy* **2024**, *232*, 121123. [CrossRef]
37. Liu, T.; Yang, Z.; Duan, Y.; Hu, S. Techno-economic assessment of hydrogen integrated into electrical/thermal energy storage in PV+ Wind system devoting to high reliability. *Energ. Convers. Manag.* **2022**, *268*, 116067. [CrossRef]
38. Baz, F.; Sharaf, M.; Aljabr, A.; Almehmadi, F.A.; Marzouk, S. Enhancing photovoltaic/thermal performance with perforated tube within cooling water duct. *Clean. Technol. Environ.* **2024**. [CrossRef]
39. Marzouk, S.A.; El-Fakharany, M.K.; Baz, F.B. Heat transfer performance of particle solar receiver: Numerical study. *Heat. Transf. Res.* **2022**, *53*, 1–19. [CrossRef]
40. Wu, M.; Sun, H.; Liu, Q.; Lv, X.; Chen, B.; Yang, M.; Song, Y. Enhancing CO_2 Sequestration Safety with Hydrate Caps: A Comparative Study of CO2 Injection Modes and Saturation Effects. *Energy* **2025**, *320*, 135044. [CrossRef]
41. Wang, Y.; Dong, H.; Ma, K.; Wang, H.; Zhang, J. Multi frequency stability optimization of integrated energy systems considering virtual energy storage characteristics of heating networks. *Appl. Therm. Eng.* **2024**, *257*, 124254. [CrossRef]
42. Hao, M.; Lan, J.; Wang, L.; Lin, Y.; Wang, J.; Qin, L. Optimized Dual-Layer Distributed Energy Storage Configuration for Voltage Over-Limit Zoning Governance in Distribution Networks. *Energies* **2024**, *17*, 1847. [CrossRef]
43. Ma, T.; Li, M.; Fan, C.; Dong, H. A novel real-time dynamic performance evaluation and capacity configuration optimization method of generation-storage-load for integrated energy system. *Appl. Energy* **2024**, *374*, 123896. [CrossRef]
44. Wang, J.L.; Yan, T.; Pan, W.G. Design and evaluation of integrated energy system combining solar energy and compressed-air energy storage. *Renew. Energy* **2024**, *232*, 121068. [CrossRef]
45. Lin, L.; Ou, K.; Lin, Q.; Xing, J.; Wang, Y. Two-stage multi-strategy decision-making framework for capacity configuration optimization of grid-connected PV/battery/hydrogen integrated energy system. *J. Energy Storage* **2024**, *97*, 112862. [CrossRef]
46. El-Fakharany, M.K.; Abo-Samra, A.A.; Abdelmaqsoud, A.M.; Marzouk, S.A. Enhanced performance assessment of an integrated evacuated tube and flat plate collector solar air heater with thermal storage material. *Appl. Therm. Eng.* **2024**, *243*, 122653. [CrossRef]
47. Zhou, J.; Wu, Y.; Zhong, Z.; Xu, C.; Ke, Y.; Gao, J. Modeling and configuration optimization of the natural gas-wind-photovoltaic-hydrogen integrated energy system: A novel deviation satisfaction strategy. *Energ. Convers. Manag.* **2021**, *243*, 114340. [CrossRef]
48. Song, Y.; Mu, H.; Li, N.; Wang, H. Multi-objective optimization of large-scale grid-connected photovoltaic-hydrogen-natural gas integrated energy power station based on carbon emission priority. *Int. J. Hydrogen Energy* **2023**, *48*, 4087–4103. [CrossRef]

energies

MDPI

Review

A Review on Phase-Change Materials (PCMs) in Solar-Powered Refrigeration Systems

Yali Guo , Chufan Liang, Hui Liu , Luyuan Gong, Minle Bao * and Shengqiang Shen

Key Laboratory of Ocean Energy Utilization and Energy Conservation of Ministry of Education, Dalian University of Technology, Dalian 116024, China
* Correspondence: mlbao@dlut.edu.cn

Abstract: Over the past few years, the combination of solar power with refrigeration technology has matured, providing a promising solution for sustainable cooling. However, a key challenge remains, namely the inherent intermittency of solar energy. Due to its uneven temporal distribution, it is difficult to ensure continuous 24 h operation when relying solely on solar energy. To address this issue, thermal energy storage technology has emerged as a viable solution. This paper presents a comprehensive systematic review of phase-change material (PCM) applications in solar refrigeration systems. It systematically categorizes solar energy conversion methodologies and refrigeration system configurations while elucidating the fundamental operational principles of each solar refrigeration system. A detailed examination of system components is provided, encompassing photovoltaic panels, condensers, evaporators, solar collectors, absorbers, and generators. The analysis further investigates PCM integration strategies with these components, evaluating integration effectiveness and criteria for PCM selection. The critical physical parameters of PCMs are comparatively analyzed, including phase transition temperature, latent heat capacity, specific heat, density, and thermal conductivity. Through conducting a critical analysis of existing studies, this review comprehensively evaluates current research progress within PCM integration techniques, methodological classification frameworks, performance enhancement approaches, and system-level implementation within solar refrigeration systems. The investigation concludes by presenting strategic recommendations for future research priorities based on a comprehensive systematic evaluation of technological challenges and knowledge gaps within the domain.

Keywords: phase-change materials; solar-powered refrigeration; renewable energy integration; thermal energy storage; energy efficiency

1. Introduction

Energy is undoubtedly one of the most critical drivers for the advancement of nations and human societies [1]. Currently, fossil fuels dominate the global energy supply, accounting for 75% of total consumption [2]. However, their use has led to substantial environmental pollution over the past several decades. In light of this, there has been a significant shift in focus toward cleaner, renewable energy sources such as solar and wind power. These energy alternatives are expected to play a crucial role in aligning societal progress with ecological preservation. At present, renewable energy contributes to approximately 30% of global energy consumption [3], and nations worldwide are actively working toward reshaping their energy infrastructures to increase the share of renewables.

As societal aspirations for elevated quality of life continue to advance, refrigeration systems have evolved from discretionary luxury goods to indispensable household necessities [4]. Statistics reveal that 40% of global food production requires refrigeration or freezing [5], making the food refrigeration sector a major consumer of energy, responsible for around 15% of global electricity consumption [6]. However, nearly 800 million people across the globe still lack reliable access to electricity [7], and many others experience inconsistent power supply. For instance, in Bangladesh, unstable voltage levels and power shortages can prevent the use of essential high-power devices, including refrigerators [8]. In this context, solar-powered refrigeration systems present an ideal solution. These systems are not only low-cost but also independent of traditional electrical grid infrastructures, making them particularly suitable for regions with unreliable power supply. Furthermore, they offer higher efficiency in utilizing renewable energy, effectively addressing these challenges. However, the intermittent nature of solar energy—its uneven distribution over time and space—poses a significant challenge, potentially leading to voltage fluctuations that could damage electrical components. By integrating energy storage technologies, such as phase-change materials (PCMs), with solar refrigeration systems, this issue can be substantially mitigated. PCMs are a cost-effective and convenient energy storage solution, making them a popular choice in the development of solar refrigeration technologies.

In recent years, the integration of PCMs into solar refrigeration systems has rapidly progressed, with a growing body of the literature exploring the subject. Notably, a 2017 review by Khan et al. provided an in-depth evaluation of the suitable PCM types, integration techniques, performance-enhancing features, and the temperature-related challenges faced by each component of a single-effect solar absorption system [9]. More recently, Patil et al. [10], in their 2024 review, focused on the stability of PCMs used in solar thermal storage systems for absorption refrigeration, offering a comprehensive overview of PCM applications within these systems. While the literature on PCM integration in solar refrigeration systems is expanding, there is still a need for further research into the specific roles and benefits of PCMs across different components of these systems.

This study endeavors to address this research deficiency through a systematic examination of contemporary solar-powered refrigeration systems and their corresponding phase-change material (PCM) applications. The investigation methodologically examines PCM integration protocols across multiple system architectures, with particular emphasis on establishing selection criteria for optimal material specifications based on thermodynamic requirements. Through comparative performance analysis, the research quantitatively evaluates energy conservation capabilities and operational efficiency metrics in PCM-enhanced solar refrigeration configurations. Furthermore, the analytical framework systematically identifies critical implementation barriers within current technological paradigms while proposing targeted optimization strategies to improve thermodynamic performance and mitigate operational constraints in practical applications.

2. Classification of Solar Refrigeration Systems

Solar refrigeration systems leverage solar energy, transforming it into thermal or electrical energy to power refrigeration processes, thereby providing cooling services to end users. System configurations are determined by the specific power input requirements and target temperature parameters of the particular cooling application. This section presents a comprehensive classification of solar refrigeration systems based on their operational principles and energy conversion mechanisms.

From a functional perspective, solar refrigeration systems can be divided into three main subsystems, which are solar energy conversion devices, refrigeration systems, and cooling load endpoints. Notably, the type of cooling load endpoint (i.e., the end user) does not have a fundamental impact on the core principles of solar refrigeration. Therefore, this paper will present a detailed and systematic classification of solar refrigeration systems based on the principles governing both solar energy conversion devices and refrigeration systems.

The solar energy conversion devices can be categorized into two primary types, namely solar thermal conversion devices and solar electric conversion devices. Refrigeration systems can further be classified into two main types, which are thermally driven refrigeration systems and electrically driven refrigeration systems. As depicted in Figure 1, the classification of solar refrigeration systems is primarily determined by the type of solar energy conversion device used, resulting in three key categories of solar thermal refrigeration systems, solar electric refrigeration systems, and solar combined refrigeration systems.

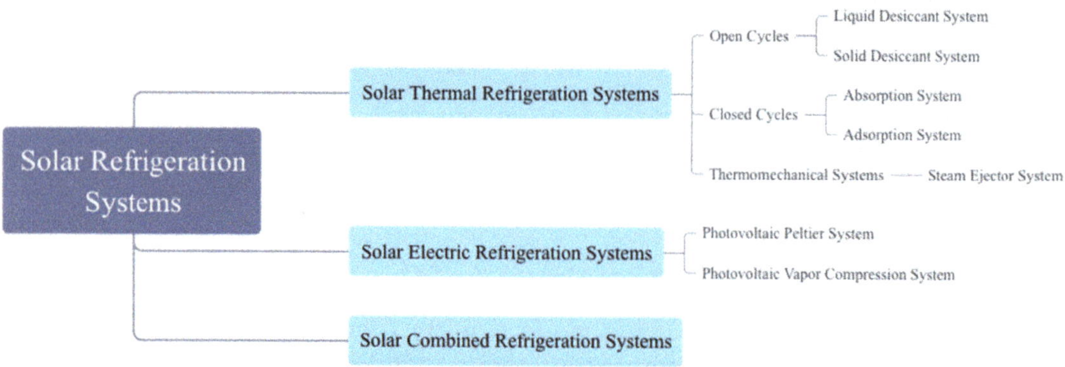

Figure 1. Energy conversion options for solar refrigeration systems.

2.1. *Solar Thermal Refrigeration*

Solar thermal refrigeration systems function by absorbing solar energy through solar thermal collectors and converting it into heat, which is then stored in a thermal storage tank. The stored heat is used to power devices that require thermal energy, and these devices utilize thermochemical or thermophysical methods to perform refrigeration. This in turn drives the operation of the refrigeration system. Based on their operational principles, solar thermal refrigeration systems can be classified into three main types, namely open adsorption cycles, closed absorption cycles, and thermomechanical refrigeration cycles.

2.1.1. Open Adsorption Cycle

The open adsorption cycle refrigeration system operates by utilizing liquid or solid desiccants to dehumidify or humidify air, effectively providing refrigeration. The system operates via two fundamental processes that are integral to its functionality. The first process, known as adsorption, involves the desiccant absorbing moisture from one airflow to another by leveraging the vapor pressure differential between the humid air and the desiccant material. When the desiccant material is dry and cold, its surface vapor pressure is lower than that of the humid air, and moisture is adsorbed onto the desiccant. During the adsorption phase, the latent heat of water vapor is released, causing an increase in temperature. As a result, the desiccant becomes wet and progresses to the second phase. The second phase, known as desorption or regeneration, involves raising the desiccant's temperature to release the absorbed moisture into the airflow. After cooling, the desiccant material can absorb moisture again. These two processes form a continuous cycle, enabling the system to transfer moisture effectively, as shown in Figure 2. The open-type adsorption cycle refrigeration system utilizing liquid desiccant consists primarily of a regulator and a regenerator. In the regulator, sprayed liquid desiccant absorbs atmospheric moisture, reducing the hot–humid air's dry-bulb temperature by 5–8 °C through dehumidification cooling. The diluted desiccant subsequently accumulates in the collection pit, corresponding to the 1→2 process in Figure 2. The regeneration phase transports this solution to the regenerator—structurally analogous to the regulator—where desiccant spraying facilitates moisture transfer to ambient air (process 2→3 in Figure 2). Finally, the reconcentrated desiccant undergoes temperature reduction via a cooling tower or heat exchanger (process 3→1), completing the thermodynamic cycle. To drive the cycle, external heat is required during the moisture release phase (from point 2 to point 3 in the figure), with heating being provided by solar energy.

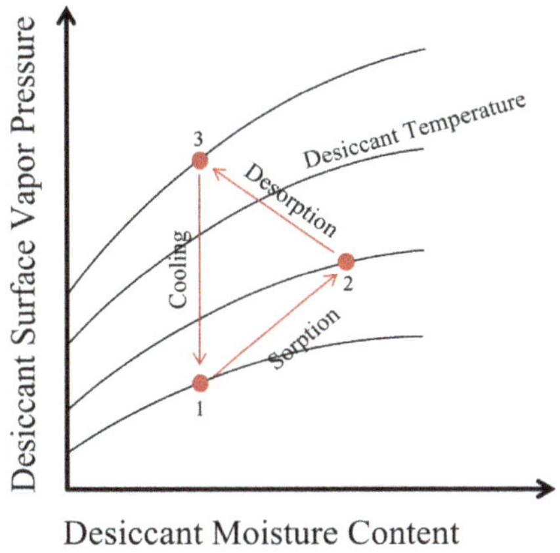

Figure 2. Process of moisture transfer by desiccant.

2.1.2. Closed Absorption Cycle

The closed absorption cycle can be classified into two types based on the nature of the absorption material, namely the liquid absorption cycle and solid adsorption cycle. The liquid absorption cycle is the most widely used absorption refrigeration cycle, while the solid adsorption cycle refers to the adsorption-based cycle.

In the liquid absorption refrigeration cycle, the system differs from traditional vapor compression refrigeration. It employs water–lithium bromide or ammonia–water as the working fluids. At low pressures, the vaporized refrigerant absorbs heat to enable cooling, while the absorbent material captures the vaporized refrigerant. In the generator, external heat sources (e.g., solar energy or industrial waste heat) drive desorption, releasing the refrigerant and preparing it for subsequent condensation and reuse. The schematic of absorption refrigeration cycle is shown in Figure 3.

Solid adsorption solar refrigeration cycles involve the use of solid adsorbents (such as zeolite, activated carbon, silica gel, etc.) combined with suitable refrigerants to form a functional refrigeration system. These adsorbents have a porous structure and, at low temperatures, absorb (or adsorb) refrigerants like a sponge. When heated, the adsorbents release (or desorb) the refrigerant. This makes the adsorption refrigeration cycle entirely driven by thermal energy, meaning the system can be powered by solar energy [11]. A comparison of solid adsorption refrigeration and liquid absorption refrigeration is shown in Table 1.

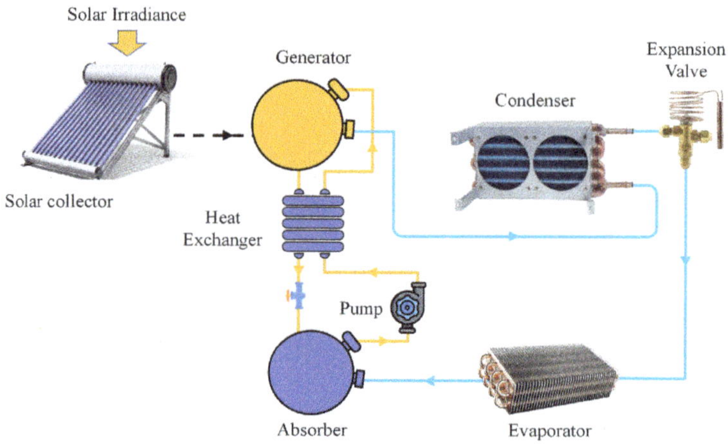

Figure 3. Schematic of absorption refrigeration cycle.

Table 1. Comparison of solid adsorption refrigeration and liquid absorption refrigeration [11–16].

Comparison Dimension	Solid Adsorption Refrigeration Systems	Liquid Absorption Refrigeration Systems
Temperature of the heat source	20–90 °C	80–150 °C
COP	0.3–0.6	0.6–1.2
Structural complexity	Simple (no pump)	Complex (requires pump)
Operating mode	Intermittent	Contiguity
Applicable scenarios	Small, decentralized	Large-scale central air conditioning
Environmental	Using natural refrigerants such as water, methanol, ammonia, etc., which are not destructive to the ozone layer	Ammonia is environmentally friendly; lithium bromide needs to be corrosion resistant
Maintenance cost	Lower	Higher (pumps, corrosion maintenance)
Technology maturity	In the pipeline	Mature and standardized

2.1.3. Thermomechanical Refrigeration System

The thermomechanical refrigeration system, commonly termed the steam ejector refrigeration system, harnesses solar energy as thermal energy using solar collectors. This thermal energy is subsequently transformed into mechanical energy via a nozzle, ultimately generating the desired cooling effect.

In a solar steam ejector refrigeration system, the solar collectors absorb solar radiation and convert it into heat energy. This thermal energy is used to heat a low-boiling-point working fluid, such as water or another suitable refrigerant, transforming it into high-pressure steam. The high-pressure steam is then expanded adiabatically through the ejector's nozzle, causing the steam's velocity to increase and its pressure to decrease. This high-speed steam creates a low-pressure region in the intake chamber of the ejector, which draws low-pressure steam from the evaporator into the ejector. The low-pressure steam mixes with the high-speed steam in the mixing chamber. The resulting mixture of gases enters the diffuser, where kinetic energy is gradually converted into pressure energy, leading to an increase in pressure. Finally, the pressurized steam enters the condenser, where it is cooled by the cooling medium and condensed back into liquid form. The liquid is partially directed through a throttling valve into the evaporator, where it absorbs heat and vaporizes, completing the refrigeration cycle. The remaining liquid is pumped back to the generator, where it is reheated and vaporized. The system is depicted in Figure 4 [17].

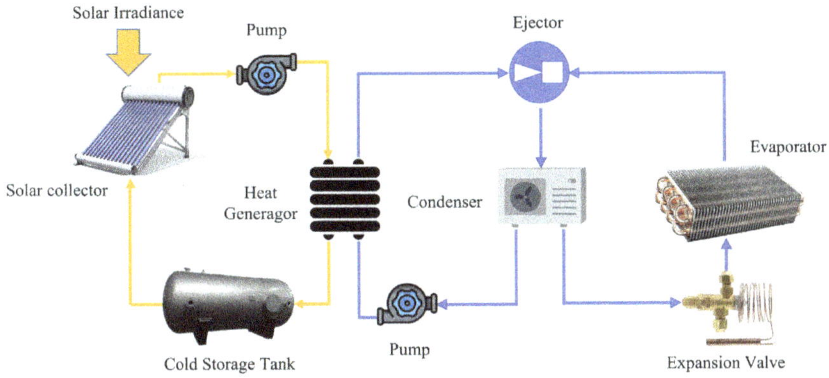

Figure 4. Schematic of steam ejector refrigeration cycle.

In an ejector refrigeration system, the circulating pump is the only moving component. The system is simpler than absorption-based systems, operates more stably, and is highly reliable. However, its main drawback is the lower refrigeration efficiency. The ejector, which is the core of this system, was invented by Sir Charles Parsons around 1901 to remove air from the condenser of a steam engine. In 1910, the ejector was used in the first steam ejector refrigeration system [18].

2.2. Solar Electric Refrigeration

Solar electric refrigeration systems operate by using solar photovoltaic panels to capture solar energy and convert it into electrical energy, which is then stored in a battery bank. The stored energy is subsequently utilized to operate refrigeration systems, including thermoelectric cooling systems and vapor compression refrigeration systems. These systems leverage the stored energy to achieve cooling through distinct thermodynamic mechanisms. As a result, solar electric refrigeration systems can be categorized into solar semiconductor refrigeration systems and solar vapor compression refrigeration systems.

Semiconductor refrigeration technology is based on the Peltier effect. When DC flows through a thermocouple composed of N-type and P-type semiconductor materials, the direction of the current entering the coupling varies, causing heat absorption and heat generation. When the current flows from N to P, the end of the P side absorbs heat, making it the cold end. Conversely, when the current flows from P to N, the end of the N side generates heat, creating the hot end [19,20].

With the development of semiconductor cooling technologies, new materials, and related advancements, scientists worldwide have been pushing for the adoption of solar semiconductor refrigeration systems in fields such as defense, industry, agriculture, and everyday life [21–23]. Ma et al. [24] designed an integrated solar semiconductor refrigeration unit for scenic areas to meet the demand for cold beverages in tourist spots. Mirand et al. [25] applied solar semiconductor refrigeration technology to design an air conditioning system for electric vehicles, achieving an efficiency of 0.5 in cooling mode and 1.72 in heating mode. Wahab et al. developed a solar semiconductor refrigeration system, finding that in comparison with solar photovoltaic vapor compression air conditioning systems, the solar semiconductor refrigeration system does not require intermediate circuits for energy conversion, has no moving or rotating parts, and does not need refrigerants. This makes it more stable, environmentally friendly, and nearly silent [26]. Song et al.

developed a solar-powered dual-function container employing thermoelectric modules, demonstrating effective temperature regulation capabilities suitable for pharmaceutical preservation in disease-endemic regions like Africa. This system serves both electrified and off-grid environments, with applications spanning automotive systems, tourism infrastructure, and medical cold chain management for vaccines and temperature-sensitive medications [27]. Solar vapor compression refrigeration systems will be detailed in the next sections.

2.3. Solar Combined Refrigeration

Solar combined refrigeration systems utilize solar energy to generate both electrical and thermal energy, which are then harnessed to power refrigeration systems driven by electricity and heat. This dual-source approach offers a more versatile solution for refrigeration, combining the benefits of solar electricity with thermal energy conversion. For instance, in 1996, Goswami introduced a novel combined system that integrates the Rankine cycle with an absorption cooling cycle, using an NH_3/H_2O mixture as the working fluid. In this system, the turbine replaces the traditional expansion valve found in absorption cycles. By expanding highly concentrated ammonia vapor to a low temperature without condensation, the turbine generates mechanical power. The low-temperature ammonia then enters the evaporator, where it vaporizes and undergoes processing in a typical absorption cycle. Despite its innovative design, the complexity of the system has limited its widespread adoption and research in the field of solar combined cooling systems [28]. This highlights the challenges faced in developing efficient, reliable, and economically viable systems that combine both thermal and electrical energy sources for refrigeration.

3. PCMs

PCMs are innovative substances that store and release thermal energy through the absorption or release of heat during phase changes. This ability to absorb or release large quantities of heat makes PCMs ideal for regulating temperatures, reducing energy consumption, and managing energy loads effectively [29]. Current energy storage technologies extend beyond PCMs to encompass electrochemical batteries, pumped hydro storage (converting electrical energy into gravitational potential energy through water elevation), flywheel systems (storing rotational kinetic energy in high-speed rotors), supercapacitors (electrostatic energy storage through surface charge adsorption on electrodes), and compressed air energy storage (CAES) (pressurized air containment in subterranean reservoirs). These systems predominantly utilize mechanical or electrochemical energy conversion mechanisms. Distinctively, PCM-based thermal energy storage employs reversible phase transitions to manage heat absorption/release cycles, characterized by non-toxic composition, ecological compatibility, simplified operational architecture, and exceptional recyclability. PCM technology remains unparalleled for thermal regulation applications, demonstrating particular efficacy in scenarios requiring low-energy-density storage, extended-duration operation, and stringent environmental compliance. The versatility of PCMs allows them to be utilized in a wide range of applications, from improving energy efficiency in buildings to enhancing cooling systems. PCMs are broadly categorized into three types, namely inorganic, organic, and eutectic phase-change materials.

Inorganic PCMs include materials such as crystallized hydrated salts, molten salts, and metals or alloys. Organic PCMs primarily consist of paraffin, acetic acid, and other organic substances, which offer certain advantages in terms of safety and ease of handling. Eutectic phase-change materials (PCMs) comprise composites of two or more organic or inorganic constituents, which can be engineered to achieve precise thermal properties through compositional adjustments. This tunability enables their application in diverse thermal management systems, ranging from energy storage to temperature regulation. This flexibility allows for precise control over important properties such as phase-change latent heat and phase-change temperature, making them highly adaptable to various energy storage needs.

The development of thermal energy storage technologies can be traced back to the Industrial Revolution, where early methods were based on relatively simple systems that relied on storing sensible heat, typically using hot water or air. These methods were limited in efficiency but provided a foundation for future advancements. Over time, as the demand for energy increased and technology advanced, thermal energy storage systems evolved, becoming more sophisticated and efficient. The introduction of PCM was a breakthrough, transforming thermal storage from sensible heat storage to latent heat storage, which is far more efficient.

Phase-change materials operate by absorbing or releasing latent heat during the phase-change process, allowing for much higher energy density compared to sensible heat storage. As a result, PCM-based thermal storage systems are capable of storing significantly more energy in the same volume. By the end of the 20th century, significant research began to explore both organic and inorganic PCMs, driven by the need for better, more efficient materials for energy storage. For example, paraffin-based organic PCMs are well-regarded for their excellent thermal stability, wide phase-change temperature range, and relatively low cost, making them a popular choice for a variety of applications. On the other hand, inorganic salts, such as nitrates, offer high latent heat capacity and are more cost-effective, making them highly attractive for large-scale thermal storage applications [30].

In recent years, PCM research has gained significant momentum. A literature review by Yataganbaba et al. found that between 1990 and 2015, there were 34,626 publications that referenced "phase change material" and 1034 that discussed "encapsulation". The research trend has become increasingly focused on improving the performance and application of PCM encapsulation, which has seen a sharp rise in the number of publications since 2000 [31]. Recent innovations in nanoparticle-enhanced phase-change materials (PCMs) have achieved significant milestones, particularly in enhancing thermal conductivity, stability, and energy storage efficiency. For instance, the integration of high-conductivity nanoparticles (e.g., silica or metal oxide nanoparticles) into PCM matrices has addressed long-standing challenges in heat transfer lag and phase segregation. Shchegolkov et al. demonstrated that incorporating carbon nanotubes and graphene into PCMs achieved over 30% thermal conductivity enhancement while enabling magnetic field-assisted thermal flow regulation. This engineered composite exhibits superior thermal modulation in critical applications, such as emergency cooling systems, effectively reducing temperatures from 100 °C to approximately 40 °C. Such innovations expand the frontiers of intelligent thermal energy storage and precision thermal management technologies [32]. Shchegolkov et al. developed a magnetically responsive thermal storage composite comprising paraffin, multi-walled carbon nanotubes, and nickel–zinc ferrite. Through ultrasonic irradiation and magnetic field processing, they systematically enhanced material properties while investigating temperature field dynamics during thermal cycling. Their experimental studies elucidated mechanisms underlying thermal storage efficiency improvements and clarified the interrelation between magnetic moment orientation and thermal transport behavior, establishing a magnetization model for predictive control. This work advances functional thermal management systems with field-programmable energy regulation capabilities [33]. Due to their high heat storage capacity—up to eight times greater than sensible heat storage [34]—PCM-based systems are now being widely adopted in various cooling applications [35]. For instance, Elsayed et al. [36] conducted experiments combining PCM with frozen products in both winter and summer. They found that under different environmental conditions, the optimal energy-saving effect was achieved when the mass ratio of frozen products to PCM was 5.0, leading to a 12.5% increase in energy saving in summer and 36.5% in winter. Rahimi et al. [37] also demonstrated that integrating phase-change cooling materials into evaporators can significantly reduce power consumption, lower the compressor's start–stop frequency, and minimize temperature fluctuations inside refrigeration units. These advancements highlight the growing potential of PCM-based systems in enhancing energy efficiency and reducing the environmental impact of cooling technologies.

Solar vapor compression refrigeration systems and solar thermal absorption refrigeration systems are two of the most widely studied and utilized solar refrigeration technologies. Given the prevalence and significance of these systems, this study focuses on two representative cases to examine the integration mechanisms of PCMs within key system components, assessing their impacts on performance metrics, energy efficiency, and thermal management optimization. This analysis aims to elucidate the role of PCMs in enhancing operational stability and scalability across diverse thermal applications.

4. PCM-Integrated Solar Vapor Compression Refrigeration System

The solar-driven vapor compression refrigeration system is a high-efficiency technology that synthesizes solar energy harvesting with conventional vapor compression cycles. By utilizing dual solar energy inputs (thermal and photovoltaic), the system optimizes renewable energy utilization for cooling and has seen substantial progress in experimental research and industrial implementation, emerging as a sustainable alternative to fossil fuel-dependent systems. For example, Wang et al. developed a thermodynamic model to optimize solar vapor compression refrigeration system performance, including the selection of suitable working fluids. They used Engineering Equation Solver (EES) software to model a low-temperature solar organic Rankine vapor compression refrigeration cycle and found that the refrigerant R123 demonstrated the best performance in terms of thermal efficiency and coefficient of performance (COP) [38]. This highlights the importance of selecting the right refrigerant to improve overall system efficiency.

In another innovative approach, Aung et al. proposed integrating solar energy directly with vapor compression systems. By using solar thermal energy to drive the vapor compression air conditioning system and incorporating components such as vacuum collectors and hot water storage tanks, the system's COP was significantly enhanced [39]. This approach not only improves efficiency but also reduces overall power consumption, making it more sustainable and cost-effective. Similarly, Hans et al. [40] designed a solar photovoltaic-driven refrigeration system that can maintain an internal temperature of 10–15 °C with a COP of 0.34, demonstrating the growing potential for solar-powered refrigeration solutions in real-world applications. Today, solar vapor compression refrigeration systems have moved beyond theoretical research and are now being applied in practical household refrigeration systems. For instance, Dhawan et al. [41] introduced a solar thermoelectric refrigeration system that uses clean solar energy to drive the compressor, resulting in a substantial reduction in power consumption—by 44–63%—for a 1.6 L household refrigerator.

Typically, these systems include key components such as photovoltaic panels, compressors, condensers, expansion valves, and evaporators. The operation follows the basic vapor compression cycle, in which the refrigerant absorbs heat in the evaporator, turning into low-temperature, low-pressure vapor, which is then compressed adiabatically in the compressor. The compression process elevates the temperature and pressure of the refrigerant vapor. The high-pressure, high-temperature vapor enters the condenser, releasing heat to the ambient environment and condensing into a high-pressure liquid state. This liquid refrigerant is subsequently expanded through the throttling valve, causing its pressure and temperature to be significantly reduced prior to entering the evaporator. After expansion, the refrigerant evaporates and absorbs heat, completing the cycle. The principle of the solar vapor compression refrigeration cycle is shown in Figure 5. However, this idealized description of the cycle is based on the Carnot cycle. In real-world systems, various irreversible factors, such as non-ideal compressor behavior and the inefficiencies associated with the expansion valve, cause deviations from the ideal cycle. These factors lead to a decrease in system efficiency. Reducing these irreversible losses is a key focus for researchers and engineers working on improving the performance of solar vapor compression refrigeration systems. By minimizing inefficiencies, we can move closer to optimizing the system for both practical and large-scale applications, making solar refrigeration systems an increasingly viable solution for sustainable cooling [42].

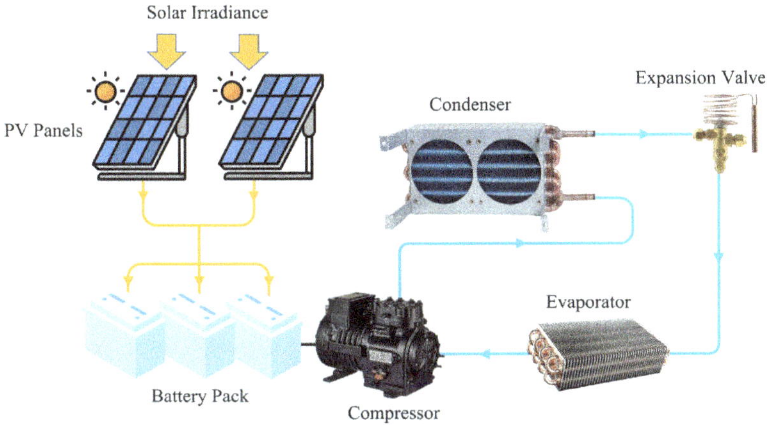

Figure 5. Schematic of solar vapor compression refrigeration cycle.

4.1. Photovoltaic Panels

Photovoltaic (PV) panels lie at the heart of modern renewable energy solutions, directly converting sunlight into electricity through the photovoltaic effect. Structurally, a PV panel consists of multiple interconnected solar cells, often made from high-efficiency materials such as monocrystalline silicon, polycrystalline silicon, or thin-film semiconductors like cadmium telluride (CdTe) and copper indium gallium selenide (CIGS). When photons in sunlight possess enough energy to surpass the bandgap of these semiconductor materials, electrons are excited from the valence band to the conduction band, resulting in the generation of direct current (DC) in an external circuit. As PV technology evolved, it became common to classify these panels into three generations based on core materials and production techniques. The first generation features traditional silicon-based solar cells; the second generation includes thin-film technologies; and the third generation encompasses emerging innovations such as organic photovoltaics, perovskite photovoltaics, and dye-sensitized solar cells. Each generation aims to enhance conversion efficiency, reduce manufacturing costs, and broaden the range of practical applications [43].

Although photovoltaic efficiency has steadily improved over the past few decades, thermal management remains a pivotal challenge. High operating temperatures can degrade cell performance, shorten component lifespans, and lower overall energy output. During the 1990s and early 2000s, researchers began paying closer attention to managing heat within PV systems [44]. As efficiency targets rose, the need to regulate temperature became even more pressing. Around the 2000s, integrating PCMs into PV panels emerged as a promising solution; by absorbing and storing excess heat when solar irradiance is high,

PCMs can help stabilize cell temperature. These efforts primarily focused on placing PCMs at the back or even within the layers of the PV panel, aiming to maintain optimal operating conditions and further boost overall energy conversion performance [45,46].

4.1.1. Effect of Photovoltaic Panel Temperature on COP

The performance of photovoltaic panels is highly temperature-sensitive, with efficiency decreasing as the surface temperature increases. This phenomenon is primarily driven by the impact on the open-circuit voltage of the panel. In practice, current photovoltaic panels only manage to convert about 10–16% of incident solar radiation into usable electrical energy. The remainder of the radiation that is not converted into electricity is absorbed by the panel, which heats it up and reduces overall efficiency [47]. On a sunny day, the surface temperature of photovoltaic panels can be as much as 40 °C higher than the surrounding ambient temperature [48]. Research has shown that for crystalline silicon photovoltaic cells, every 1 °C increase in panel temperature above 25 °C results in a 0.4–0.65% drop in efficiency [49–52]. Boudhina et al. (2019) systematically analyzed PV module performance across varying ambient temperatures, establishing comparative temperature coefficient benchmarks. Their findings confirm that PV cells exhibit a negative temperature coefficient—quantifying a percentage power loss per 1 °C rise—typically ranging from −0.2% to −0.5%. This parameter indicates that the PV module output decreases by 0.2–0.5% per 1 °C temperature elevation, critically informing photovoltaic system efficiency optimization under thermal stress conditions [53]. Given that solar panels typically operate at temperatures between 60 and 80 °C, this results in a power generation efficiency of only about 10% [54].

These factors highlight the critical need for effective temperature management strategies to enhance the efficiency of photovoltaic systems. As the photovoltaic industry continues to evolve, managing heat within solar panels has become a priority. Cooling technologies engineered to mitigate these challenges are broadly classified into active and passive cooling systems. In contrast to active systems—which depend on external energy inputs and typically incorporate mechanical components—passive cooling systems operate without auxiliary energy input and require minimal maintenance. This distinction underscores the growing preference for passive systems in sustainable thermal management applications. Among passive cooling methods, the integration of PCMs into photovoltaic systems—known as PV-PCM systems—has proven to offer superior cooling performance and more stable temperature regulation. Xiao et al. (2022) demonstrated that a stearic acid–lauric acid composite (phase-change temperature: 42.9 °C) applied as a PCM to photovoltaic panels reduced peak operational temperatures from 61.6 °C to 48.9 °C, significantly enhancing thermal regulation efficiency [55]. The most significant advantage of PV-PCM systems lies in the latent heat storage capabilities of PCMs. These materials can absorb large amounts of heat without experiencing a significant increase in temperature, making it possible to maintain the photovoltaic panel at an optimal operational temperature range for extended periods. This passive method of cooling ensures that the panel operates efficiently without requiring external energy or mechanical components, which not only improves the longevity and performance of the system but also makes it more environmentally friendly and cost-effective.

4.1.2. Methods of Integrating PCM with Photovoltaic Panels

Phase-change material (PCM)-integrated photovoltaic panels leverage latent heat absorption to stabilize module temperatures within the 25–40 °C high-efficiency conversion range, effectively curbing power loss from thermal degradation. This integration achieves 20–30% peak temperature reduction, mitigating encapsulant aging while extending module lifespan by 10–15%. The system's passive thermal regulation operates energy-autonomously via solid–liquid phase transitions, providing diurnal thermal buffering that suppresses microcrack propagation induced by thermal cycling. Such thermal stress management enhances operational stability and reliability under extreme temperature differentials. Efficiently managing the heat generated by PV panels is a key challenge in maintaining high energy output and prolonging panel lifespan. One promising approach involves incorporating PCMs directly onto the back surface of PV panels, enabling a more effective transfer of heat away from the solar cells. One of the methods is where the PCM is encapsulated and placed between the PV cells and the collection back panel. In this arrangement, the PCM absorbs excess heat during periods of high solar irradiance, preventing the PV panel from reaching excessively high temperatures that would otherwise reduce conversion efficiency [56,57].

A distinctive feature of the PCM-PV setup is its ability to maintain a near-constant temperature at the phase-change point while the material melts. As soon as the PV panel's temperature drops below the melting point, the PCM begins to solidify, gradually releasing stored heat in the process. The phase transition thermal inertia of PCM maintains near-constant thermal equilibrium during solidification exothermic processes, preserving its characteristic phase-change temperature plateau. This thermal stabilization mechanism prevents the photovoltaic efficiency losses typically associated with temperature fluctuations, as the PCM's phase transition plateau avoids detrimental temperature elevation during energy discharge. This cyclical absorption and release of heat helps stabilize the panel temperature, ultimately boosting the system's overall power output [58]. Beyond that, the stored latent heat can be harnessed for other thermal management purposes within the system, offering additional flexibility in design [59–61].

Recent studies have further validated the benefits of integrating PCMs into PV systems. Hasan et al. [62] performed a parametric investigation examining the influence of various PCM types and thicknesses on temperature regulation. Their findings revealed that this technique could reduce the panel's temperature by up to 18 °C within just 30 min. In a more recent study, Ahmadi et al. embedded PCMs within a thermally conductive foam layer attached to a PV panel. This design successfully lowered panel temperatures by 6.8%, resulting in a 14% increase in energy conversion efficiency [63]. Similarly, Fayaz et al. [61] demonstrated—using both numerical simulations and experiments—that a PVT system employing polyethylene glycol as its phase-change material exhibited a 12.91% and 12.75% performance improvement over a conventional PVT setup without PCM. Refaey et al. (2024) demonstrated that integrating aluminum fins into phase-change material (PCM) encapsulation significantly improves photovoltaic thermal management. Their experimental framework evaluated three PCM compositions and fin geometries, identifying RT42 (a Rubitherm Technologies organic phase-change material) paired with triangular fin arrays as the optimal configuration. This combination reduced PV module peak operating temperatures by 14.5 °C (from 74.7 °C to 60.2 °C) while elevating power conversion efficiency from 17.1% to 18.5%. The triangular fin design enhanced heat dissipation through an increased surface-area-to-volume ratio, facilitating efficient thermal energy transfer from PV modules to the PCM during phase transition cycles [64]. Al Arni et al. (2024) engineered a gradient concentration three-layer PCM system incorporating graphene nanoparticles, where particle density decreased progressively from the photovoltaic interface layer to the base substrate. Encapsulated within high-thermal-conductivity aluminum shells, this configuration achieved a 5 °C operational temperature reduction (46 °C to 41 °C) in photovoltaic modules alongside a 0.4% efficiency enhancement (16.2% to 16.6%). Experimental validation demonstrated that the graded nanoparticle distribution optimized interfacial heat flux while maintaining material stability during phase transition cycles [65]. Collectively, these approaches highlight the growing appeal of PCM-based heat management strategies in PV technology. By effectively flattening temperature spikes and leveraging latent heat storage, PCM integration contributes not only to enhanced efficiency but also to improved system stability and a potentially longer operational life for photovoltaic installations.

4.1.3. Selection of PCM

Selecting an appropriate PCM is crucial for effective thermal management, particularly in solar-powered applications such as photovoltaic (PV) cooling. Broadly, PCMs fall into three categories, namely organic, inorganic, and eutectic. Organic PCMs, like paraffin, typically offer stable chemical properties and a flexible range of melting points but can have lower latent heat capacities and undergo significant volume changes during phase transitions [58]. Inorganic PCMs, often in the form of salt hydrates (e.g., $CaCl_2{\cdot}6H_2O$), boast higher thermal conductivity and greater latent heat but can suffer from stability issues, including incongruent melting and supercooling. Eutectic PCMs combine two or more substances, aiming to fine-tune phase-change temperatures and latent heats; however, their application in PV panel cooling remains relatively unexplored [66,67]. Table 2 summarizes and compares the typical thermophysical properties of the three types of PCMs.

Table 2. Comparison of thermal properties of three types of PCMs [68,69].

Thermophysical Parameters	Organic PCMs	Inorganic PCMs	Eutectic PCMs
Latent heat (kJ/kg)	120–250	150–400	100–300
Thermal conductivity (W/m·K)	0.1–0.3	0.5–1.5	0.5–8.5
Phase transition temperature (°C)	−20–150	20–800	Can be adjusted by changing the ratio of components

Historically, researchers have focused on two key properties when selecting a PCM, namely phase-change temperature and latent heat. The phase-change temperature must align with the operating range of the system—60–80 °C for typical solar PV panels [54]—to ensure that the PCM effectively stores and releases energy under real-world conditions. Latent heat capacity determines how much thermal energy can be stored or released during the phase transition, directly influencing the effectiveness of temperature regulation. In laboratory settings, differential scanning calorimetry (DSC) is commonly used to evaluate these parameters, providing insights into a material's suitability for integration into thermal storage systems.

Inorganic PCMs, particularly those involving water and salts, offer high latent heat and robust thermal conductivity but face two persistent challenges. First, incongruent melting can lead to irreversible melt–freeze cycles if the salt hydrate lacks sufficient water of crystallization. Thickeners can mitigate this to some extent, but complete prevention is difficult [70]. Second, pronounced supercooling occurs when nucleation during freezing is hindered. Although adding nucleating agents alleviates this, the phenomenon is not fully eliminated [71]. These issues have largely deterred researchers from incorporating inorganic salts into PV-PCM systems, with most investigations focusing only on fundamental melting–freezing behaviors rather than practical integration.

In contrast, organic PCMs—notably paraffin-based compounds—have gained widespread adoption in photovoltaic cooling applications owing to their chemical inertness, negligible supercooling tendencies, and design flexibility. These attributes enable reliable thermal regulation while maintaining long-term operational consistency in solar energy systems. Paraffin is a petroleum-derived mixture of saturated hydrocarbons with phase-change temperatures that can be tailored by varying chain length. Studies have consistently shown paraffin's effectiveness in cooling PV panels and boosting efficiency. For example, Ahmadi et al. integrated paraffin (melting at 37 °C) into a PV system, reducing the panel's temperature by 6.8% and improving power output by 14% [63]. Table 3 compares the stability of the three types of PCMs.

Table 3. Stability comparison of three types of PCMs [72,73].

Parameters	Organic PCMs	Inorganic PCMs	Eutectic PCMs
Cycling stability	Latent heat decay <5% after 500 cycles	Susceptible to phase separation and subcooling, up to 20% degradation after 100 cycles	Phase separation inhibited by encapsulation or thickening agents, <10% decay over 300 cycles
Chemical stability	Non-corrosive, acid- and alkali-resistant	Partially hydrated salts tend to corrode metal containers	Reduced corrosion risk through cladding
Thermal stability	Decomposition temperature >200 °C	Decompose easily at high temperatures	Can be adjusted by changing the ratio of components

Nonetheless, paraffin's low thermal conductivity remains a notable limitation, prompting extensive research into thermal conductivity enhancements. Wu et al. [74,75] investigated the addition of copper nanoparticles at various weight percentages, finding improvements of up to 14.2% (solid phase) and 18.1% (liquid phase) in thermal conductivity at 2 wt% copper. A subsequent study comparing copper and aluminum nanoparticles identified copper as more effective; a 1 wt% copper addition reduced heating and cooling times by over 30%, with negligible impacts on latent heat or phase-change temperature. These findings underscore the potential for nanoparticle doping to address paraffin's inherent conductivity shortcomings, making organic PCMs an even more viable option for real-world, large-scale PV cooling applications. Table 4 summarizes and compares the PCM types, compositions, and their thermophysical parameters used in the above articles.

4.2. Condenser

The condenser is a critical component in vapor compression refrigeration systems, designed to transform a high-temperature, high-pressure refrigerant vapor into a high-pressure liquid while releasing heat to the surroundings. In most setups, air serves as the cooling medium. Without forced convection, the rising air temperature around the condenser lowers the available temperature gradient, diminishing overall heat exchange efficiency. Furthermore, lowering the condensing temperature can significantly improve the system's COP, as a reduced discharge temperature and pressure translate into lower compressor power consumption. Studies show that each 1 °C reduction in condenser inlet water temperature can lift a chiller's COP by about 2–5% [76]. By integrating PCMs into the condenser, the refrigerant exchanges heat with a substance capable of absorbing and releasing thermal energy at a (relatively) fixed temperature, thereby stabilizing the heat absorption side and helping maintain a beneficial temperature differential for condensation.

One challenge, however, is that embedding PCMs within a heat exchanger inherently adds thermal resistance, initially hindering heat transfer. Yet, a PCM with a suitably low phase-change temperature can offset these losses by creating a larger temperature difference, and any accumulated heat can dissipate when the compressor is idle. However, to solve the problem of additional thermal resistance caused by PCMs, the most direct and effective way is to improve the thermal conductivity of PCMs and PCM packaging or to combine heat pipe technology to improve the heat transfer rate and also to reduce the thermal resistance by changing the PCM integration method. In practice, combining PCMs with the condenser often leads to marked gains in system performance. For instance, Wang et al. [77–79] conducted comprehensive optimization studies on PCM integration in a vapor compression system. In their initial phase, they compared the placement of the PCM (latent heat of 220 kJ/kg, phase-change temperature of 21 °C) between the compressor and condenser versus between the condenser and expansion valve. The latter arrangement raised the system's COP by 8% and delivered a 10 °C temperature drop at the expansion valve—superior to the 5 °C cooling achieved in the former setup. Building on these insights, they developed a dynamic model featuring a liquid refrigerant flash mechanism to predict COP gains, which correlated well with experimental data but showed limited accuracy when significant subcooling or superheating occurred. To augment thermal stability, the researchers integrated a PCM-based storage tank strategically positioned between the evaporator and compressor units, thereby enhancing operational reliability. By carefully regulating refrigerant inlet temperatures, this configuration reduced overall operating temperatures and shaved peak heat loads. Notably, employing a composite PCM arrangement (linking heat exchangers in both the main and bypass loops) enabled the active control of condenser pressure/temperature, even in low-ambience conditions. This strategy effectively maintained minimal condensing temperatures, significantly improving operational stability and achieving meaningful energy savings—a testament to the importance of PCM spatial optimization in dynamic load scenarios. Cheng et al. took a different approach by engineering a PCM that blends paraffin/high-density polyethylene with graphite powder (GP) or expanded graphite (EG). The addition of EG multiplied the material's thermal conductivity fourfold [80]. Incorporating this PCM into a household refrigerator condenser allowed it to absorb heat during compressor operation and release that heat when the compressor was off [81]. Experimental measurements recorded a 6.3 °C drop in outlet temperature and a 2.3 °C reduction at the condenser's center, boosting the COP by 12%. However, the system's compressor start–stop frequency increased, raising concerns about potential compressor wear. A subsequent dynamic numerical model validated these observations, predicting up to a 19% improvement in the COP [82]. Abdulaziz et al. (2025) demonstrated that integrating a phase-change material (PCM) unit as a pre-cooling module upstream of air conditioning condensers significantly enhances system efficiency. Their hybrid methodology combining numerical simulations and experimental validation under varied summer conditions revealed an 8.6% annual energy consumption reduction and a peak coefficient of performance (COP) of 5.63 compared to conventional configurations. This innovation leverages the PCM's thermal buffering capacity to precondition condenser inlet air, thereby optimizing heat rejection processes in HVAC systems [83]. Ismail et al. (2024) developed a dual-PCM configuration incorporating distinct phase-change materials on air conditioning evaporator and condenser components. Their methodology combined ANSYS (ANSYS fluent 2020 R2)-based 2D transient numerical modeling with experimental validation, revealing seasonal energy savings of 11.8% (summer) and 12.8% (winter) compared to conventional systems. This hybrid thermal management approach optimized heat exchange dynamics through targeted PCM deployment at critical refrigeration cycle interfaces [84]. While this condenser–PCM integration has already been practically applied in small domestic refrigerators, its scalability to larger systems remains uncertain. Recent research on larger-scale implementations is sparse, likely due to the risk of intensified compressor cycling. Moving forward, deeper investigations into modeling, PCM property enhancement, and full-scale system validation are essential to unlock the potential of PCM-integrated condensers in diverse refrigeration environments, ensuring both efficiency gains and operational reliability. Table 5 systematically presents a comprehensive summary of the PCM types incorporated into condenser components, detailing their key thermophysical characteristics along with their corresponding thermophysical properties documented in existing studies.

Table 4. PCMs integrated onto PV panels.

Research Study	Type	PCM	Density (Solid)	Density (Liquid)	Specific Heat Capacity (Solid)	Specific Heat Capacity (Liquid)	Thermal Conductivity (Solid)	Thermal Conductivity (Liquid)	Latent Heat of Fusion	Melting Point	Performance
			[kg/m^3]	[kg/m^3]	[kJ/kg·K]	[kJ/kg·K]	[W/m·K]	[W/m·K]	[kJ/kg]	[°C]	
[55]	Organic	Mixture of stearic acid and lauric acid	/	/	/	/	/	/	/	42.9 °C	Reduced maximum temperature of PV plate from 61.6 °C to 48.9 °C
[61]	Organic	A44	805	805	2.15	/	/	0.18	242.0	44.0	12.91% (numerical simulation) and 12.75% (experimental) increase in maximum electrical efficiency with 1000 W/m^2 radiation.
[63]	Organic	PA	805	805	/	/	/	0.4	124.6	47.0	A 6.8% lower temperature and 14% higher power generation efficiency at 1100 W/m^2 radiation.
[64]	Organic	Wax (RT42)	880	770	/	2	0.2	0.2	165.0	38–43	At a radiation intensity of 850 W/m^2, the power generation is increased to 737.4 W-h/m^2 and the PV plate temperature is reduced to 60.2 °C, which is 19.4% lower than that of the uncooled plate.
	Organic	Wax	860	/	/	2	/	0.2	160.0	32–38	
	Organic	Wax (RT31)	/	760	/	2	0.2	0.2	165.0	29–34	
[65]	Organic	Wax (RT26)	880	760	/	2	/	0.2	180	25–26	The electrical efficiency of the PV plate increased to 16.1% within 100 min, and the temperature of the PV plate reduced by 5 °C compared to the single-layer PCM.
[62]	Organic	Wax (RT20)	880	/	2.0	/	0.2	0.2	140.3	21.2	Maximum temperature reduction of 14 °C at 1000 W/m^2 radiation for 30 min.
	Organic	Mixture of capric–lauric acid	880	863	/	/	0.139	0.139	171.9	20.8	Maximum temperature reduction of 16.5 °C at 1000 W/m^2 radiation for 30 min.
	Organic	Mixture of capric–palmitic acid	883	840	/	/	0.143	0.143	196.1	22.3	Maximum temperature reduction of 16.5 °C at 1000 W/m^2 radiation for 30 min.
	Organic	Commercial blend (SP22)	1490	1430	2.5	/	0.6	0.6	182.0	23.0	Maximum temperature reduction of 14 °C at 1000 W/m^2 radiation for 30 min.
	Inorganic	$CaCl_2·6H_2O$	1710	/	1.4	/	1.090	1.090	213.1	29.2	Maximum temperature reduction of 18 °C at 1000 W/m^2 radiation for 30 min.

Table 5. PCMs integrated in the condenser.

Research Study	Type	PCM	Density (Solid)	Density (Liquid)	Specific Heat Capacity (Solid)	Specific Heat Capacity (Liquid)	Thermal Conductivity (Solid)	Thermal Conductivity (Liquid)	Latent Heat of Fusion	Melting Point	Performance
			[kg/m^3]	[kg/m^3]	[kJ/kg·K]	[kJ/kg·K]	[W/m·K]	[W/m·K]	[kJ/kg]	[°C]	
[81]	Organic	SSPCM	957	/	/	/	0.31	/	133.1	44.0	A 12% higher COP and 6.5 °C lower condensing temperature
[85]	Organic	RT35HC	880	/	2000	/	0.2	/	240.0	35.0	Reduced temperature fluctuations; the COP of the system averages about 1.82
[86]	Organic	SP24E	1500	1400	/	/	0.6	/	180.0	25.0	The COP was increased by 50% with a power saving of 6.85%
[83]	Inorganic	SP31	1350	1250	2	2	0.5	0.5	210.0	30.0	Annual energy consumption reduced by up to 8.6% with a maximum COP of 5.63
[84]	Inorganic	SP24E (summer)	1500	1400	2	2	0.5	0.5	180.0	25.0	An 11.8% reduction in energy consumption
	Inorganic	SP11gel (winter)	1330	1320	2	2	0.5	/	155.0	12.5	Reduction in energy consumption by 12.8 per cent
[87]	Inorganic	SP24E	1500	1400	/	/	0.6	/	180.0	25.0	Energy consumption is reduced by about 9.8% to 11.2%

4.3. Evaporator

The evaporator plays a central role in refrigeration systems, where it serves as the primary site for heat absorption. As low-pressure liquid refrigerant flows through the evaporator, it transitions into a vapor phase, extracting heat from the surrounding environment and thus lowering the ambient temperature to achieve cooling [88]. The evaporation temperature is a key determinant of both the system's power consumption and its COP. If the evaporator runs at excessively high temperatures, the cooling effect diminishes. Conversely, if it runs at too low a temperature, the compressor's pressure ratio increases, heightening energy consumption and undermining efficiency. Crucially, small increases in the evaporation temperature can enhance the COP by reducing the compressor's workload and boosting the refrigerant's heat absorption capacity—studies indicate that each 1 °C rise in evaporator outlet temperature can raise a chiller's COP by 1.5–3% [76]. Moreover, in applications like cold cabinets, the evaporation temperature can significantly influence both temperature gradients and fluctuations within the cabinet itself.

A viable strategy for stabilizing evaporator operation entails the integration of phase-change materials (PCMs). These materials absorb and release latent heat during phase transitions, effectively mitigating thermal fluctuations. Researchers have investigated diverse PCM configurations—including plates, granular beds, and tubular geometries—to optimize interfacial surface area and enhance effective heat exchange. Such innovations aim to balance energy storage capacity with thermal responsiveness in dynamic operating conditions. Bista et al. [89] found that adding PCMs to the evaporator steadied temperature fluctuations and provided a buffer against changing thermal loads, leading to improved operational stability. By slowing the rate at which heat is transferred to the refrigerant, the PCM also tempers the evaporator's temperature, allowing for a steadier and more continuous evaporation process. This, in turn, reduces temperature swings throughout the system, enhancing response time and increasing control precision [90].

From the viewpoint of solar-driven refrigeration, using PCMs in the evaporator can help capture and store more heat consistently, thus supplying greater cooling potential. An experimental study by Maiorino et al. [91] demonstrated that introducing PCMs into an evaporator significantly reduced internal temperature gradients and product temperature fluctuations in a cooled cabinet while extending the compressor's off cycle. In various conditions—ranging from ambient temperatures and heat loads to differing PCMs—household refrigerators equipped with PCM-based evaporators have shown COP improvements between 2% and 74% [92]. Msell et al. integrated a PCM with a phase-change temperature of 2–4 degrees Celsius into an evaporator and experimentally tested the performance of the evaporator in charging and discharging modes of operation, which showed that the evaporator was able to efficiently store and release the cooling capacity in the PCM, which significantly improved the performance of the system and resulted in energy savings [93]. Table 6 summarizes information including the types of PCMs used in the literature mentioned above and their thermophysical parameters.

Table 6. PCMs integrated into the evaporator plate.

Research Study	Type	PCM	Density (Solid)	Density (Liquid)	Specific Heat Capacity (Solid)	Specific Heat Capacity (Liquid)	Thermal Conductivity (Solid)	Thermal Conductivity (Liquid)	Latent Heat of Fusion	Melting Point	Performance
			[kg/m^3]	[kg/m^3]	[kJ/kg·K]	[kJ/kg·K]	[W/m·K]	[W/m·K]	[kJ/kg]	[°C]	
[93]	Organic	Wax	880	770	/	/	0.20	0.20	/	5.0	COP increased by 8–10%.
[94]	Organic	Corn oil esters	/	/	/	/	/	/	/	−15.0	CCOP increased by 6% and compressor power consumption reduced by 4%.
[86]	Organic	PCM10HC	/	/	/	/	0.2	/	200.0	12.0	The COP increased by 13.5% and the power saving was 3.9%.
[95]	Organic	Diethylene glycol	1200	1320	4.86	/	/	/	247.0	−10.0	PCM stores cold during the day and releases it at night.
[96]	Inorganic	H_2O	917	1000	2.01	4.18	2.2	0.58	334.0	0	When the compressor stops working, the PCM is able to utilize its latent heat of phase change to maintain a low cabinet temperature.
[97]	Inorganic	H_2O	917	1000	2.01	4.18	2.2	0.56	333.6	0	Cooling capacity is stored during the nighttime trough hours and released during the daytime peak hours, thereby taking advantage of peak and valley tariff differences and reducing operating costs.

As a rule of thumb, the PCM's phase-change temperature should be set 5–10 °C above the evaporation temperature, especially in sub-zero applications where solid–liquid PCMs are favored for their high latent heat, minimal supercooling, adequate thermal conductivity, and stable chemistry. Li et al. offered a comparative analysis of several PCMs suitable for air conditioning systems; the graph compares four common phase-change materials (PCMs) across six critical parameters, namely thermal conductivity, volumetric latent heat, density, corrosion potential, supercooling tendency, and phase separation behavior. Salt hydrates and co-crystals exhibit superior volumetric energy storage capacity, enhanced thermal transfer properties, inherent flame resistance, and cost-effectiveness. However, these materials demonstrate the three following significant limitations: pronounced phase separation during thermal cycling, substantial supercooling effects, and corrosive interactions with metallic containment systems, collectively resulting in compromised cyclic stability. Paraffin waxes and fatty acids represent organic phase-change materials that demonstrate distinct advantages over inorganic counterparts, including high chemical inertness, structural stability, and recyclability. These organic compounds exhibit minimal supercooling tendency while maintaining freedom from phase separation and corrosive behavior. However, their practical application is constrained by relatively low thermal conductivity and reduced phase-change enthalpy. Refrigerant-based hydrates present unique thermodynamic characteristics, existing in gaseous states under atmospheric conditions and requiring pressurized containment for hydrate formation. Notably, their fusion enthalpy significantly exceeds that of low-temperature eutectic salts, paraffinic compounds, and fatty acid systems. (refer to Figure 6) [98].

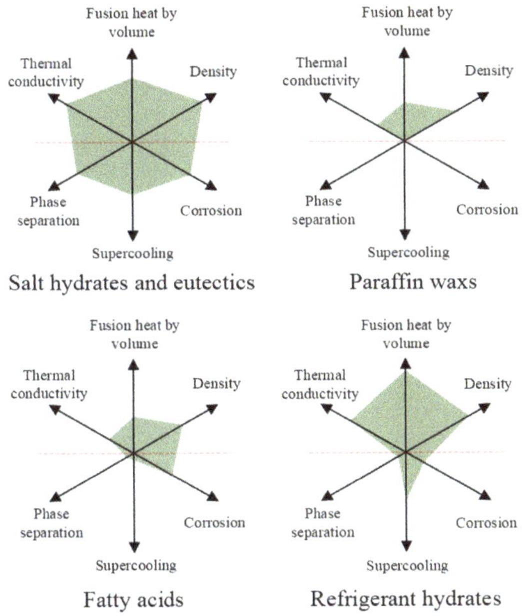

Figure 6. Comparison of thermal properties of different PCMs [98].

Despite these gains, PCM integration in evaporators faces some hurdles. Chief among them is the slow charging and discharging inherent to many PCMs, largely due to their low thermal conductivity, which can lead to delayed system responses. Proposed solutions involve the incorporation of microchannel arrays or porous architectures to augment the effective heat transfer surface area, as well as the integration of thermally conductive nanostructured additives (e.g., metallic nanoparticles or graphene) into the PCM matrix. Recent findings suggest that composite PCM/metal systems can markedly enhance thermal conductivity and expedite heat transfer [99,100]. Another concern is the volumetric change that occurs during phase transitions, potentially causing structural stress or deformation over extended operation. Advanced encapsulation—utilizing metallic encapsulation shells or polymeric enclosures—effectively mitigates leakage and thermal stress, whereas tailoring the PCM's intrinsic material properties to minimize volume expansion provides a foundational solution. Looking ahead, refining these approaches and exploring new PCM formulations will be pivotal to harnessing the full potential of PCM-enhanced evaporators, particularly in the context of dynamic load environments and solar-driven cooling applications [101,102].

5. PCM-Integrated Solar Thermal Absorption Refrigeration System

Absorption refrigeration cycles hinge on four key components—an evaporator, an absorber, a generator, and a condenser—and typically unfold in five distinct stages. First, in the evaporator, the refrigerant (often water or ammonia) absorbs heat and evaporates, thereby cooling the surrounding environment. The resultant refrigerant vapor is then drawn into the absorber, where it is taken up by a weaker solution—commonly water–lithium bromide or ammonia–water—raising the system pressure. This diluted solution is subsequently pumped to the generator, where external heat (which may be waste heat or solar-derived) drives out the refrigerant, forming high-pressure vapor. That vapor progresses to the condenser, where it discharges its latent heat and liquefies, and finally, the liquid refrigerant undergoes throttling through an expansion valve before returning to the evaporator to repeat the cycle [103,104].

Among the strengths of absorption refrigeration systems is a relatively high COP, spurring considerable research interest. Xu et al. [105] devised a solar absorption refrigeration system that employed variable mass energy transformation and storage (VMETS) technology, attaining COPs of 0.7525 (air-cooled condenser) and 0.7555 (water-cooled condenser). Others have compared the fundamental performance of vapor compression and absorption cycles. One notable advantage of photovoltaic vapor compression systems lies in their reduced land footprint, cutting initial capital investment. By contrast, solar thermal absorption systems rely on solar collectors, whose required area is shaped by collector efficiency and attainable high temperatures. If concentrated solar collectors are adopted, higher collector outlet temperatures become feasible, minimizing the physical footprint and boosting practicality. Moreover, absorption refrigeration often pairs well with solar energy, and single-effect solar absorption systems have attracted growing attention in recent years for their relatively high efficiency and cost-effectiveness [16,106].

In such setups, the generator's external heat source can stem from waste heat or solar-heated fluids. Once heated, the low-pressure refrigerant solution frees high-pressure refrigerant vapor that passes to the condenser [107]. Notably, each component in a solar absorption refrigeration system—the solar collector, generator, absorber, condenser, and evaporator—stands to benefit from PCM integration. By absorbing and releasing latent heat, PCMs can buffer thermal fluctuations and elevate system performance. In the following sections, we explore how PCMs can be integrated with each component, discuss suitable material selections, detail the impact on individual component efficiency, and evaluate the overall effect on system performance.

5.1. Solar Collector

Within solar-driven absorption refrigeration systems, the solar collector constitutes the critical component, distinguishing the system through its unique energy conversion role.

The collector harnesses incident solar radiation, converting it into usable thermal energy that subsequently powers the generator to initiate the refrigeration cycle. Because solar energy supply fluctuates throughout the day, it is difficult to provide a continuous stable heat source for the generator solely from direct sunlight. Consequently, integrating thermal storage is imperative. By embedding PCMs into the collector, any surplus thermal energy can be stored during periods of strong sunlight. Later, when solar irradiance weakens or after sunset, the PCM can release the stored heat to sustain the generator's operation. This strategy not only improves system performance but also maximizes the utilization of available solar energy. As previously discussed, we will not repeat the basic principles and classifications of solar collectors here.

The concept of combining PCMs with solar thermal collectors dates back to the late 1980s. Serale et al. [108] investigated a flat-plate collector enhanced with a PCM, reporting an 8% gain in efficiency compared to traditional heat storage methods and a 20–40% rise in annual average thermal conversion efficiency. As with other PCM-integrated systems, a careful selection of phase-change temperature and latent heat is key. The solar collector's working temperature range dictates which PCM is most suitable, and references often provide typical output temperatures for different collector types [109].

Brancato et al. [110] studied PCMs with melting points of 80–100 °C in solar thermal absorption systems. Although these materials offered high latent heat, they also encountered issues like supercooling and incongruent melting. Agyenim et al. [111,112] explored phase-change temperatures in the 100–130 °C range for use in a LiBr-H_2O absorption system operating at 120 °C. Erythritol, with a 117 °C phase-change temperature, emerged as their choice due to its stable phase transition and high latent heat. Because its melting point was 3 °C lower than the generator temperature, it provided a sufficient temperature differential for heat exchange. Their subsequent tests, examining four container designs (no fins, circular fins, longitudinal fins, multi-tube) and three different PCMs (erythritol, RT100, and magnesium chloride hexahydrate), highlighted that a high latent heat and a phase-change temperature aligned with the system requirements, which were the primary selection criteria—supercooling concerns were secondary if latent heat was sufficiently high. Chopra, K. et al. integrated phase-change materials (PCMs) into heat pipe evacuated tube solar collectors (ETCs) by charging each tube with 2.25 kg of SA-67. Their experimental results demonstrated a 32.73–37.56% increase in daily thermal efficiency for PCM-enhanced ETCs, along with sustained high water temperature maintenance during nocturnal periods or low solar radiation intervals [113]. Naghavi et al. and Abokersh et al. incorporated phase-change materials (PCMs) into evacuated tube collector (ETC) arrays, demonstrating significant heat loss reduction, elevated system thermal efficiency, and enhanced operational stability under fluctuating irradiance conditions [114,115].

Currently, one key limitation for PCM-integrated solar collectors is limited adaptability to diverse or rapidly changing weather conditions. Researchers recommend several optimization strategies. For example, advanced thermal regulation strategies and adaptive control systems enable coordinated heat collection and storage, ensuring stable system operation. Innovative heat exchanger configurations enhance thermal coupling between the PCM and solar collector, thereby improving overall energy conversion efficiency. These advancements collectively address transient thermal mismatches and elevate system reliability under variable solar irradiance conditions. Additionally, modular system designs allow for easier adaptation and scaling in different climates, ensuring that the solar collector's integration with PCMs remains robust and practical [116–118]. Table 7 summarizes the information including the types of PCMs integrated on solar collectors and their thermophysical parameters used in the literature mentioned above.

Table 7. PCMs integrated into solar collector.

Research Study	Type	PCM	Density (Solid)	Density (Liquid)	Specific Heat Capacity (Solid)	Specific Heat Capacity (Liquid)	Thermal Conductivity (Solid)	Thermal Conductivity (Liquid)	Latent Heat of Fusion	Melting Point	Performance
			[kg/m^3]	[kg/m^3]	[kJ/kg·K]	[kJ/kg·K]	[W/m·K]	[W/m·K]	[kJ/kg]	[°C]	
[112]	Organic	Erythritol	1480	1300	1.4	1.4	0.7	0.3	339.8	117.7	The performance of the heat collector can be significantly improved.
[113]	Organic	SA-67	1100	1190	2.01	2.47	1.1	1.2	244.2	67.1	Thermal efficiency upgraded from 55.46% to 87.80.
[114]	Organic	Wax	990	916	2.76	2.48	0,35	0.17	174.0	64.0	Systems with integrated PCMs exhibit higher thermal efficiency at different water flow rates and are insensitive to changes in water flow rate.
[115]	Organic	Wax	920	795	2.3	2.3	0.21	0.21	189.0	58–62	System efficiency increased by approximately 14%.
[116]	Organic	Caprylic acid and lauric acid	2000	2000	2	2	0.2	0.3	130.0	21.0	In the active heat release mode, the ratio of heat release to heat storage increased by 30.12%.

5.2. Absorber

Absorption refrigeration systems rely on external heat sources to drive the cooling process, utilizing the interaction between the absorbent and refrigerant to achieve the desired cooling effect. However, because these systems often operate with fluctuating heat inputs, incorporating PCMs can help buffer these variations, significantly improving both thermal storage and release efficiency. The absorber plays a critical role in the absorption refrigeration cycle, with the falling film absorber being the most widely used design. In this type of absorber, refrigerant vapor is absorbed by a thin film of solution flowing over horizontal tubes [119], and the heat transfer rate is directly proportional to the flow rate of the solution [120,121].

The absorption process in the absorber is a combination of chemical and physical heat exchange, and it is an exothermic process. As the refrigerant vapor is absorbed, the temperature of the absorber increases. Therefore, to prevent overheating, heat must be released to the environment. In the early designs of absorbers, this heat dissipation was addressed by using cooling water to absorb the excess heat. However, PCM integration within the absorber enables the capture and storage of surplus thermal energy, which can subsequently be redirected to auxiliary heat-dependent equipment, thereby enhancing overall solar energy utilization efficiency. This approach aligns with circular energy management principles, reducing waste and maximizing renewable resource exploitation.

Given the inherent instability of the heat source in absorption refrigeration systems, integrating PCMs allows the system to better cope with fluctuations in thermal load. The PCM helps to absorb the excess heat when the system experiences high load and releases stored heat when the load decreases, ensuring the system remains stable and operational [122]. Ji et al. [123] found that by incorporating PCMs into the absorber, temperature fluctuations in the absorbent are significantly reduced, allowing the system to operate efficiently across a wider temperature range and improving the overall reliability of the system.

5.3. Generator

In an absorption refrigeration system, the generator's principal job is to heat the absorbent solution so that refrigerant vapor can be separated. In a LiBr-H_2O absorption refrigeration cycle, for instance, hot water or steam elevates the temperature of the diluted lithium bromide solution, progressively vaporizing the aqueous refrigerant. This phase-change process is critical for regenerating the concentrated absorbent solution and sustaining the refrigeration cycle. This process must also build up enough pressure for the vapor to flow into the condenser, functioning similarly to a compressor in traditional vapor compression systems. With advances in solar thermal technology, more efficient heat exchangers are now employed to drive this step in solar absorption systems. Common strategies for supplying heat include using thermal storage tanks, which store the necessary heat and release it as needed.

As the generator functions inherently as a heat exchanger, integrating PCMs can offer distinct advantages for thermal energy storage in solar-driven absorption systems. This integration facilitates load shifting and stabilizes thermal output during intermittent solar availability, enhancing system reliability. For instance, Ponshanmugakumar et al. [124] designed a vertically oriented generator with an integrated PCM section for an NH_3-H_2O solar absorption refrigeration system, facilitating forced convective boiling. They built a TRNSYS model to simulate a 1 m long coaxial generator, interlinked with a hot water storage tank and a boiler. Their research showed that when cooling loads are high or solar irradiance is low, the PCM substantially reduces the reliance on auxiliary heat sources. In a separate study, Lorente et al. conducted numerical simulations of a cylindrical storage tank filled with a paraffin PCM, featuring a vertically arranged spiral heating tube. By adjusting the spiral's radius, they assessed how geometry affected the melting fraction and temperature distribution. Lorente et al. [125] underscored the value of combining numerical modeling with experimental work to advance generator design.

Although the concept of using PCMs in heat exchangers is not new [126], applying it specifically to the generator in solar absorption systems remains relatively novel. In theory, selecting a phase-change material (PCM) with a phase-change temperature and latent heat capacity well aligned with the generator's operational parameters enables the PCM to act as an auxiliary thermal buffer, thereby stabilizing the generator's thermal output. Prior investigations have primarily focused on structural innovations; going forward, researchers should also seek or develop new PCMs and explore ways to enhance the thermal properties of existing ones. This dual approach—combining structural optimization with material innovation—promises further improvements in the overall efficiency and reliability of solar absorption refrigeration systems. Table 8 summarizes the information including the types of PCMs integrated on the generator and their thermophysical parameters used in the literature mentioned above.

Table 8. PCMs integrated into generator.

Research Study	Type	PCM	Density (Solid)	Density (Liquid)	Specific Heat Capacity (Solid)	Specific Heat Capacity (Liquid)	Thermal Conductivity (Solid)	Thermal Conductivity (Liquid)	Latent Heat of Fusion	Melting Point	Performance
			[kg/m^3]	[kg/m^3]	[kJ/kg·K]	[kJ/kg·K]	[W/m·K]	[W/m·K]	[kJ/kg]	[°C]	
[124]	Organic	Erythritol	1480	1300	1.4	1.4	0.7	0.3	339.8	117.7	By optimizing the use of PCMs, the overall performance of the system can be improved and the system can achieve an average annual solar utilization of 0.58.
[125]	Organic	Wax	800	800	1.25	/	0.2	0.2	125	30	The use of vertical spiral heaters significantly improves the melting efficiency of PCMs.
[127]	Organic	RT90HC	850	/	2.0	/	0.2	/	170.0	91.0	Electricity consumption reduced by 6.2%, and heat delivery time increased by 27.8%.
[128]	Organic	RT82	950	/	2.0	2.0	0.2	0.2	176.0	82.0	A 170% increase in COP.
	Organic	RT100	/	/	1.8	2.4	0.2	0.2	168.0	105.0	
	Inorganic	$MgCl_2 \cdot 6H_2O$	1570	1450	2.3	2.6	0.70	0.57	167.0	118.0	

6. Conclusions

This paper has thoroughly surveyed the use of PCMs in solar refrigeration systems, discussing the criteria for selecting PCMs, techniques for integrating PCMs with various system components, and the resulting effects on overall performance. By highlighting both the hurdles and promising avenues in this field, it provides a roadmap for future research and practical solutions to enhance system reliability, efficiency, and cost-effectiveness. From the analysis presented, the following primary conclusions emerge:

- Phase-change temperature and latent heat are the most critical parameters for PCM selection. Ideally, the PCM should have a phase-change temperature that aligns with the system's operating range and a high latent heat to maximize energy storage.
- Choosing the right PCM depends on whether a component is intended to absorb or release heat. For evaporators, using a PCM with a phase-change temperature above the evaporation temperature ensures sufficient heat absorption; for condensers, a phase-change temperature below the condensation temperature aids effective heat release.
- In addition to phase-change temperature and latent heat, supercooling can significantly affect a system's performance and must be considered during PCM selection and design.
- Microencapsulation technology enhances the heat transfer surface area of inorganic phase-change materials (PCMs) while mitigating their inherent phase separation challenges. Concurrently, the incorporation of nucleating agents effectively addresses the pronounced supercooling phenomenon in these materials by promoting controlled crystallization during phase transition processes.
- Organic PCMs, often chosen for their low supercooling, remain prevalent despite lower thermal conductivity. Inorganic PCMs feature higher thermal conductivity but may suffer from more pronounced supercooling. Currently, organic PCMs are still predominant in practical applications.
- Much of the existing literature focuses on improving the thermal conductivity of PCMs by introducing nanoparticles or employing microencapsulation techniques. However, these methods can be expensive, limiting their economic feasibility.
- PCMs experience volumetric changes during phase transitions, potentially causing deformation or structural stress on storage tanks and containers, thereby reducing their service life.
- Although a direct contact integration of PCMs is broadly validated in research, actual integration strategies remain relatively narrow, suggesting the need for innovative designs to optimize heat transfer and encapsulation.
- While adding PCMs raises initial costs, it confers significant advantages, including enhanced energy efficiency, better performance under fluctuating weather conditions, and partial mitigation of the inherent intermittency of solar energy.
- Experimental and numerical investigations on phase-change material integration across multiple components within the same solar cooling system remain insufficiently explored in the current literature.
- Although integrating PCMs in both the condenser and evaporator enhances the temperature gradient for heat transfer and elevates system performance, inherent limitations in existing integration approaches frequently amplify thermal resistance, leading to significant energy dissipation.
- The integration of phase-change materials with solar collectors remains relatively uncommon in current practice, with existing implementations often necessitating solution pump operation that introduces additional electrical power consumption.

- The incorporation of phase-change materials (PCMs) in generator units presents significant theoretical advantages, as their superior thermal exchange properties effectively mitigate energy losses, thereby enhancing overall system efficiency.

7. Outlook

- While organic PCMs dominate current applications, their widespread use stems largely from the stability challenges that plague inorganic counterparts—particularly phase separation. Nonetheless, many inorganic salt solutions outshine organics in terms of thermal performance, thanks to their higher thermal conductivity. Bridging the gap in phase separation and other drawbacks could unlock significant potential for inorganic PCMs.
- Research on eutectic materials remains comparatively sparse, presenting a promising area for innovation. Further investigation into eutectic mixtures could broaden the range of operating temperatures and provide specialized solutions for diverse refrigeration and thermal storage needs.
- Embedding nanoparticles in PCMs can markedly boost thermal conductivity but often at a high cost. Future studies should aim to develop more cost-effective additives, balancing affordability with the desired performance gains.
- Current PCM integration methods across various systems are relatively limited and uniform. Exploring novel designs—potentially involving new heat exchanger geometries or hybrid approaches—could further improve heat transfer efficiency and reliability.
- The existing one-size-fits-all integration strategies may not accommodate varying operational or climatic demands. Modular or detachable PCM systems could adapt to shifting loads or seasonal changes, enhancing both system resilience and practical utility. A representative application involves integrating phase-change material (PCM) behind photovoltaic panels, where automated regulation enables heat absorption during peak operating temperatures, followed by detachment upon complete phase transition to facilitate thermal dissipation.
- Phase transition in PCMs is typically accompanied by volumetric changes that can induce mechanical stress on storage containers, shortening their lifespan. Ongoing research should focus on reducing or managing this expansion, whether through advanced encapsulation techniques or by modifying the PCM's intrinsic properties.
- The majority of current experimental and modeling efforts address small-scale solar refrigeration units, such as household refrigerators or commercial freezers. Extending PCM integration to large-scale or industrial refrigeration systems represents an important frontier with the potential to substantially increase energy savings and system stability on a bigger stage.
- Future research should prioritize experimental and numerical investigations into concurrent PCM integration across multiple system components to enhance integration methodology flexibility while simultaneously advancing solar cooling system performance.

Funding: The authors are grateful for the support of the National Natural Science Foundation of China (no. 51936002), the China Postdoctoral Science Foundation (2024M750315), and the Foreign Expert Project (S20240314).

Conflicts of Interest: The authors declare no conflict of interest.

References

1. Mondal, M.A.H.; Denich, M. Assessment of renewable energy resources potential for electricity generation in Bangladesh. *Renew. Sustain. Energy Rev.* **2010**, *14*, 2401–2413.
2. Yüksel, I. Global warming and renewable energy sources for sustainable development in Turkey. *Renew. Energy* **2008**, *33*, 802–812.

3. International Energy Agency (IEA). *World Energy Outlook 2022*; IEA: Paris, France, 2022.
4. Tabor, H. Use of solar energy for cooling purposes. *Sol. Energy* **1962**, *6*, 136–142.
5. James, S.; James, C. The food cold-chain and climate change. *Food Res. Int.* **2010**, *43*, 1944–1956.
6. Coulomb, D. Refrigeration and cold chain serving the global food industry and creating a better future: Two key IIR challenges for improved health and environment. *Trends Food Sci. Technol.* **2008**, *19*, 413–417.
7. International Energy Agency (IEA). *Electricity 2024*; IEA: Paris, France, 2024.
8. UNESCAP. *Share of Total Energy Consumption from Renewable Sources in Bangladesh from 2005 to 2015*; Bangladesh Statistics: Dhaka, Bangladesh, 2017.
9. Khan, M.M.A.; Saidur, R.; Al-Sulaiman, F.A. A review for phase change materials (PCMs) in solar absorption refrigeration systems. *Renew. Sustain. Energy Rev.* **2017**, *76*, 105–137.
10. Patil, B.; Salunke, N.; Diware, V.; Ar, S.R.; Ansari, K.B. Stability assessment of emerging phase change materials for solar thermal storage in absorption refrigeration: A review. *Int. J. Green Energy* **2024**, *22*, 253–280. [CrossRef]
11. Anyanwu, E.E. Review of solid adsorption solar refrigerator I: An overview of the refrigeration cycle. *Energy Convers. Manag.* **2003**, *44*, 301–312.
12. Totla, N.B.; Arote, S.S.; Gaikwad, S.V.; Jodh, S.P.; Kattimani, S.K. Comparison of the performances of NH_3-H_2O and LiBr-H_2O vapour absorption refrigeration cycles. *Int. J. Eng. Res. Appl.* **2016**, *6*, 8–13.
13. Xu, X.; Li, Y.; Yang, S.; Chen, G. A review of fishing vessel refrigeration systems driven by exhaust heat from engines. *Appl. Energy* **2017**, *203*, 657–676.
14. Nikbakhti, R.; Wang, X.; Hussein, A.K.; Iranmanesh, A. Absorption cooling systems—Review of various techniques for energy performance enhancement. *Alex. Eng. J.* **2020**, *59*, 707–738.
15. Wouagfack, P.A.N.; Tchinda, R. Finite-time thermodynamics optimization of absorption refrigeration systems: A review. *Renew. Sustain. Energy Rev.* **2013**, *21*, 524–536. [CrossRef]
16. Sarbu, I.; Sebarchievici, C. Review of solar refrigeration and cooling systems. *Energy Build.* **2013**, *67*, 286–297. [CrossRef]
17. Huang, B.J.; Chang, J.M.; Petrenko, V.A.; Zhuk, K.B. A solar ejector cooling system using refrigerant R141b. *Sol. Energy* **1998**, *64*, 223–226. [CrossRef]
18. Gosney, W.B. *Principle of Refrigeration*; Cambridge University Press: Cambridge, UK, 1982.
19. Tassou, S.A.; Lewis, J.; Ge, Y.; Hadawey, A.; Chaer, I. A review of emerging technologies for food refrigeration application. *Appl. Therm. Eng.* **2010**, *30*, 263–276. [CrossRef]
20. Cao, Z.H. A New Refrigeration Method—A Discussion on Semiconductor Refrigeration Technology. *Clean Air Cond. Technol.* **2017**, *1*, 106–107.
21. Xie, B.B.; Zhang, L. Comprehensive Study on the Operating Conditions of Semiconductor Refrigeration Systems. *Shandong Ind. Technol.* **2015**, *68*. [CrossRef]
22. Kobayashi, E.; Watabe, Y.; Hao, R.; Ravi, T.S. High Efficiency Heterojunction Solar Cells on n-type Kerfless Monocrystalline Silicon Wafers by Epitaxial Growth. *Appl. Phys. Lett.* **2015**, *106*, 96. [CrossRef]
23. Richter, A.; Hermle, M.; Glunz, S.W. Reaction of the Limiting Efficiency for Crystalline Silicon Solar Cells. *IEEE J. Photovohaics* **2013**, *3*, 1184–1191. [CrossRef]
24. Ma, X.Y.; Zheng, D.Y.; Zhu, X.J. Design and Experimental Analysis of Solar Semiconductor Refrigeration Box. *Sci. Technol. Innov.* **2017**, *32*, 145–146.
25. Miranda, A.G.; Chen, T.S.; Hong, C.W. Feasibility study of a green energy powered thermoelectric chip based air conditioner for electric vehicles. *Energy* **2013**, *59*, 633–641.
26. Abdul-Wahab, S.A.; Elkamel, A.; Al-Damkhi, A.M.; Al-Habsi, I.A.; Al-Rubai'Ey', H.S.; Al-Battashi, A.K.; Al-Tamimi, A.R.; Al-Mamari, K.H.; Chutani, M.U. Design and experimental investigation of portable solar thermoelectric refrigerator. *Renew. Energy* **2009**, *34*, 30–34.
27. Song, C.; Zhu, J.; Yang, X. Innovative Design and Experiments of a Semiconductor Cooling and Heating Box Driven by Solar Energy. *IOP Conf. Ser. Mater. Sci. Eng.* **2019**, *677*, 032108.
28. Tamm, G.; Goswami, D.Y.; Lu, S.; Hasan, A.A. Theoretical and experimental investigation of an ammonia-water power and refrigeration thermodynamic cycle. *Sol. Energy* **2004**, *76*, 217–228.
29. Tang, L.; Zeng, D.; Ling, Z. Research progress on phase change cold storage materials and system applications. *Chem. Ind. Prog.* **2023**, *42*, 4322–4339.
30. Li, S.K.; Lin, Y.; Pan, F. Research progress in thermal energy storage and conversion technology. *Energy Storage Sci. Technol.* **2022**, *11*, 1551–1562.
31. Yataganbaba, A.; Ozkahraman, B.; Kurtbas, I. Worldwide trends on encapsulation of phase change materials: A bibliometric analysis (1990–2015). *Appl. Energy* **2017**, *185*, 720–731.
32. Shchegolkov, A.; Shchegolkov, A.; Demidova, A. The Use of Nanomodified Heat Storage Materials for Thermal Stabilization in the Engineering and Aerospace Industry as a Solution for Economy. *MATEC Web Conf.* **2018**, *224*, 03012.

33. Shchegolkov, A.; Shchegolkov, A.; Dyachkova, T.; Semenov, A. The heat storage material based on paraffin-modified multilayer carbon nanotubes with Nickel-zinc ferrite. *IOP Conf. Series Mater. Sci. Eng.* **2017**, *312*, 012023.
34. Tatsidjodoung, P.; Le, P.N.; Luo, L. A review of potential materials for thermal energy storage in building applications. *Renew. Sustain. Energy Rev* **2013**, *18*, 327–349.
35. Oro, E.; Gracia, A.; Castell, A.; Farid, M.M.; Cabeza, L.F. Review on phase change materials (PCMs) for cold thermal energy storage applications. *Appl. Energy* **2012**, *99*, 513–533.
36. Elsayed, M.; Mansour, M.S.; Eid, M.; Abdel-Raouf, M. Thermal behavior of frozen products paired with PCM during winter and summer power outages. *Case Stud. Therm. Eng.* **2024**, *53*, 103956.
37. Rahimi, M.; Ranjbar, A.A.; Hosseini, M.J. Experimental Investigation on PCM/Fin Slab Incorporation in a Evaporator Side of a Household Refrigerator. *Energy Rep.* **2023**, *10*, 407–418.
38. Wang, L.B.; Liu, L.N.; Li, H.S.; Bu, X.B.; Gong, Y.L. Optimization of Working Fluid and Operation Condition for Organic Rankine Vapor Compressor Refrigeration System Driven by Solar Energy. *Adv. NR Energy* **2017**, *5*, 386–393.
39. Aung, Z.T.; Mon, M.; Nu, S.S. Theoretical study on energy-saving of steam compression air conditioning system using solar thermal energy. In Proceedings of the 13th International Conference on Electrical Engineering/Electronics, Computer, Telecommunications and Information Technology, Chiang Mai, Thailand, 28 June–1 July 2016.
40. Hans, R.; Kaushik, S.C.; Manikandan, S. Experimental study and analysis on novel thermo-electric cooler driven by solar photovoltaic system. *Appl. Sol. Energy* **2016**, *52*, 205–210.
41. Dhawan, S.; Sinha, R.; Chaturvedi, S.; Parvez, Y.; Haq, A.U. Development and Performance Analysis of an Automated Solar-Powered Thermoelectric Refrigeration System. *Appl. Sol. Energy* **2023**, *59*, 226–238.
42. Richard, G.; Peter, E.; Bill, B. Thermodynamics. In *Mechanical Engineering Systems*; IIE Core Textbooks Series; Butterworth-Heinemann: Oxford, UK, 2001; pp. 7–111.
43. Neha, K.; Sanjay, K.S.; Sanjay, K. A comparative study of different materials used for solar photovoltaics technology. *Mater. Today Proc.* **2022**, *66*, 3522–3528.
44. Green, M.A.; Emery, K.; Hishikawa, Y.H.; Warta, W.; Dunlop, E. Solar cell efficiency tables (version 49). *Prog. Photovolt. Res. Appl.* **2016**, *25*, 3–13.
45. Yu, T.W.K.P.; Lam, K.H.; Chan, L.S.H. Thermal management in photovoltaic cells using phase change materials. *Renew. Energy* **2011**, *36*, 2085–2091.
46. Lee, D.J.C.; Barlow, D.D.W.K.; Chan, C.K.L. Performance improvement of solar cells using phase change materials for temperature control. *Energy Procedia* **2014**, *57*, 1822–1831.
47. Rosenthal, A.H.; Gonçalves, B.P.; Beckwith, J.A.; Gulati, R.; Compere, M.D.; Boetcher, S.K.S. Phase-change material to thermally regulate photovoltaic panels to improve solar to electric efficiency. In Proceedings of the ASME 2015 International Mechanical Engineering Congress and Exposition, IMECE 2015, Houston, TX, USA, 13–19 November 2015; Volume 57502, p. V08BT10A00.
48. Huang, M.J. Two Phase Change Material with Different Closed Shape Fins in Building Integrated Photovoltaic System Temperature Regulation. In Proceedings of the World Renewable Energy Congress-Sweden, Linköping, Sweden, 8–13 May 2011; Volume 57, pp. 2938–2945.
49. Biwole, P.H.; Eclache, P.; Kuznik, F. Improving the Performance of Solar Panels by the Use of Phase-Change Materials. In Proceedings of the World Renewable Energy Congress, Linköping, Sweden, 8–13 May 2011; pp. 2953–2960.
50. Radziemska, E.; Klugmann, E. Thermally affected parameters of the current-voltage characteristics of silicon photocell. *Energy Convers* **2002**, *43*, 1889–1900.
51. Krauter, S. Increased electrical yield via water flow over the front of photovoltaic panels. *Sol. Energy Mater. Sol. Cells* **2004**, *82*, 131–137.
52. Krauter, S.; Araújo, R.G.; Schroer, S.; Hanitsch, R.; Salhi, M.J.; Triebel, C.; Lemoine, R. Combined photovoltaic and solar thermal systems for facade integration and building insulation. *Sol. Energy* **1999**, *67*, 239–248.
53. Boudhina, N.; Jebali, R. Effects of Temperature on the Efficiency of Photovoltaic Modules: A Comparative Study. *Energy Rep.* **2019**, *5*, 797–804.
54. Radziemska, E. Performance Analysis of a Photovoltaic-Thermal Integrated System. *Int. J. Photoenergy* **2009**, *2009*, 732093.
55. Xiao, L.; Fu, X.; Zhou, H.; Qi, W.; Zhang, X.; Meng, W. Design and Experimental Study of PV/TEG/PCM Thermal Control Solar Hybrid Power Generation System. *J. Chuxiong Norm. Univ.* **2022**, *37*, 15–19.
56. Ma, T.; Zhao, J.; Li, Z. Mathematical modelling and sensitivity analysis of solar photovoltaic panel integrated with phase change material. *Appl. Energy* **2018**, *228*, 1147–1158.
57. Sharma, S.; Tahir, A.; Reddy, K.; Mallick, T.K. Performance enhancement of a Building-Integrated Concentrating Photovoltaic system using phase change material. *Sol. Energy Mater. Sol. Cell* **2016**, *149*, 29–39.
58. Maghrabie, H.M.; Mohamed, A.S.A.; Fahmy, A.M.; Samee, A.A. Performance Augmentation of PV Panels Using Phase Change Material Cooling Technique: A Review. *SVU-Int. J. Eng. Sci. Appl.* **2021**, *2*, 1–13. [CrossRef]

59. Malvi, C.S.; Dixon-Hardy, D.W.; Crook, R. Energy balance model of combined photovoltaic solar-thermal system incorporating phase change material. *Sol. Energy* **2011**, *85*, 1440–1446.
60. Kazemian, A.; Hosseinzadeh, M.; Sardarabadi, M.; Passandideh-Fard, M. Experimental study of using both ethylene glycol and phase change material as coolant in photovoltaic thermal systems (PVT) from energy, exergy and entropy generation viewpoints. *Energy* **2018**, *162*, 210–223. [CrossRef]
61. Fayaz, H.; Rahim, N.A.; Hasanuzzaman, M.; Nasrin, R.; Rivai, A. Numerical and experimental investigation of the effect of operating conditions on performance of PVT and PVT-PCM, Renew. *Energy* **2019**, *143*, 827–841.
62. Hasan, A.; McCormack, S.J.; Huang, M.J. Evaluation of Phase Change Materials for Thermal Regulation Enhancement of Building Integrated Photovoltaics. *Sol. Energy* **2010**, *84*, 1601–1612. [CrossRef]
63. Ahmadi, R.; Monadinia, F.; Maleki, M. Passive/active photovoltaic-thermal (PVT) system implementing infiltrated phase change material (PCM) in PS-CNT foam. *Sol. Energy Mater. Sol. Cells* **2021**, *222*, 110942. [CrossRef]
64. Refaey, H.; Abdo, S.; Saidani-Scott, H.; El-Shekeil, Y.; Bendoukha, S.; Barhoumi, N.; Abdelrahman, M. Thermal Regulation of Photovoltaic Panels Using PCM with Multiple Fins Configuration: Experimental Study with Analysis. *Therm. Sci. Eng. Prog.* **2024**, *49*, 102457. [CrossRef]
65. Al Arni, S.; Mahdi, J.M.; Abed, A.M.; Hammoodi, K.A.; Hasan, H.A.; Homod, R.Z.; Ben Khedher, N. Novel Multi-Layer Nano-Modified PCM Configuration for Efficient Thermal Management of Photovoltaic-Thermal Systems. *J. Energy Storage* **2024**, *103*, 114352. [CrossRef]
66. Hasan, A.; McCormack, S.J.; Huang, M.J.; Norton, B. Energy and cost saving of a photovoltaic-phase change materials (PV-PCM) system through temperature regulation and performance enhancement of photovoltaics. *Energies* **2014**, *7*, 1318–1331. [CrossRef]
67. Browne, M.C.; Norton, B.; McCormack, S.J. Heat retention of a photovoltaic/thermal collector with PCM. *Sol. Energy* **2016**, *133*, 533–548. [CrossRef]
68. Venkitaraj, K.P.; Suresh, S. Experimental study on thermal and chemical stability of pentaerythritol blended with low melting alloy as possible PCM for latent heat storage. *Exp. Therm. Fluid Sci.* **2017**, *88*, 73–87. [CrossRef]
69. Sarcinella, A.; Cunha, S.; Aguiar, J.; Frigione, M. Thermo-chemical characterization of organic phase change materials (PCMs) obtained from lost wax casting industry. *Sustainability* **2024**, *16*, 7057. [CrossRef]
70. Ryu, H.W.; Woo, S.W.; Shin, B.C.; Kim, S.D. Prevention of supercooling and stabilization of inorganic salt hydrates as latent heat storage materials. *Sol. Energy Mater. Sol. Cells* **1992**, *27*, 161–172. [CrossRef]
71. Sharma, A.; Tyagi, V.V.; Chen, C.R.; Buddhi, D. Review on thermal energy storage with phase change materials and applications. *Renew. Sustain. Energy Rev.* **2009**, *13*, 318–345. [CrossRef]
72. Raheem Junaidi, M.A. Influence of advanced composite phase change materials on thermal energy storage and thermal energy conversion. *J. Therm. Anal. Calorim.* **2025**, 1–22. [CrossRef]
73. Kamaraj, D.; Senthilkumar, S.R.R.; Ramalingam, M.; Vanaraj, R.; Kim, S.-C.; Prabakaran, M.; Kim, I.-S. A review on the effective utilization of organic phase change materials for energy efficiency in buildings. *Sustainability* **2024**, *16*, 9317. [CrossRef]
74. Wu, S.Y.; Wang, H.; Xiao, S.; Zhu, D.S. An investigation of melting/freezing characteristics of nanoparticle-enhanced phase change materials. *J. Therm. Anal. Calorim.* **2012**, *110*, 1127–1131. [CrossRef]
75. Wu, S.; Zhu, D.; Zhang, X.; Huang, J. Preparation and melting/freezing characteristics of Cu/paraffin nanofluid as phase-change material (PCM). *Energy Fuels* **2010**, *24*, 1894–1898. [CrossRef]
76. Jiang, M.H. The influence of operating conditions on the performance of chiller units. *Refrig. Technol.* **2021**, *41*, 78–82.
77. Wang, F.; Maidment, G.; Missenden, J.; Tozer, R. The novel use of phase change materials in refrigeration plant. Part 1: Experimental investigation. *Appl. Therm. Eng.* **2007**, *27*, 2893–2901. [CrossRef]
78. Wang, F.; Maidment, G.; Missenden, J.; Tozer, R. The novel use of phase change materials in refrigeration plant. Part 2: Dynamic simulation model for the combined system. *Appl. Therm. Eng.* **2007**, *27*, 2902–2910. [CrossRef]
79. Wang, F.; Maidment, G.; Missenden, J.; Tozer, R. The novel use of phase change materials in refrigeration plant. Part 3: Pcm for control and energy savings. *Appl. Therm. Eng.* **2007**, *27*, 2911–2918. [CrossRef]
80. Cheng, W.L.; Zhang, R.M.; Xie, K.; Liu, N.; Wang, J. Heat conduction enhanced shapestabilized paraffin/HDPE composite PCMs by graphite addition: Preparation an thermal properties. *Sol. Energy Mater. Sol. Cells* **2010**, *94*, 1636–1642. [CrossRef]
81. Cheng, W.L.; Mei, B.J.; Liu, Y.N.; Huang, Y.H.; Yuan, X.D. A novel household refrigerator with shape-stabilized PCM (Phase Change Material) heat storage condensers: An experimental investigation. *Energy* **2011**, *36*, 5797–5804.
82. Cheng, W.L.; Yuan, X.D. Numerical analysis of a novel household refrigerator with shape-stabilized PCM (phase change material) heat storage condensers. *Energy* **2013**, *59*, 265–276.
83. Alasiri, A.; Nasser, M. Comparative analysis of PCM configurations for energy-efficient air conditioning systems: A case study in Riyadh, Saudi Arabia. Case Stud. *Therm. Eng.* **2025**, *65*, 105691.
84. Ismail, M.; Hassan, H. Enhancing Air Conditioning System Performance via Dual Phase Change Materials Integration: Seasonal Efficiency and Capsulation Structure Impact. *Int. J. Thermophys.* **2024**, *45*, 115.

85. Cavargna, A.; Mongibello, L.; Iasiello, M.; Bianco, N. Analysis of a Phase Change Material-Based Condenser of a Low-Scale Refrigeration System. *Energies* **2023**, *16*, 3798. [CrossRef]
86. Said, M.A.; Hassan, H. Impact of Energy Storage of New Hybrid System of Phase Change Materials Combined with Air-Conditioner on Its Heating and Cooling Performance. *J. Energy Storage* **2021**, *36*, 102400. [CrossRef]
87. Said, M.A.; Hassan, H. An experimental work on the effect of using new technique of thermal energy storage of phase change material on the performance of air conditioning unit. *Energy Build.* **2018**, *173*, 353–364.
88. Yu, H. Modeling and Simulation of Evaporators in HVAC Systems. Ph.D. Thesis, Shandong University, Jinan, China, 2006. [CrossRef]
89. Bista, S.; Hosseini, S.E.; Owens, E.; Phillips, G. Performance improvement and energy consumption reduction in refrigeration systems using phase change material (PCM). *Appl. Therm. Eng.* **2018**, *142*, 723–735.
90. Salah, B.; Sifeddine, R. Phase Change Materials for Enhancing Heat Transfer in Evaporator Systems. *Energy Convers. Manag.* **2015**, *93*, 267–276.
91. Maiorino, A.; Del Duca, M.G.; Mota-Babiloni, A.; Greco, A.; Aprea, C. The thermal performances of a refrigerator incorporating a phase change material. *Int. J. Refrig.* **2019**, *100*, 255–264.
92. Joybari, M.M.; Haghighat, F.; Moffat, J.; Sra, P. Heat and Cold Storage Using Phase Change Materials in Domestic Refrigeration Systems: The State-of-the-Art Review. *Energy Build.* **2015**, *106*, 111–124.
93. Mselle, B.D.; Vérez, D.; Zsembinszki, G.; Borri, E.; Cabeza, L.F. Performance Study of Direct Integration of Phase Change Material into an Innovative Evaporator of a Simple Vapour Compression System. *Appl. Sci.* **2020**, *10*, 4649. [CrossRef]
94. Arsana, M.E.; Temaja, I.W.; Widiantara, I.B.G.; Sukadana, I.B.P. Corn oil phase change material (PCM) in frozen food cooling machine to improve energy efficiency. *J. Phys. Conf. Ser.* **2020**, *1450*, 012107.
95. Ghorbani, B.; Mehrpooya, M. Concentrated Solar Energy System and Cold Thermal Energy Storage (Process Development and Energy Analysis). *Sustain. Energy Technol. Assess.* **2020**, *37*, 100607.
96. Jokiel, M.; Sevault, A.; Banasiak, A.; NÆSS, E. Cold Storage Using Phase Change Material in Refrigerated Display Cabinets: Experimental Investigation. In Proceedings of the PCM2021 13th IIR, Phase Change Materials and Slurries for Refrigeration and Air Conditioning Conference, Vicenza, Italy, 1–3 September 2021; pp. 1–13.
97. Cheralathan, M.; Velraj, R.; Renganarayanan, S. Performance analysis on industrial refrigeration system integrated with encapsulated PCM-based cool thermal energy storage system. *Int. J. Energy Res.* **2007**, *31*, 1398–1413.
98. Li, G.; Hwang, Y.; Radermacher, R. Review of cold storage materials for air conditioning application. *Int. J. Refrig.* **2012**, *22*, 108–120.
99. Zhang, Y.; Li, Y.; Zhang, W. A review of phase change materials and their applications in energy storage systems. *Energy* **2012**, *47*, 98–108.
100. Chen, H.; Zhao, X.; Li, Y. Enhancement of heat transfer in phase change material based thermal storage using graphene nanomaterials. *Int. J. Heat Mass Transf.* **2016**, *98*, 869–876.
101. Zhang, Q.; Lu, J.; Wang, H.; Han, X. Phase change materials for thermal energy storage: A review. *Renew. Sustain. Energy Rev.* **2012**, *16*, 2247–2257.
102. Zhu, Y.; Shen, Z.; Xia, Y.; Bao, W. Development and performance of encapsulated phase change material for thermal energy storage. *Renew. Energy* **2015**, *73*, 60–66.
103. Ullah, K.R.; Saidur, R.; Ping, H.W.; Akikur, R.; Shuvo, N. A review of solar thermal refrigeration and cooling methods. *Renew. Sustain. Energy Rev.* **2013**, *24*, 499–513.
104. Boopathi, R.V.; Shanmugam, V. A review and new approach to minimize the cost of solar assisted absorption cooling system. *Renew. Sustain. Energy Rev.* **2012**, *16*, 6725–6731.
105. Xu, S.M.; Huang, X.D.; Du, R. An investigation of the solar powered absorption refrigeration system with advanced energy storage technology. *Sol. Energy* **2011**, *85*, 1794–1804.
106. Shirazi, A.; Pintaldi, S.; White, S.D.; Morrison, G.L.; Rosengarten, G.; Taylor, R.A. Solarassisted absorption air-conditioning systems in buildings: Control strategies and operational modes. *Appl. Therm. Eng.* **2016**, *92*, 246–260. [CrossRef]
107. Wang, S.G.; Wang, R.Z. Recent developments of refrigeration technology in fishing vessels. *Renew. Energy* **2005**, *30*, 589–600. [CrossRef]
108. Serale, G.; Baronetto, S.; Goia, F.; Perino, M. Characterization and Energy Performance of a Slurry PCM-based Solar Thermal Collector: A Numerical Analysis. *Energy Procedia* **2014**, *48*, 223–232.
109. Kalogirou, S.A. Solar thermal collectors and applications. *Prog. Energy Combust. Sci.* **2004**, *30*, 231–295.
110. Brancato, V.; Frazzica, A.; Sapienza, A.; Freni, A. Identification and characterization of promising phase change materials for solar cooling applications. *Sol. Energy Mater. Sol. Cells.* **2017**, *160*, 225–232.
111. Agyenim, F.; Smyth, M.; Eames, P. A review of phase change material energy storage; selection of materials suitable for energy storage in the 100–130 °C temperature range. In Proceedings of the World Renewable Energy Congress (WREC 2005), Aberdeen, UK, 22–27 May 2005; pp. 384–389.

112. Agyenim, F. The use of enhanced heat transfer phase change materials (PCM) to improve the coefficient of performance (COP) of solar powered LiBr/H_2O absorption cooling systems. *Renew. Energy* **2016**, *87*, 229–239.
113. Chopra, K.; Pathak, A.K.; Tyagi, V.V.; Pandey, A.K.; Sari, A.; Sarid, A. Thermal Performance of Phase Change Material Integrated Heat Pipe Evacuated Tube Solar Collector System: An Experimental Assessment. *Energy Convers. Manag.* **2020**, *203*, 112205.
114. Naghavi, M.S.; Ong, K.S.; Badruddin, I.A.; Mehrali, M.; Silakhori, M.; Metselaar, H.S.C. Theoretical Model of an Evacuated Tube Heat Pipe Solar Collector Integrated with Phase Change Material. *Energy* **2015**, *91*, 911–924. [CrossRef]
115. Abokersh, M.H.; El-Morsi, M.; Sharaf, O.; Abdelrahman, W. An Experimental Evaluation of Direct Flow Evacuated Tube Solar Collector Integrated with Phase Change Material. *Energy* **2017**, *139*, 1111–1125.
116. Li, Z.; Ge, Z.; Jing, A. Thermal management of PCM-based solar thermal collectors: System design and optimization. *Renew. Energy* **2018**, *120*, 407–417.
117. Yousef, A.; Sharma, Q.; Tahir, K. PCM-based thermal energy storage systems for solar collectors: A review on performance, heat transfer, and integration. *Renew. Sustain. Energy Rev.* **2015**, *48*, 423–437.
118. Li, A.; Duan, S.P.; Han, R.B.; Wang, C.Y. Investigation on the Dynamic Thermal Storage/Release of the Integrated PCM Solar Wall Embedded with an Evaporator. *Renew. Energy* **2022**, *200*, 1506–1516.
119. Srikhirin, P.; Aphornratana, S.; Chungpaibulpatana, S. A review of absorption refrigeration technologies. *Renew. Sustain. Energy Rev.* **2001**, *5*, 343–372.
120. Jeong, S.; Sang, K.; Koo, K.K.; Ziegler, F. Heat transfer performance of a coiled tube absorber with working fluid of ammonia/water/discussion. *ASHRAE Trans.* **1998**, *104*, 1577.
121. Matsuda, A.; Choi, K.H.; Hada, K.; Kawamura, T. Effect of pressure and concentration on performance of a vertical falling-film type of absorber and generator using lithium bromide aqueous solutions. *Int. J. Refrig.* **1994**, *17*, 538–542.
122. Zhang, H.; Zhang, S.; Wu, W.; Wang, S. Performance enhancement of absorption refrigeration system integrated with phase change material. *Appl. Therm. Eng.* **2014**, *63*, 306–315.
123. Ji, X.; Liu, G.; Li, Q.; Wu, J.; Xiao, R. Performance and volume expansion of PCM-based thermal storage systems for absorption refrigeration. *Energy* **2016**, *97*, 448–457.
124. Ponshanmugakumar, S.B.A.; Deepak, P.; Sivaraman, H.; Vignesh, K.R. Numerical investigation on vertical generator integrated with phase change materials in vapour absorption refrigeration system. *Appl. Mech. Mater.* **2015**, *766–767*, 468–473. [CrossRef]
125. Lorente, S.; Bejan, A.; Niu, J.L. Constructal design of latent thermal energy storage with vertical spiral heaters. *Int. J. Heat Mass Transf.* **2015**, *81*, 283–288. [CrossRef]
126. He, Q.; Zhang, W. A study on latent heat storage exchangers with the hightemperature phase-change material. *Int. J. Energy Res.* **2001**, *25*, 331–341. [CrossRef]
127. Migla, L.; Bogdanovics, R.; Lebedeva, K. Performance Improvement of a Solar-Assisted Absorption Cooling System Integrated with Latent Heat Thermal Energy Storage. *Energies* **2023**, *16*, 5307. [CrossRef]
128. Soliman, A.S.; Zhu, S.; Xu, L.; Dong, J.; Cheng, P. Design of an H_2O-LiBr absorption system using PCMs and powered by automotive exhaust gas. *Appl. Therm. Eng.* **2021**, *191*, 116881. [CrossRef]

energies

MDPI

Article

Performance Enhancement of a Solar Air Heater Equipped with a Tree-like Fractal Cylindrical Pin for Drying Applications: Tests Under Real Climatic Conditions

Chotiwut Prasopsuk, Kittiwoot Sutthivirode and Tongchana Thongtip *

Thermal and Fluid Laboratory (TFL), Department of Teacher Training in Mechanical Engineering, King Mongkut's University of Technology North Bangkok, 1518 Phacharat 1 Rd., Bang Sue, Bangkok 10800, Thailand; s6202017910012@email.kmutnb.ac.th (C.P.); kittiwoot.s@fte.kmutnb.ac.th (K.S.)
* Correspondence: tongchana.t@fte.kmutnb.ac.th; Tel.: +66-2-5552000

Abstract: This paper reports the improved thermal and drying performance of a solar air heater powered by real solar irradiance and equipped with a tree-like fractal-based cylindrical pin (SAH-TFCP) as a turbulator for drying applications. The main purpose of this work is to demonstrate the SAH-TFCP's improvement potential based on its thermal and drying performance as compared with a conventional solar air heater based on a flat-plate absorber (SAH-FP). The test was implemented based on solar time from 8:30 to 17:30 under Thailand's climatic conditions at a latitude angle of 14° and a longitude angle of 100°. Turmeric slices were used to evaluate the SAH's drying performance. The thermal efficiency, moisture content wet basis (MC_{wb}), drying rate (DR), and drying efficiency were measured as parameters of interest to assess the improvement potential of the SAH-TFCP over the SAH-FP. The results indicate that the SAH-TFCP provides better thermal and drying performance than the SAH-FP. A higher flow rate yields a higher thermal efficiency and a greater improvement potential. The improvement potential is around 44–85%. The drying efficiency of the SAH-TFCP is always higher than that of the SAH-FP and has an improvement potential of 32–44%, depending on the airflow rate.

Keywords: solar air heater; tree-like fractal; cylindrical pin; solar dryer; improvement potential

1. Introduction

Utilizing renewable energy sources is an efficient way to address the fossil fuel shortage in the near future. At the same time, it can also mitigate the environmental impacts of burning fossil fuels to generate useful thermal energy or electricity [1]. The most widely used renewable source is solar energy because of its abundance, especially in tropical countries. Hence, many researchers have focused on methods for converting solar irradiance into useful energy, such as wall-mounted solar heating for houses [2]; heating through stored thermal energy via solar power [3]; and drying applications via solar air heaters, indirect solar air dryers [4], hybrid SAHs [5], and forced convection solar dryers [6].

In the food preservation and herb production industry, drying is an important process for removing water from products (moisture content). This process mainly requires thermal energy to perform the water removal process via evaporation. In large-scale agricultural drying, the thermal energy used in the drying process mostly comes from electrical air heaters and heat pumps. Winans et al. [7] focused on the drying process via an electrical heater, while Cop et al. [8] and Hu et al. [9] discussed the drying of agricultural products via heat pumps. Their drying machines require quite high electrical energy per unit mass of dried product. This means that the increasing need for drying comes with an increased energy demand, which leads to problems such as primary energy shortages and environmental impacts. Therefore, many researchers have proposed alternative technologies to reduce the energy consumption of agricultural drying. One of the most suitable technologies is the solar air heater (SAH). The SAH can convert solar irradiance into useful thermal energy (hot air supplied for the drying process). This has encouraged many researchers to design and propose the development of solar air heaters for agricultural drying applications, as proposed by Afzal et al. [10], Karwa et al. [11], and Singh, V.P et al. [12]. They aimed to make energy usage more sustainable and to mitigate the environmental impacts of using fossil energy.

Jongpluempiti et al. [13] developed and tested a solar air heater with a conventional flat-plate absorber and investigated the thermal performance and hot air supplied when drying tofu with the spray-drying method. Their prototype solar air heater was tested using real solar irradiance. Y. He et al. [14] performed a numerical investigation of a solar air heater with a flat-plate absorber. Their main focus was on its convective heat transfer performance. They concluded that the key performance parameter in producing hot air is the convective heat transfer mechanism, which is significantly dependent on the Nusselt number, Reynolds number, and Pranthl number. Boulemtafes-Boukadoum et al. [15] also investigated the performance of a solar air heater based on heat transfer and aimed to demonstrate the influence of the convective heat transfer coefficient on energy conversion efficiency. Their findings indicated that the energy conversion of the solar air heater significantly depends on the airflow rate, the gap between the absorber plate and glass cover (aspect ratio), and its ability to generate a turbulent flow (or recirculation) of the air stream flowing through the absorber plate. The solar air heater based on the conventional flat-plate absorber had relatively low energy efficiency due to the lower turbulent flow of the air stream (laminar boundary layer flow) because the air stream flows over a smooth surface. This results in poor heat transfer performance and consequently results in poor drying performance, as supported by [16]. This has encouraged some researchers to propose an alternative design technique to enhance the heat transfer performance by means of installing a turbulator in the absorber plate. Previous research that focused on solar air heaters equipped with a turbulator is summarized in Table 1.

Table 1. Summary of research focusing on the turbulator-enhanced performance of SAHs.

Research Purpose	References (Authors)	System Details	Operating Condition	Significant Findings
Turbulator-enhanced heat transfer performance of SAH	M.T. Baissi et al., 2020 [17]	Experiments with LCD baffles 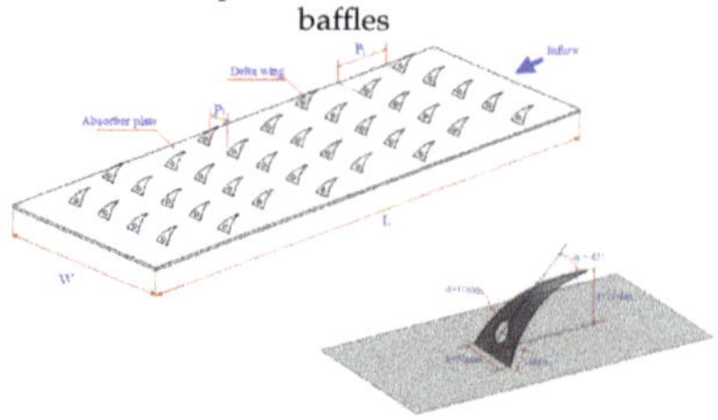	Re = 2500–12,000 $e/H = 0.8$ $P_l/e = 3, 4, 5$ $P_t/e = 0.6, 0.8, 1$	$Nu/Nu_0 = 5.13–6.94$ $f/f_0 = 29–43.41$ TEF = 1.46–2.26

Table 1. *Cont.*

Research Purpose	References (Authors)	System Details	Operating Condition	Significant Findings
Turbulator-enhanced heat transfer performance of SAH	A. Saravanan et al., 2021 [18]	Experiment with C-finned 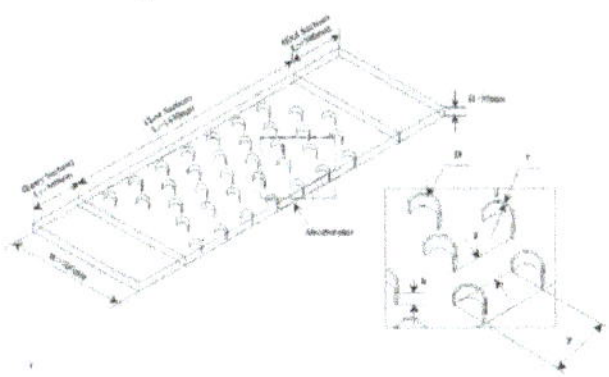	Re = 3000–24,000 h/H = 0.3–3.8 h/d_p = 1.5–3 P/g = 3.2–3.8	Max enhancement Nu = 2.67 f = 5.34 TEF = 1.64
	R.K. Ravi et al., 2018 [19]	Multi-V-shaped rib-roughness-based simulation 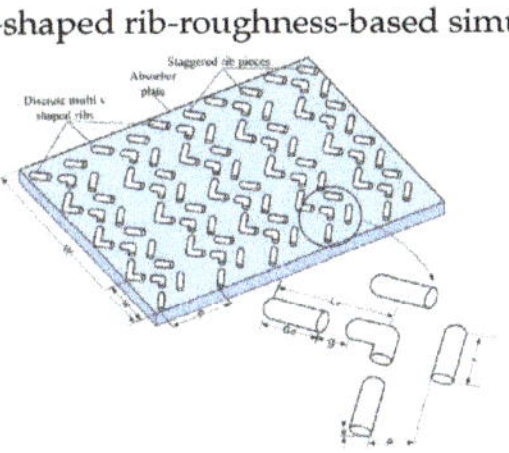	Re = 2000–20,000 $p/e = 10$ $e/D = 0.043$ W/w = 5–8	Max improvement Nu = 4.52 f = 3.13 TPP = 3.2–3.8
	Tabish Alam et al., 2001 [20]	Conical protrusion rib roughness 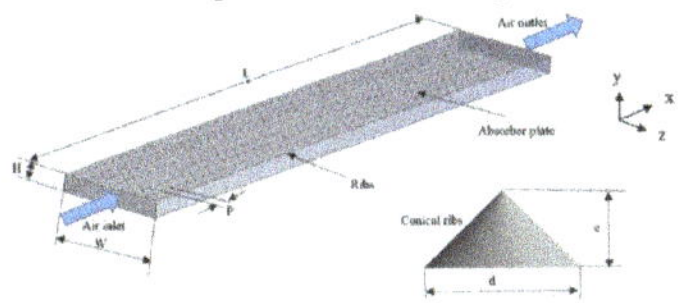	Re = 13,520–21,780 e/D = 0.020–0.044 p/e = 6–12 I = 600–1000 W/m^2	η_{th} = 70.92% at $e/D = 0.0289$ $p/e = 10$
	A.S. Yadav et al., 2022 [21]	Numerical with semicircular ribs 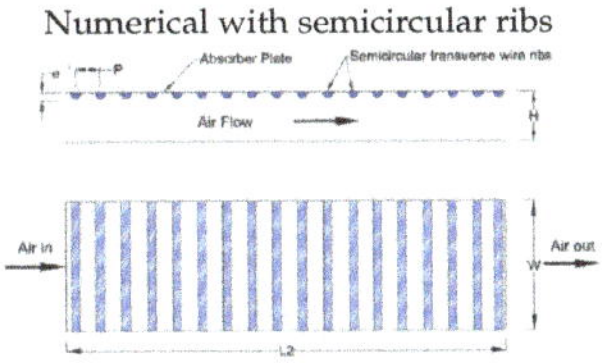	(RNG)k-ε model Re = 3800–18,000 RRP = 5–16.67 RRH = 0.03–0.06 Pr = 0.71	MaxTHPP = 1.98 at Re = 15,000 RRP = 10 RRH = 0.06
Turbulator-enhanced drying performance	P. J. Bezbaruah et al., 2020 [22]	Plus-shaped perforated baffles 	Mass flow rate = 0.009, 0.011 kg/s	Highest η_{th} = 84.3% in PPSCDB condition Drying efficiency: 21.6%
	A. Hassan et al., 2022 [23]	V-groove absorber 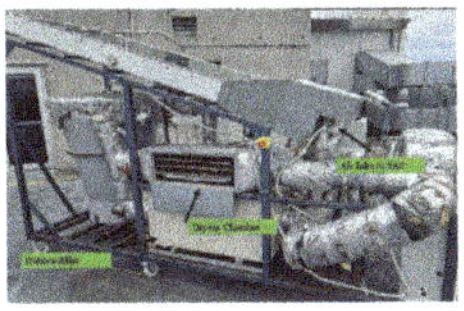	Mass flow rate = 0.021–0.061 kg/s	Maximum thermal efficiency : 88.8% $\dot{m}$ = 0.041 kg/s higher moisture removal

Table 1. *Cont.*

Research Purpose	References (Authors)	System Details	Operating Condition	Significant Findings
Turbulator-enhanced drying performance	E. Tefera et al., 2023 [24]	SAH with aluminum can for red pepper drying 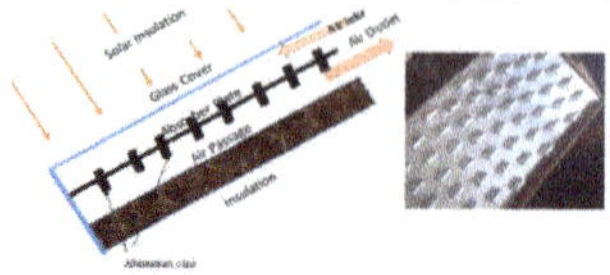	Blowing rate = 0.0383 Drying time/day = 9 h Crop = red pepper	$\eta_c = 79\%$ $\eta_d = 46\%$ (greenhouse drying system)
	S.A. Mohammed et al., 2024 [25]	Fin on absorber plate 	Free Convection Drying time per day = 10 h Sample: orange slice	Maximum η_c = 65% is coating 15% silicon model and 18% moisture content
	R. Rani et al., 2023 [26]	Semicircular loop solar collector (SLSC) 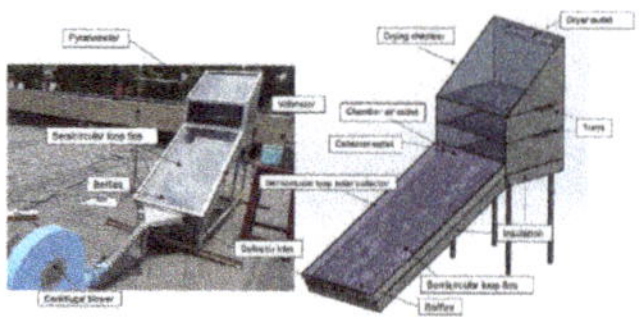	Simulation was used for FE approach Compared with FPSC and SLSC Sample: Apple slice	SLSC time used to dry (10% wb) was less than FPSC: 2.5 h.
Empirical tests of solar drying	E.K. Akpinar. 2010 [27]	The thin-layer open sun drying 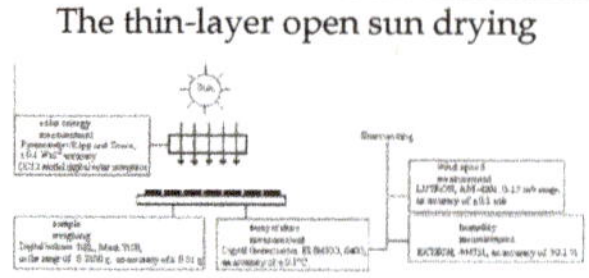	Sample: mint leaves $T_{ambient} = 30 - 46.4\ °C$ $I = 561.5 - 939\ W/m^2$	Moisture content Open sun drying $MR = 0.09$, time 23,400 s
	V. Patel et al., 2023 [28]	Gooseberry slice, open sun-dried 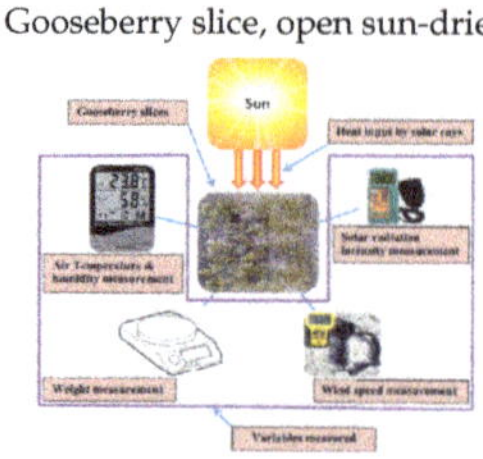	Sample: gooseberry Drying time: 10:00 a.m.–5:00 p.m.	M_w: 87.81% to 5.72%- Time taken: 23 h Effective diffusivity: $5.406 \times 10^{-11}\ m^2s^{-1}$
	C. Prajapati et al., 2023 [29]	Cabinet-type solar dryer 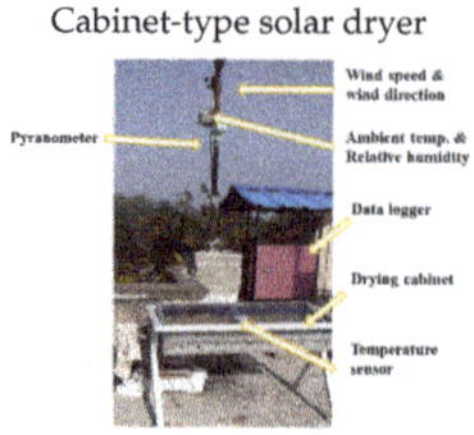	Sample: gooseberry Drying time: 9.00–17.00 Total time: 35 h of drying	Drying efficiency for open sun is 12.09% MR = 0.06 open sun

The surveyed literature listed in Table 1 shows that the thermal and drying efficiency of solar air heaters for drying applications can be increased by installing a turbulator. Hence, energy usage becomes more efficient. A key way to improve performance is to intensify the turbulent flow of the air stream as it flows through the absorber plate. A higher turbulence intensity can promote better convective heat transfer. Previous research that supports this was conducted by Naef A.A. Qasem et al. [30] and N. Jayranaiwachira et al. [31]. Their experiments showed that a higher convective heat transfer coefficient is obtained when a turbulator is installed on a smooth surface; however, it comes with a larger pressure drop across the absorber, which increases the energy consumption of an air blower or fan [32]. Therefore, the trade-off between enhanced heat transfer and a higher pressure drop has been studied in detail by researchers to investigate the value of using a turbulator [33]. Similarly, Sureandhar et al. [34] studied the arc-rib fin turbulator arrangement installed in an absorber plate with a fixed arc-rib fin configuration (e/D_e = 0.0422) and one with varying sizes (e/D_e = 0.0422–0.0541). The results showed that the varied arc-rib fin significantly improved the heat transfer process and provided better thermo-hydraulic characteristics than the fixed arc-rib fin. The friction factors of the varied arc-rib and fixed arc-rib are not much different; hence, the thermal enhancement factor of the varied arc-rib fin is higher than that of the fixed one.

Some researchers have applied fractal theory to formulate specific turbulator shapes to enhance heat transfer. Zhang [35] studied heat transfer augmentation with novel truncated ribs designed using fractal theory. Their aim was to propose new turbulator shapes (coral, elliptical, cylindrical, etc.) with sizes that could be varied. This resulted in heat transfer enhancement, which they investigated in detail. They found numerically and experimentally that the fractal turbulator could increase the Nusselt number and the convective heat transfer coefficient with the same flow Reynolds number. A slight increase in the pressure drop was found when using the fractal turbulator, which resulted in a higher thermal enhancement factor as compared with other turbulators [36]. However, the available literature has few studies investigating a fractal installed in a solar air heater. Yu et al. [37] proposed a fractal technique to improve an energy storage system with tree-like fractal fins. Zhang et al. [38] employed fractal theory to create a tree-like fractal formation, the aim of which was to improve heat transfer performance. They concluded that the fractal can be used as a tool for heat transfer enhancement due to its shape. The fractal technique makes the large-scale design of a heat exchanger easier and more efficient [39].

However, to the best of the authors' knowledge, fractal theory applied to a solar air heater for drying applications has not been reported in the available literature. This is a major research gap that is addressed in this work. It is expected that a solar air heater equipped with a fractal working as a turbulator will have significant improvements in both thermal and drying performance. This is because it can improve convective heat transfer between the air stream and absorber, which results in better energy conversion (converting solar radiation into thermal energy). However, a reasonably accurate case study must first be performed based on solar time to assess the performance improvement potential. This will be useful as a base case for designing further SAHs equipped with fractal-shaped pins for larger-scale solar drying.

As explained earlier, this work investigates the performance of a solar air heater equipped with a tree-like fractal-based cylindrical pin (SAH-TFCP) with the aim of demonstrating the improvement potential of the thermal and drying performance as compared with the conventional solar air heater based on a flat-plate absorber (SAH-FP). The test was implemented based on solar time from 8:30 to 17:30 under Thailand's climatic conditions at a latitude angle of 14° and a longitude angle of 100°. Turmeric slices were used as samples to evaluate the drying performance. The thermal efficiency, moisture content wet basis (MC_{wb}), drying rate (DR), and drying efficiency were measured as the parameters of interest for analyzing the improvement potential of the SAH-TFCP over the SAH-FP. The test results demonstrate the feasibility of using the TFCP to improve the thermal and drying performance. The major contribution of this work is that it provides a reference case for improving solar drying performance. The novelty of this work is characterized below:

- A tree-like fractal-based cylindrical pin was designed and constructed to be installed in the absorber plate of the solar air heater and work as a turbulator. This has not been reported in the available literature for solar drying applications.
- The improvement potential of the SAH-TFCP over the conventional SAH was demonstrated in tests under real climatic conditions.
- The performance of the two systems under real solar irradiance is detailed and evaluated.
- Thermal and drying performance is discussed and compared to indicate the advantages of using a tree-like fractal.

This work is composed of four sections. The research problem and literature review demonstrating the research gap are presented in Section 1. The test setup and the construction details of the SAH equipped with/without the tree-like fractal-based cylindrical pin (SAH-TFCP) are provided in Section 2, which also describes the data collected and the details of system control. Section 3 provides a discussion of the results based on thermal and drying performance to demonstrate the improvement potential of the SAH-TFCP over the conventional SAH. Concluding remarks on the significant findings of this work are summarized in Section 4.

2. Test Setup of Solar Air Heater

This research aims to investigate the thermal and drying performance enhancement of a solar air heater for drying applications by installing a tree-like fractal-based cylindrical pin in the absorber plate (SAH-TFCP). The thermal and drying performance indicators of the SAH-TFCP are compared with those of the conventional flat-plate solar air heater (SAH-FP). An empirical test of the solar-driven air heater under Thailand's climatic conditions was implemented to analyze the improvement potential of the SAH-TFCP over the SAH-FP. Turmeric was used as the sample product to demonstrate the drying performance. The test was implemented based on real solar time, as this demonstrates the real thermal and drying performance of the solar-driven SAH. The test stand unit was located at a latitude angle of 14° and a longitude angle of 100° (Lopburi province, Thailand). During the test, the airflow rate and solar irradiance were measured as the variable parameters to investigate the thermal performance and drying efficiency. This section provides the details of the test setup of the SAH (Section 2.1) and the detailed development of the tree-like fractal-based cylindrical pin, which was installed in the absorber of the SAH (Section 2.2). The data collected during empirical tests are described in Section 2.3.

2.1. Construction and Test Setup of Solar Air Heater for Drying Applications

Figure 1 shows the solar-driven air heater test unit, which is mainly composed of an absorber plate, a drying chamber, and measuring devices. The absorber plate was designed to be easily exchanged so that the performance of the SAH with and without the TFCP can be fairly compared using the same major hardware. The airflow through the absorber was supplied using an air blower, in which the flow rate can be precisely regulated by an inverter.

The operation of the SAH for drying turmeric begins when solar irradiance falls on the absorber plate. At the same time, ambient air can enter the absorber (point 1) via the air blower. As the air stream flows through the absorber, convective heat transfer takes place, resulting in hot air produced at the outlet of the absorber (point 2). This hot airflow then enters the drying chambers to perform the drying process (points 4–6) of the agricultural product (turmeric slices). The water contained within the turmeric slices (moisture content) is evaporated from their surface as hot air flows through the drying chamber. Thereafter, air with higher humidity (higher moisture) leaves the drying chamber at point 7 to complete the operation. The temperature at the points of interest, the airflow rate, and the weight loss of the dried product are important parameters for this investigation.

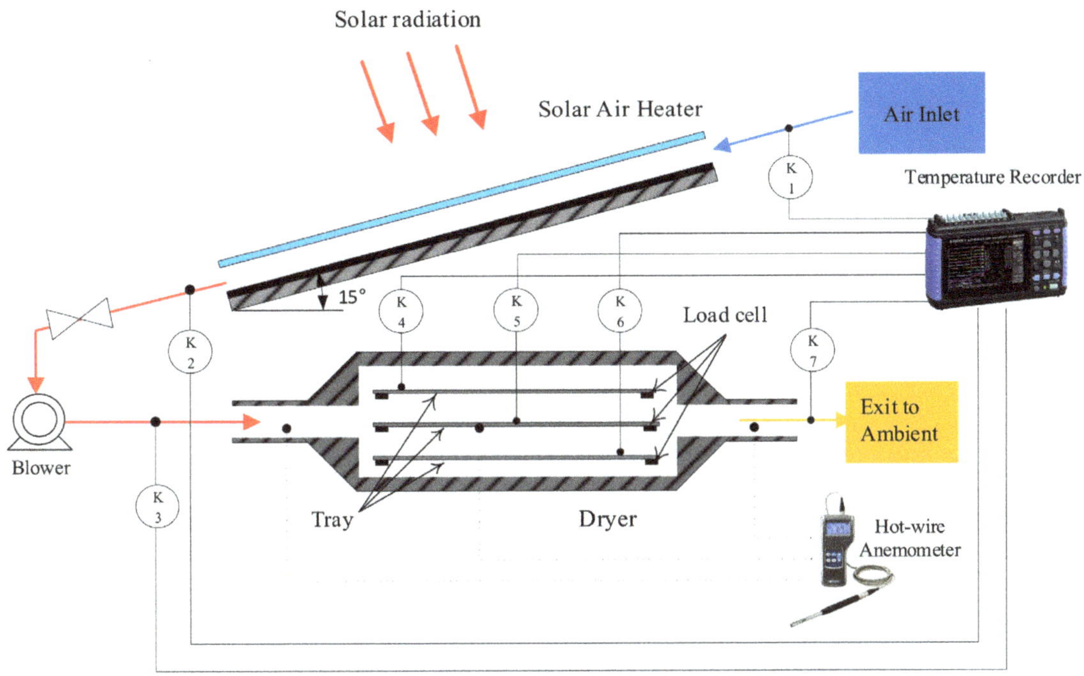

Figure 1. Schematic of the experimental setup.

The absorber plate was constructed based on ASHARE 93-77 recommendations [40]. The dimensions (W × L × H) are 150 × 75 × 4 cm, as shown in Figure 2 and Table 2. The absorber plate was fabricated from aluminum alloy (AA-1060) with a 3 mm thickness. This aluminum alloy was commercially available in Thailand market. Smart wood was placed behind the absorber plate to support the test stand. The absorber plate was painted matte black to maximize its ability to absorb solar irradiance (high absorbability). A neoprene foam insulator was used to prevent heat loss from the absorber plate to its surroundings, and a polycarbonate plate was placed behind the insulator to support the test stand. A glass cover with a 3 mm thickness was installed over the absorber plate to form an airflow gap to generate hot air via a convective heat transfer mechanism. The gap height is 40 mm, which corresponds to an aspect ratio (r, W/H) of around 19; this r value optimizes the trade-off between the pressure drop and heat transfer performance, as supported by [41].

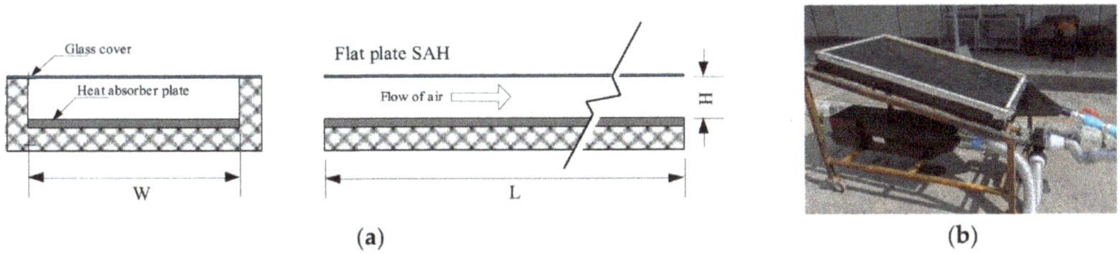

Figure 2. The geometrical drawing of the absorber plate of the SAH. (**a**) The sectional view of the absorber (**b**) Photograph of the test unit.

Table 2. Dimensions of the SAH.

Parameters	Values	Units
Heat absorber plate		
Length (L)	1500	mm
Width (W)	750	mm
Height (H)	40	mm
Thickness	1.5	mm
Emissivity	0.05	
Absorptivity	0.95	
Drying Chamber		
Length	1000	mm
width	300	mm
Height	270	mm
Drying tray		
Length	600	mm
Width	250	mm
Thickness	1.5	mm

Based on the criteria recommended in [42], the drying chamber was designed so that the agricultural products can be dried by supplied hot air. This makes agricultural products and herbs more economical. The dimensions are shown in Table 2. The hot air from the absorber enters the drying chamber at point (3) by means of an air blower and discharges to the ambient environment at point (7). In this work, turmeric slices were selected to demonstrate the drying performance of the SAH. They were prepared with approximately 3 mm thickness, as shown in Figure 3.

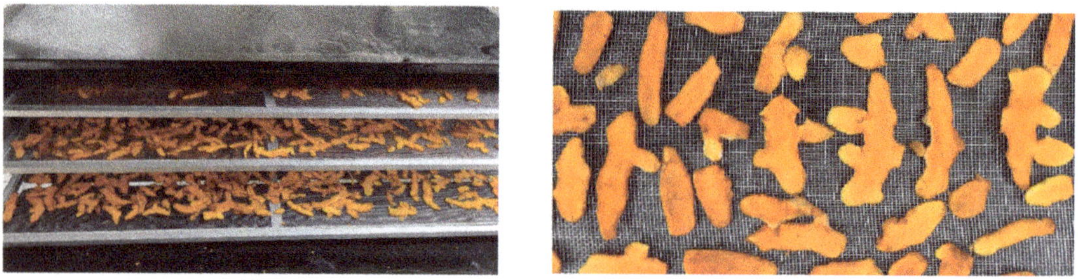

(a) (b)

Figure 3. The drying chamber and trays. (**a**) Drying samples on the dryer tray, (**b**) Turmeric slices as samples.

The air blower is electrically driven by an AC motor (220 V, 0.35 kW), and its flow rate can be regulated precisely by an inverter. This allows the airflow rate to remain constant at the desired point. An empirical test was implemented at three different mass flow rates: 0.016, 0.034, and 0.05 kg/s. The mass flow rate was the only parameter that varied during the tests. These mass flow rates were proven to provide acceptable performance for this particular absorber size in previous work by the authors of [43], where it was proven that an acceptable thermal enhancement factor (trade-off between heat transfer enhancement and the increased pressure drop (higher energy consumption)) can be obtained with mass flow rates of 0.016–0.05 kg/s. In addition, the investigated range of mass flow rates is consistent with the published work by Abushanab et al. [44].

To maintain the mass flow rate at a certain value, a hot-wire anemometer was used to measure the air velocity within a constant-area duct installed at the absorber inlet. The constant-area duct has a diameter of 45 mm and a length of 600 mm. The air velocity can be varied by fine-tuning the inverter frequency until the air flowing through the constant-area duct reaches a constant velocity, and hence, the mass flow rate can be calculated via Equation (4). Solar irradiance, ambient temperature, and relative humidity vary with the daytime period, which is a common manner of field testing of solar drying. Such parameters are beyond the control of the researcher. Therefore, in this work, the performance of the SAH was evaluated using a practical solar drying application.

Uncertainty Analysis

During the test, temperatures at the points of interest (points 1–7) were measured using a thermocouple probe (K-type) with an uncertainty of ±0.25 °C, and the real-time temperature was recorded using a data logger (Hioki-LR8431). The air velocity was detected using a hot-wire anemometer with an uncertainty of 2.0% of the read value. The solar irradiance falling on the absorber plate was monitored via a pyrometer. The weight of the sample was recorded using a digital load cell (uncertainty of 1.0%). All measuring devices and their uncertainties are listed in Table 3.

Table 3. List of instruments applied during experiments. (**a**) The measuring devices with their uncertainties; (**b**) the uncertainties of the parameters of interest (during the test).

(**a**)

Parameters	Uncertainty
Solar irradiance	±2.50%
Air mass flow rate	±2.75%
Air temperature	±0.52%
Moisture content	±2.25%
Thermal efficiency	±1.75%
Drying efficiency	±1.85%

(**b**)

Parameter	Equipment/Instrument	Model	Accuracy
Solar intensity	Pyrometer	Lutron, model SPM-116SD	±2.5%
Air velocity	Hot wire anemometer	Fluke 925	±2.0%
Relative humidity	Hygrometer	Climomaster, model 6501	±2.0%
Data collector	Data logger	Hioki-LR8431	±1.5 °C
Temperature	Thermocouples	K-Type	±0.5%
Weight	Load cell	Strain gauge	±1.0%

The uncertainty of the experimental setup was analyzed based on the method described in [45]. Let the independent variables be $a_1\ a_2, \ldots, a_n$ and let the dependent variable be Y. Thus,

$$Y = f(a_1,\ a_2, \ldots\ a_n) \tag{1}$$

The uncertainty is calculated as

$$\delta Y = \sqrt{\left(\frac{\delta Y}{\delta a_1}\delta a_1\right)^2 + \left(\frac{\delta Y}{\delta a_2}\delta a_2\right)^2 + \ldots + \left(\frac{\delta Y}{\delta a_n}\delta a_n\right)^2} \tag{2}$$

where δY is the uncertainty in the function Y, $\frac{\delta Y}{\delta a_n}$ is the partial derivative of the corresponding a_i, and δa_i is the uncertainty in the parameter.

The results for the uncertainties in the parameters of interest are tabulated in Table 3—(b).

2.2. Construction of Tree-like Fractal-Based Cylindrical Pin (TFCP)

The major purpose of this work is to compare the thermal and drying performance of the solar-driven air heater equipped with/without a tree-like fractal (TFCP). The conventional absorber plate (flat plate) and that equipped with the TFCP (SAH-TFCP) were constructed for empirical solar drying tests. The dimensions of the conventional flat plate are 1.5 × 0.75 m^2, and its thickness is 1.5 mm. It was fabricated from aluminum alloy AA-1060. The absorber plate of the SAH-TFCP was designed to have similar dimensions to the conventional plate, but the tree-like fractal-based cylindrical pin is placed on the top of the absorber plate, as shown in Figure 4. As a result of installing the fractal in the absorber plate, heat transfer enhancement is made possible by the fractal working as a turbulator to facilitate the heat transfer mechanism. Its improvement potential will be addressed in Section 3.

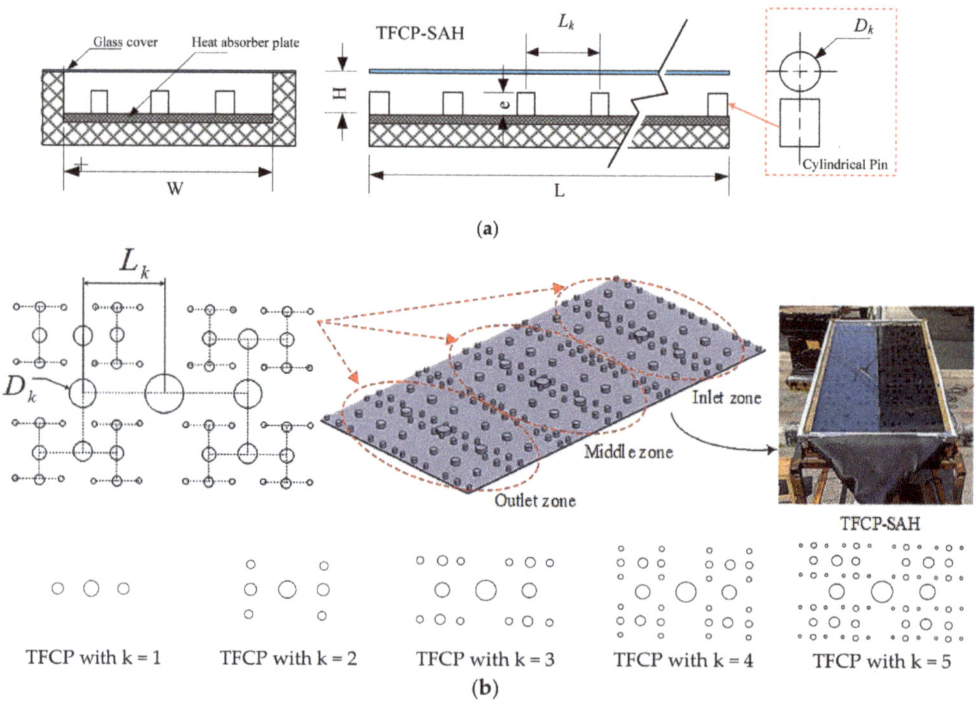

Figure 4. The construction of the SAH-TFCP. (**a**) Details of SAH-TFCP, (**b**) The formulation of the TFCP with an increase in branching order.

To indicate the location for installing the tree-like fractal and the diameter of the cylindrical pin, the fractal formula is used, which is stated in a tree-like fractal equation. This formula was previously proposed by C. Xia et al. [46]:

$$L_k = L_0 N^{\frac{-k}{\Delta_L}} \tag{3}$$

where L_0 is the length of the initial branches of the fractal.

$$D_k = D_0 N^{\frac{-k}{\Delta_D}} \tag{4}$$

Here, D_0 is the length of the initial diameter of the pin. According to Murray's law, the optimal values of parameters Δ_D and Δ_L provide the minimum flow resistance.

In constructing the TFCP, a bifurcation angle of 90° was used to create the cylindrical pin fractal (pin diameter and position), as shown in Figure 4b. The branching order (k) of the fractal is key to identifying the cylindrical pin's installation location and dimensions, calculated using Equations (3) and (4). A branching order of 5 (k = 5) was used to construct the fractal [47]. In the initial design, the absorber was separated into three zones, as shown in Figure 2. An initial length of 250 mm (L_0 = 250 mm) and an initial diameter of 50 mm (D_0 = 50 mm) were used to calculate the fractal TFCP for each zone. The branching order is key to generating the fractal TFCP. The formulation of the TFCP with increasing branching order (k = 1–5) is indicated in Figure 4b.

In this work, the height ratio (e/H) was fixed at 0.5 throughout the experiment (the cylindrical pin fractal height is 20 mm). The absorber equipped with the TFCP was fabricated as shown in Figure 4a. The TFCP was made of aluminum alloy AA-1060, which is similar to the absorber plate.

The installation of a tree-like fractal can enhance heat transfer performance because the air stream flowing through the fractal becomes more turbulent. This can improve heat transfer performance by eliminating the boundary layer flow (laminar sub-layer flow). The authors implemented CFD simulations to visualize the flow pathline with/without the TFCP and determine whether the TFCP could provide a more turbulent flow, as described in [48]. The simulation was implemented at an airflow velocity of 5 m/s as the inlet boundary condition. The velocity contour representing the flow physics with/without the TFCP is schematically shown in Figure 5. It is clearly seen that the air stream flowing through the absorber equipped with the TFCP is much more turbulent, and the recirculation zone of the air stream is established. This can enhance the convective heat transfer performance due to an increased Nusselt number [49]. In other words, the fractal (TFCP) works as a turbulator to increase the turbulence of the air stream flow. The use of a tree-like fractal based on a cylindrical shape for solar drying applications has not been reported in the available literature. This is proposed and discussed in this work.

2.3. Data Collected and Determination of Performance Parameters

2.3.1. Thermal Performance Parameter

During the test, solar irradiance is absorbed by the absorber plate (via short-wave radiation), and then, the absorbed heat is transferred to the air as it flows through the air gap between the absorber plate and the glass cover. The heat transfer rate can be calculated using Equation (3) [50]:

$$Q = \dot{m} C_p (T_{out} - T_{in}) \tag{5}$$

where

$$\dot{m} = \rho A V \tag{6}$$

Meanwhile, the absorbed heat from the solar irradiance falling on the absorber with a certain surface area can be estimated using Equation (6):

$$Q_{solar} = I_{solar} A_{absorb} \tag{7}$$

Therefore, the thermal efficiency [51] of the SAH can be determined by

$$\eta_{th} = \left(\frac{Q_{out}}{Q_{in}}\right) \times 100\% \quad (8)$$

where Q_{out} is the useful energy generated by the SAH, and Q_{in} is the solar irradiance falling on the absorber plate.

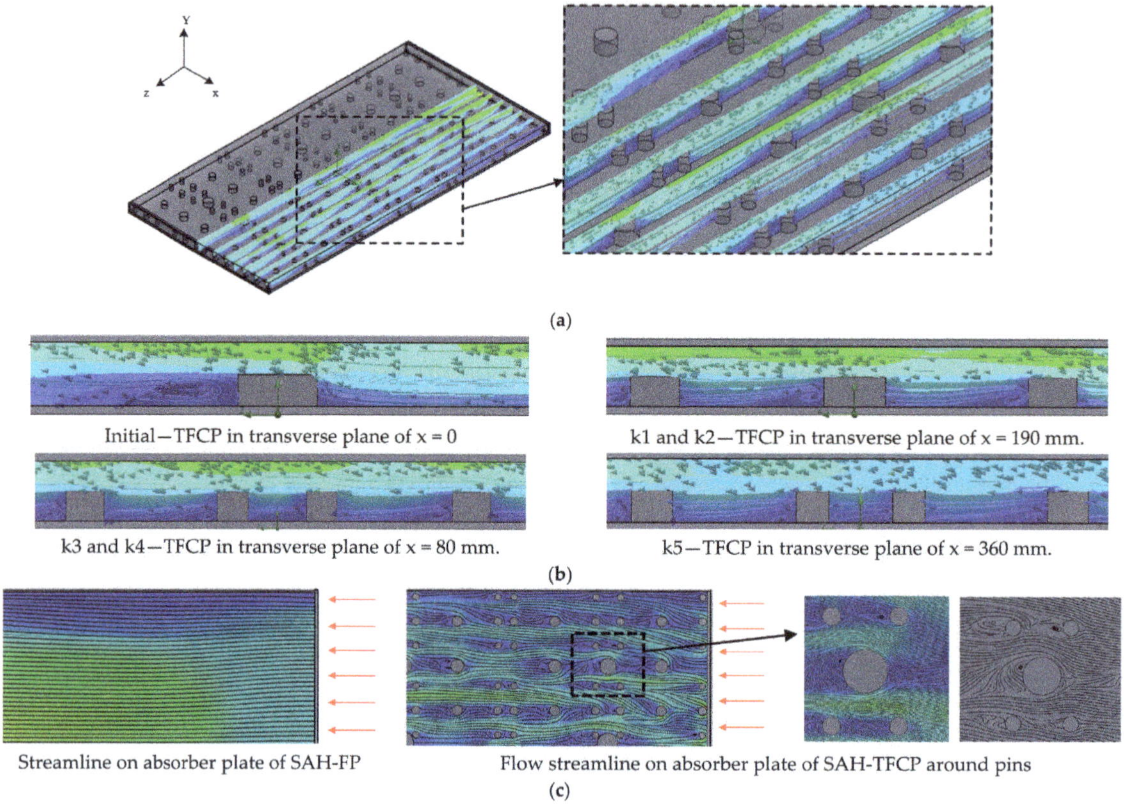

Figure 5. Comparison of typical flow physics of the air stream flowing through SAH-FP and SAH-TFCP. (**a**) Flow contour of air stream at various sections, (**b**) Velocity contours in transverse plane of x (different sections), (**c**) Velocity streamline on absorber plate of SAH.

Since this paper reports thermal performance based on the real-time operation of the SAH, thermal efficiency is key to demonstrating the improvement in the performance of the SAH equipped with/without the TFCP.

2.3.2. Drying Performance Parameters

The objective of the drying process is to remove water from the product (moisture content). This can be indicated by the moisture content's wet basis, as stated in Equation (9) [52]:

$$MC_w = \left(\frac{m_i - m_d}{m_i}\right) \times 100\% \quad (9)$$

where MC_w is the moisture content (wet basis), m_i is the initial product weight (g) (before drying), and m_d is the weight of the dried product (g).

The drying rate is recognized as a significant parameter for indicating drying performance improvement and can be determined using Equation (10) [53]:

$$DR = \left(\frac{m_0 - m_f}{\Delta t}\right) \tag{10}$$

where m_0 is the initial mass of the product (before drying), m_f is the final mass of the product (after drying), and Δt is the time interval for drying (hour).

The parameter that indicates the overall energy efficiency for solar drying applications is the drying efficiency. It can be determined using Equation (11) [54].

$$\eta_{dry} = \frac{Q_{dry}}{Q_{solar} + W_{blower}} = \frac{\dot{m}_w h_{fg}}{(I_{solar} A_{abs}) + W_{blower}} \tag{11}$$

Additionally,

$$m_w = m_m \frac{\left(MC_i - MC_f\right)}{(100 - MC_i)} \tag{12}$$

The aforementioned parameters were determined to analyze the performance improvement of the solar air heater equipped with the TFCP (SAH-TFCP) for drying applications. This is discussed in the next section.

3. Test Results and Discussion

This paper aims to demonstrate the thermal and drying performance of a solar air heater equipped with/without a TFCP when the system is driven by real solar irradiance under Thailand's climatic conditions. The main purpose is to demonstrate the improvement potential of a solar air heater equipped with a tree-like fractal (SAH-TFCP) as compared with the conventional flat-plate absorber (SAH-FP), which is not reported in the available literature. The tests were implemented within a daytime period of 8.30–17.30 (drying time of 9 h). The air temperature, humidity, and solar irradiance at the points of interest during the operating time were measured so that the thermal and drying performance of the SAH could be analyzed comprehensively. Moreover, the initial weight of the turmeric slices and their real-time weight were recorded (every 15 min) to assess the drying performance. Subsequently, the thermal and drying performance of the SAH-TFCP was compared with that of the SAH-FP (conventional SAH) to determine the improvement potential of the SAH-TFCP.

3.1. *Comparison of Thermal Performance of SAH-TFCP and SAH-FP*

In comparative experiments, the airflow rate of the SAH was held constant in each test case (this was the only control variable parameter of the investigations) by carefully adjusting the rotational speed of the blower. To determine the desired mass flow rate, a hot-wire anemometer installed at the inlet of the absorber was used to measure the air velocity within the constant-area duct. The air mass flow rate was then determined using Equation (6). The initial weight of turmeric was recorded, and real-time weight was monitored via a digital load cell and recorded using a data acquisition unit. The test started at 8.30 and finished at 17.30 (9 h) under clear-sky test conditions. The environmental conditions during the test periods are depicted in Figure 6. The parameters of interest were recorded every 15 min via the data acquisition unit. Additionally, comparative tests were implemented using different airflow rates ranging from 0.016 to 0.05 kg/s. The test results of the SAH are depicted in Figure 7.

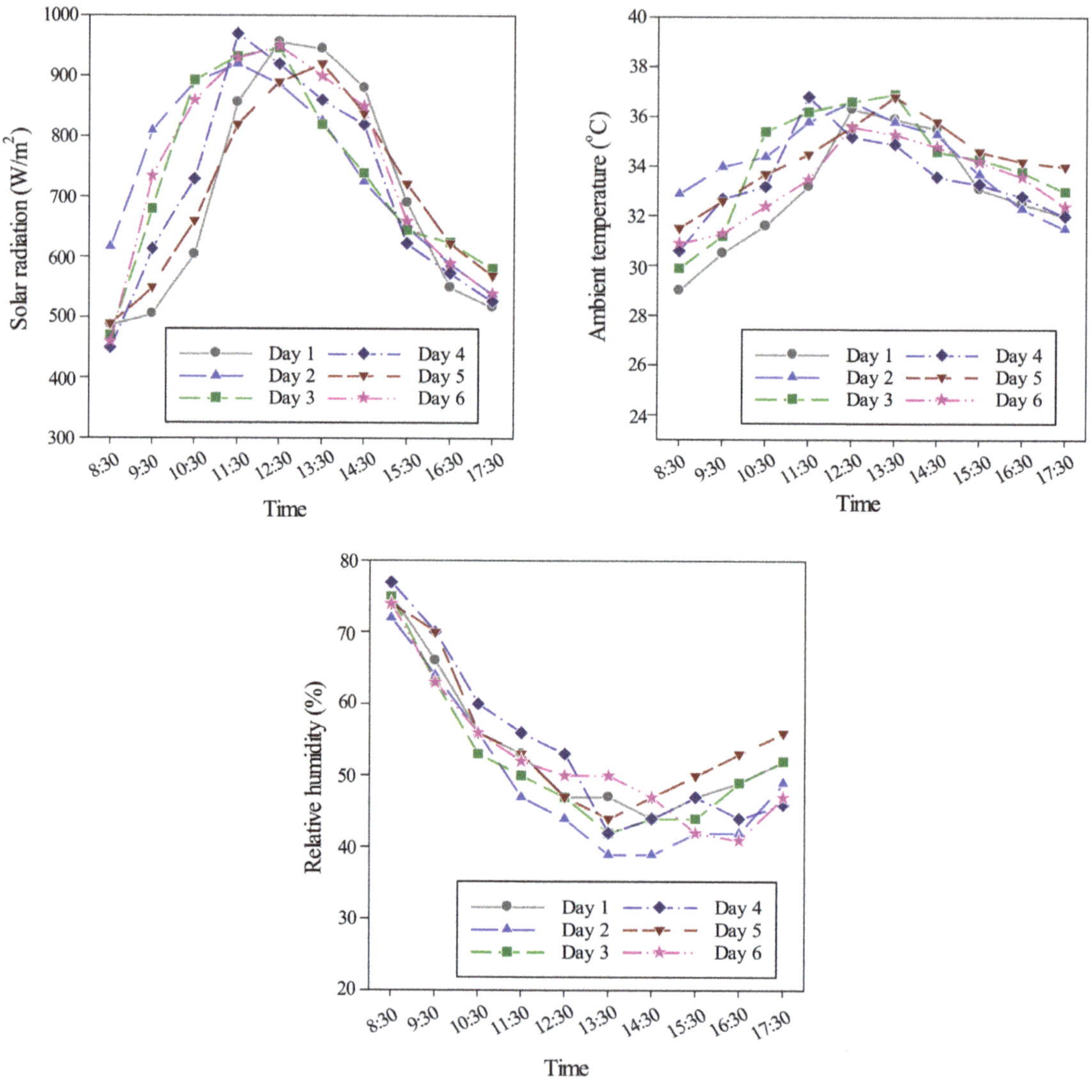

Figure 6. Environmental conditions during the test period.

Figure 7a–f show the ambient air temperatures at the absorber inlet (or T_{amb}), the absorber outlet ($T_{hot\text{-}air}$), the inlet/outlet of the drying chamber, and variations in solar irradiance with time. These results are also depicted for different airflow rates (0.016–0.050 kg/s). It can be seen that for a mass flow rate of 0.016 kg/s (Figure 7a,b), the SAH-TFCP can produce a much higher air temperature at the outlet of the absorber than the SAH-FP. The highest possible $T_{hot\text{-}air}$ produced by the SAH-TFCP is 57.3 °C with solar irradiance of around 985 W/m^2. Meanwhile, that produced by the SAH-FP is 48.5 °C with solar irradiance of around 980 W/m^2. This indicates that a higher hot-air temperature ($T_{hot\text{-}air}$) is made possible by the SAH-TFCP. This results in a better drying process (discussed later in the next section).

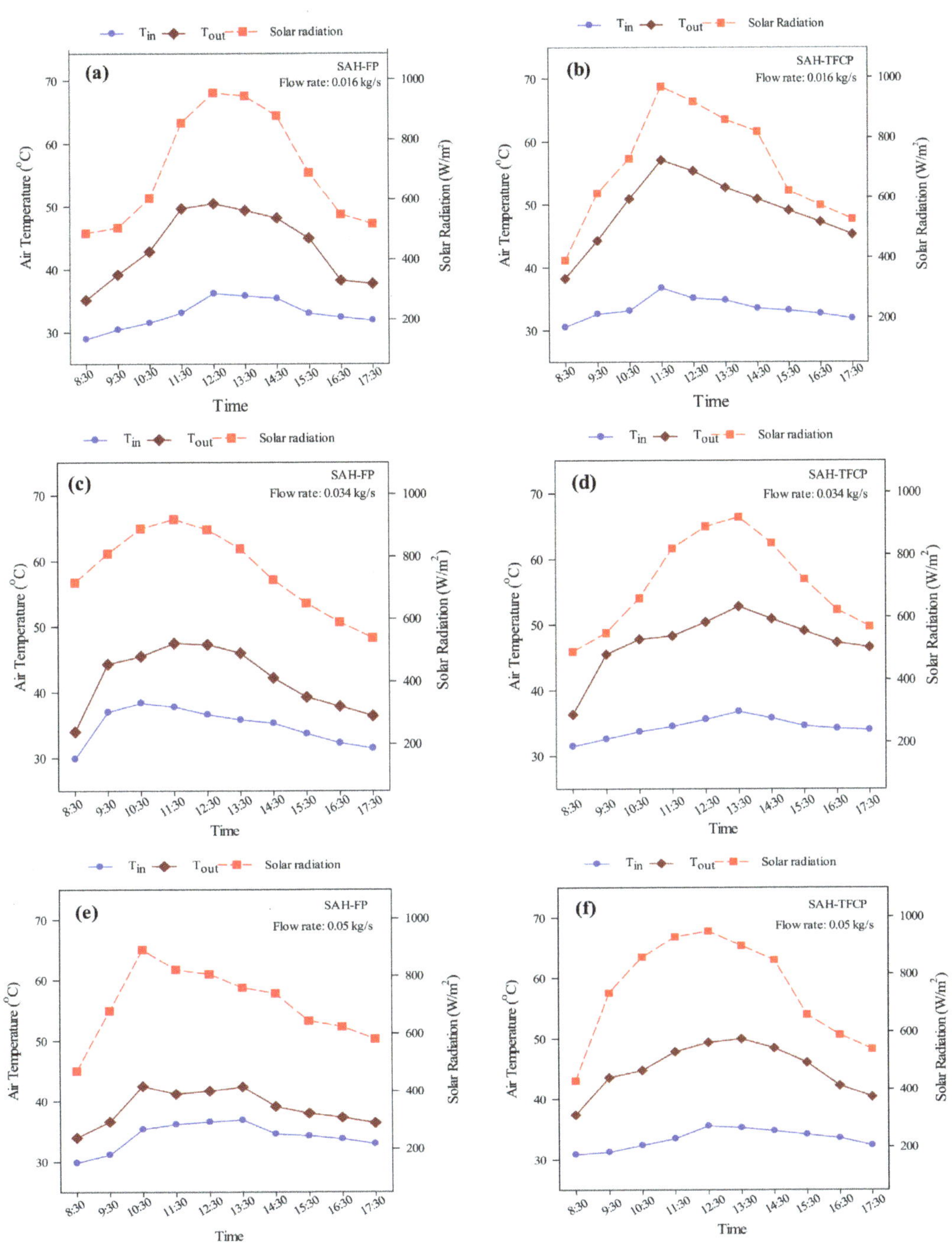

Figure 7. (**a**,**b**) Test results at the airflow rate of 0.016 kg/s; (**c**,**d**) Test results at the airflow rate of 0.034 kg/s; (**e**,**f**). Test results at the airflow rate of 0.05 kg/s.

The results for mass flow rates of 0.034 and 0.05 kg/s, as shown in Figure 7c,d and Figure 7e,f, respectively, show that the air temperature at the absorber outlet increases significantly with increasing solar irradiance and vice versa. In other words, a higher

$T_{hot\text{-}air}$ is produced when solar irradiance increases and vice versa. Increasing $T_{hot\text{-}air}$ leads to a higher drying rate and better drying performance (which will be discussed in the next section). A higher $T_{hot\text{-}air}$ is the result of better heat transfer via the convection process due to the working absorber temperature being increased with solar irradiance. It is interesting to see that the $T_{hot\text{-}air}$ produced by the SAH-TFCP is always higher than that produced by the SAH-FP. Since the air velocity is fixed for a particular test case, a larger temperature differential between the absorber surface and the air is the result of an increased heat transfer rate. As a result, a higher air temperature at the absorber outlet is produced. However, the test was conducted at different solar times (day-by-day variations), and hence, fair comparisons of the thermal efficiency of the two absorbers (demonstrating the trade-off between useful thermal energy and solar irradiance) must be employed for accurate results. Comparisons of the thermal efficiency of the two systems and the percent improvement are shown in Figure 8.

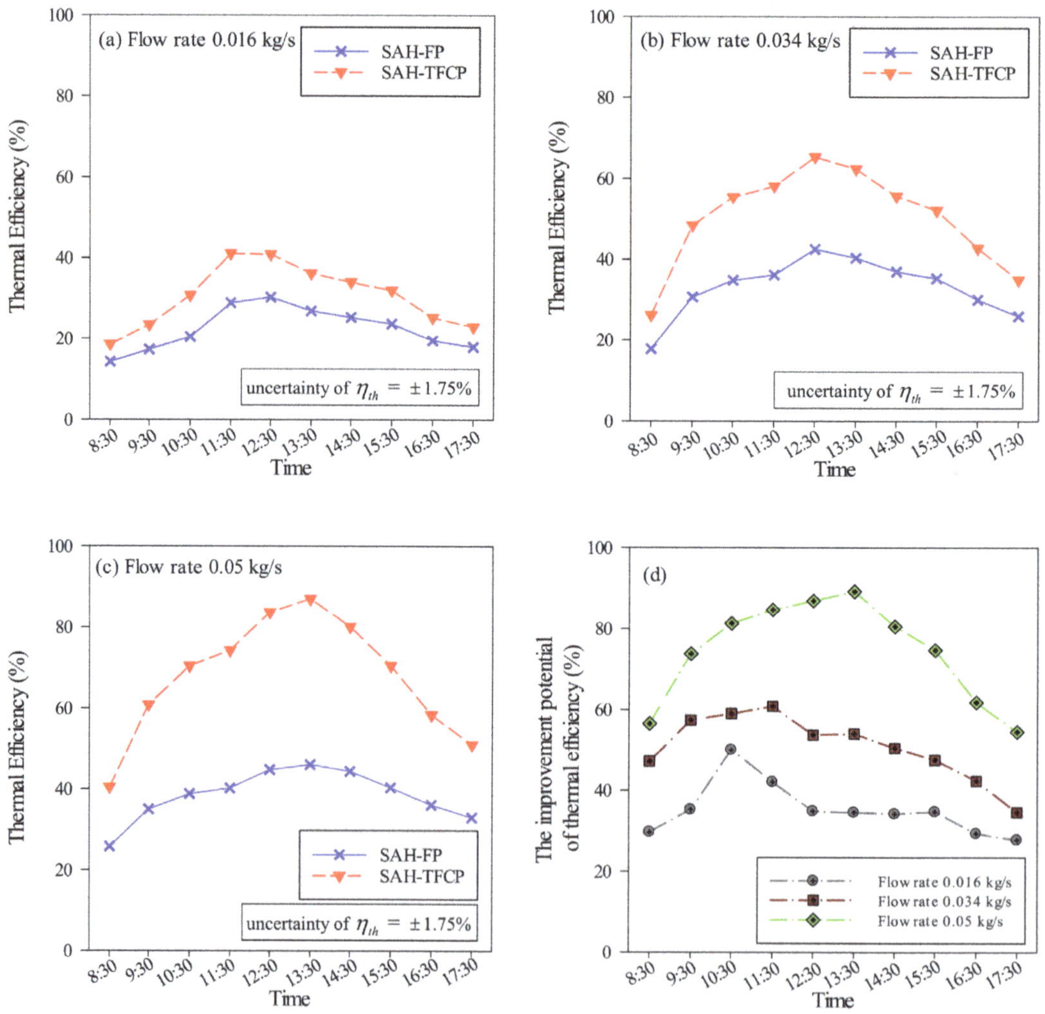

Figure 8. Thermal efficiency of the SAH-TFCP compared with the SAH-FP at different air flow rates of (**a**) 0.016 kg/s (**b**) 0.034 kg/s (**c**) 0.05 kg/s (**d**) Improvement potential of thermal efficiency of SAH.

It is clearly seen in Figure 8a–c that, for a given airflow rate, the thermal efficiency of the SAH-TFCP is higher than that of the SAH-FP throughout the test period. The improvement potential (%IMP) is found to vary with the airflow rate and solar time, and a higher improvement potential is found when the airflow rate is increased, as shown in Figure 8d.

As previously shown in Figure 7a–f, the air temperature at the absorber outlet ($T_{hot\text{-}air}$) produced by the SAH-TFCP is much higher than that produced by the SAH-FP throughout the test period. Notably, the two systems were tested with the same airflow rate (same flow Reynolds number), and thermal efficiency was used as a key parameter for comparison. This shows that heat transfer performance is significantly enhanced after the cylindrical fractal (TFCP) is installed in the absorber plate. The key to the heat transfer enhancement is that the TFCP can generate more turbulence in the air stream that flows through the absorber plate (as previously shown in Figure 5). A higher turbulence intensity can enhance the convective heat transfer mechanism, in which the Nusselt number increases (or the convective heat transfer coefficient increases) even when the flow Reynolds number is identical.

It is also found that when the airflow rate increases (or the flow Reynolds number increases), the improvement potential increases (please see Figure 8d). This is due to the fact that the convective heat transfer coefficient increases with the flow Reynolds number, as supported by [55]. In this case, a more turbulent air stream is achieved with a higher flow velocity at the absorber plate. As a result, the SAH-TFCP can produce a higher improvement potential when the airflow rate is increased. This analysis has demonstrated that installing the TFCP in the absorber plate can greatly enhance the heat transfer mechanism of the solar air heater, resulting in improved thermal performance.

It is also seen that the air temperature at the absorber outlet ($T_{hot\text{-}air}$) decreases when the airflow rate is increased (seen in Figure 7). This is because a higher convective heat transfer coefficient is obtained due to being operated with a higher mass flow rate (or higher air velocity). In this case, the Nusselt number (or convective heat transfer coefficient) is increased due to an increase in the flow Reynolds number when the air mass flow rate is increased. However, a lower temperature differential between the absorber surface and the airflow causes $T_{hot\text{-}air}$ to decrease. This is a common convective heat transfer mechanism that occurs when operating with a higher convective heat transfer coefficient, as supported by [56].

The above analysis has shown that the thermal performance improvement of the solar air heater is made possible by installing the tree-like fractal in the absorber. The key to this heat transfer enhancement is the more turbulent flow of the airflow stream. The results reported in this section were obtained under Thailand's climatic conditions, and the SAH was powered by real-time solar irradiance. This can serve as a reference case for further improving and validating the design model of the SAH. This will benefit the drying performance, which will be discussed in the next section.

3.2. Comparative Drying Performance of SAH-TFCP with SAH-FP

This section demonstrates the drying performance improvement of the SAH-TFCP as compared with the SAH-FP. The impact of the airflow rate on the drying performance is explored and discussed. In this test, the hot air produced at the absorber outlet was supplied to the drying chamber to produce dried turmeric. In this case, drying samples and turmeric slices were used to evaluate the drying performance. The percent decrease in moisture content (wet basis), drying rate, and drying efficiency are used as drying performance indicators to demonstrate the improvement potential of the SAH-TFCP in the drying process. The test was implemented based on real solar irradiance and carried out from 8:30 to 17:30 (9 h drying time).

Figure 9 depicts the moisture content wet basis (MC_W) over the drying period at various airflow rates (Figure 9a–c). The results clearly show that for a given airflow rate, the MC_W with the SAH-TFCP is always lower than that with the SAH-FP for the same drying time. This means that the sample dries faster when the SAH-TFCP is used. A lower MC_W with the same drying time reflects better water removal from the drying sample, meaning that the water in the sample (moisture content) is removed faster when using the SAH-TFCP. In the first four hours of the drying period, the MC_W of both systems drops sharply with the same trend. This is because the drying sample still contains quite high moisture, resulting in a higher evaporation rate of water at the surface of the drying sample. Notably, the MC_W obtained with the SAH-TFCP is always higher than that with the SAH-FP, which reflects a higher water evaporation rate. After four hours of drying, the MC_W decreased slightly with the drying time in all cases.

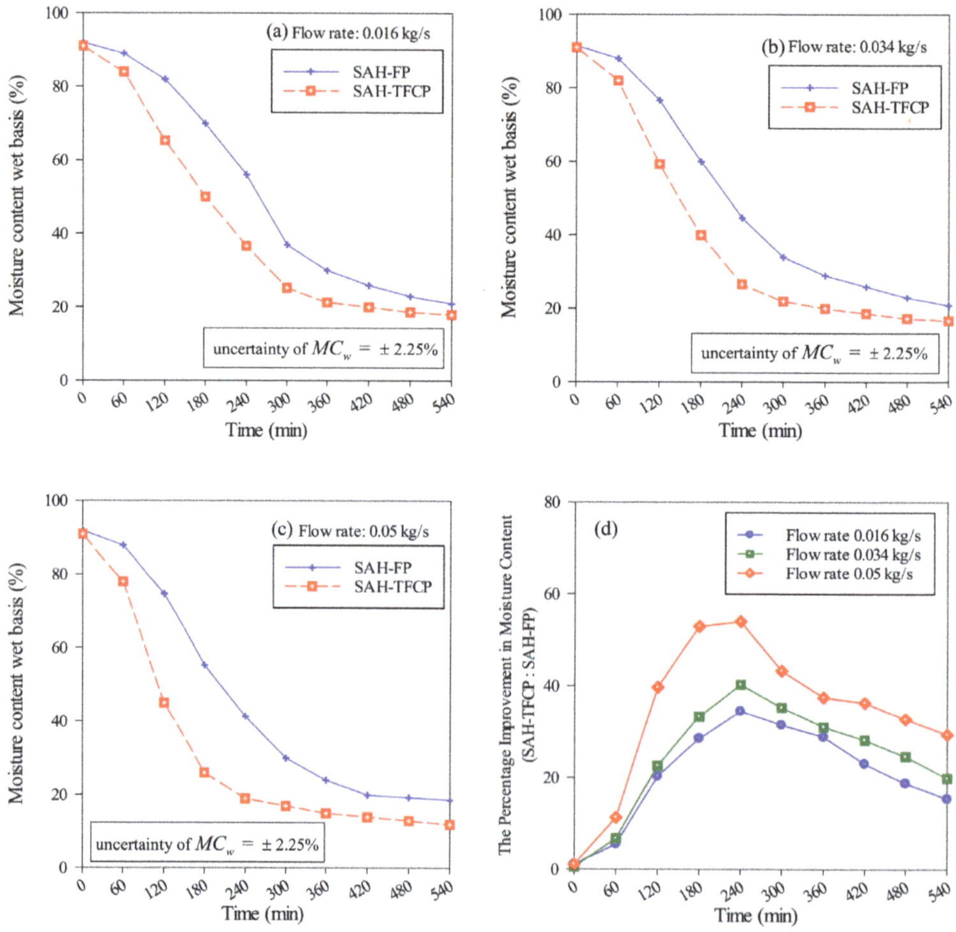

Figure 9. Moisture content produced by the SAH-TFCP compared with that produced by the SAH-FP at different air mass flow rates of (**a**) 0.016 kg/s (**b**) 0.034 kg/s (**c**) 0.05 kg/s (**d**) Improvement potential of the produced moisture content.

Figure 9d shows that in the first four hours of drying, the percent improvement in MC_w increases with the drying time until it reaches its maximum value after 240 min of drying. Then, the percent improvement of MC_w decreases continuously until the drying period ends. In addition, for the same drying time, a higher percent improvement in MC_w was found when the airflow rate was increased. From the test, the maximum percent improvement of MC_w is as follows: 54.03% at a flow rate of 0.05 kg/s, 40.29% at a flow rate of 0.034 kg/s, and 34.52% at a flow rate of 0.016 kg/s.

For a given airflow rate, the faster water removal from the drying sample is caused by enhanced heat transfer via the SAH-TFCP (by producing a higher $T_{hot\text{-}air}$), as discussed in the previous section. In this case, a higher air temperature yields a higher water evaporation rate at the surface of the sample (due to a larger temperature differential between the hot air and the sample surface). This causes a sharp decrease in MC_w with the same drying time. It is noticeable in Figure 9d that operating the SAH at a higher airflow rate increases the improvement potential. However, the previous section has shown that $T_{hot\text{-}air}$ (seen in Figure 7) decreases significantly when the airflow rate is increased. Hence, the airflow rate seems to have a greater impact on the moisture removal rate than the drying temperature ($T_{dry} = T_{hot\text{-}air}$). A possible reason for this is that a higher airflow rate can accelerate the water evaporation rate more than a higher drying temperature can because the water in the sample (turmeric slice) is dehydrated by the evaporation process. Hence, a higher velocity of air flowing over the sample surface can yield a higher water evaporation rate at the sample surface, resulting in a lower MC_w with the same drying time. This is because the airflow has a relatively low relative humidity.

The above results demonstrate the benefit of using a tree-like fractal as the turbulator for performance enhancement in drying applications based on the moisture removal rate. However, to provide a different perspective on drying performance, the drying rate and drying efficiency during the drying period are demonstrated in Figures 10 and 11. The drying rate involves latent heat transfer based on the evaporation of water in the drying sample. A higher latent heat transfer rate yields a higher drying rate. The drying efficiency represents the efficiency of energy usage, and hence, a higher drying efficiency is preferable.

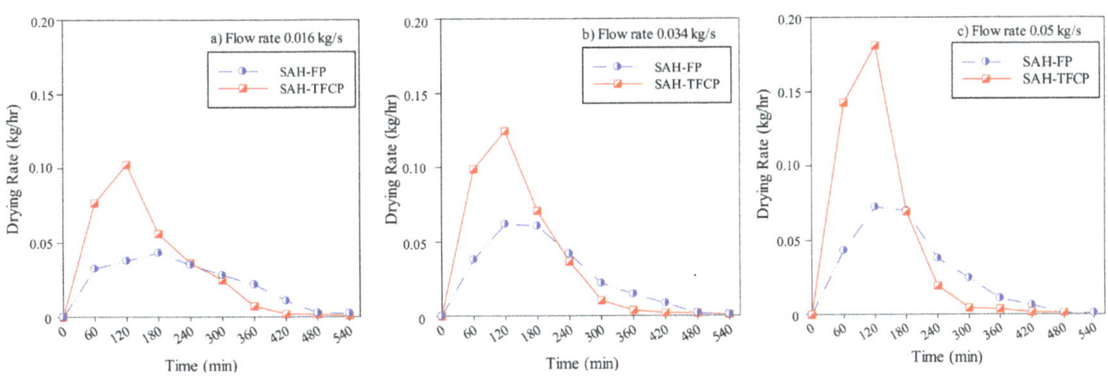

Figure 10. Comparisons of the drying rate at different flow rates.

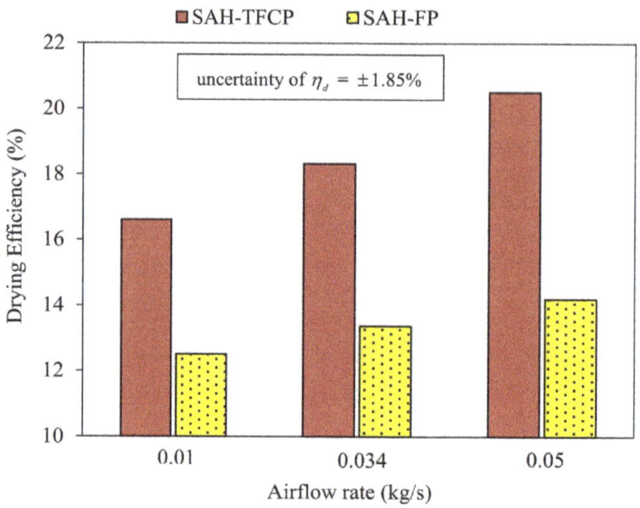

(**a**)

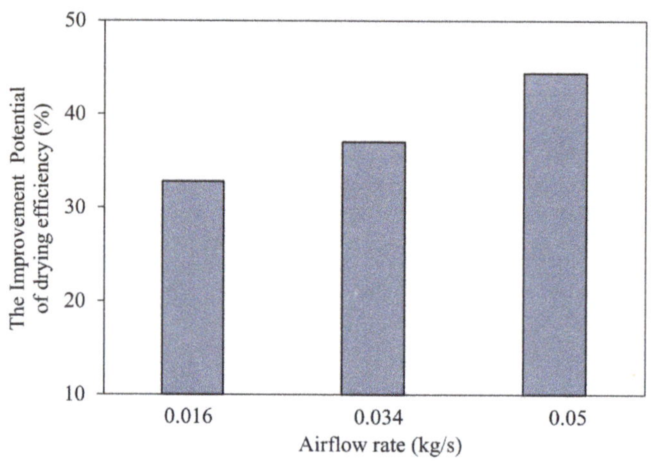

(**b**)

Figure 11. The drying efficiency and improvement potential at different airflow rates. (**a**) The drying efficiency at different airflow rates, (**b**) The improvement potential at different airflow rates.

It is seen in Figure 10 that at a fixed airflow rate, the drying rate of the SAH-TFCP is much higher than that of the SAH-FP in the first two hours. This is due to a higher moisture removal rate, as discussed earlier for MC_W. However, after three hours, the drying rate of the SAH-TFCP becomes lower than that of the SAH-FP. The reason is that the moisture content of the product (after three hours) dried by the SAH-TFCP is lower than that dried by the SAH-FP, and the surface area of the dried product for water evaporation is reduced significantly (seen in Table 4). These are the reasons that the drying rate of the SAH-TFCP is lower than that of the SAH-FP after three hours.

Table 4. The drying process of turmeric slices using the SAH-TFCP at a 0.05 kg/s mass flow rate.

Time of the Drying Process				
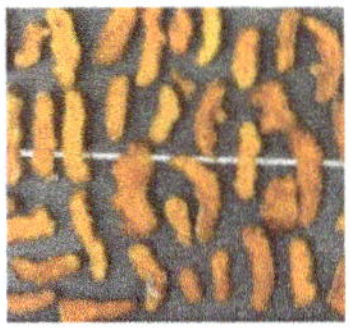	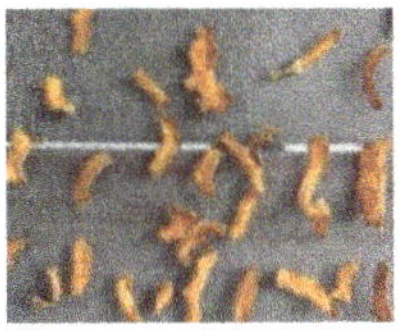	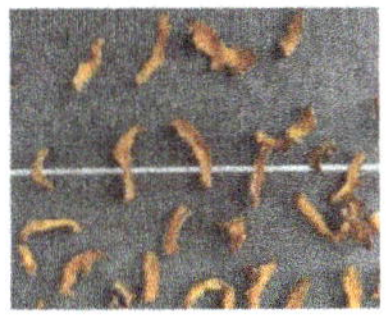		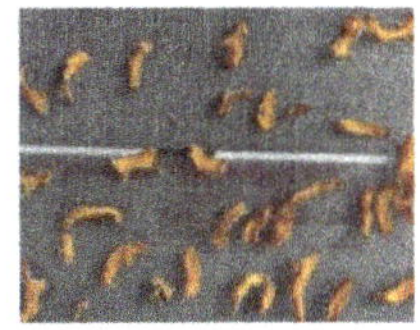
Start Drying	2 h	4 h	6 h	9 h

Figure 11a shows that at a fixed airflow rate, the drying efficiency obtained using the SAH-TFCP is much higher than that using the SAH-FP. It also shows that the drying efficiency of the two systems increases with an increase in the airflow rate. The SAH-TFCP always has a higher drying efficiency throughout the investigated range of mass flow rates. The improvement potential (or percent improvement) is depicted in Figure 11b. The improvement in drying efficiency is the result of heat transfer performance enhancement by the TFCP, which works as a turbulator to increase the turbulence of the air stream flowing through the absorber plate, which results in a higher $T_{hot\text{-}air}$ at a fixed flow rate. In addition, when increasing the airflow rate, the turbulent flow of the air stream is more intense because the air velocity is increased as it flows through the fractal. This results in a higher heat transfer rate, which results in better energy conversion. As a result, it has a high potential to have better drying performance, which is indicated by the drying efficiency improvement shown in Figure 11b.

The results in this section show that the drying performance, in terms of the drying rate and drying efficiency, can be improved by installing the tree-like fractal-based cylindrical pin in the absorber plate of the solar air heater. A significant drying performance improvement is obtained with solar energy radiation under Thailand's climatic conditions. Hence, the results reported in this section can be used as a base reference case for further developing a more accurate mathematical model to predict the drying performance of the SAH operating with a turbulator.

4. Conclusions

The improvement potential of a solar air heater equipped with a TFCP was investigated under Thailand's climatic conditions. The thermal and drying performance of the SAH-TFCP was compared with that of the SAH-FP (conventional one) for analysis. Turmeric slices were used as drying samples for the drying performance assessment. The tests were conducted at different airflow rates ranging from 0.016 to 0.050 kg/s under real solar irradiance. The findings of this work can be summarized as follows:

- At a fixed airflow rate, the temperature of hot air ($T_{hot\text{-}air}$) produced by the SAH-TFCP is much higher than that produced by the SAH-FP. This is due to convective heat transfer enhancement via the tree-like fractal-based cylindrical pin.
- The temperature of hot air from both systems decreases when increasing the airflow rate; however, the $T_{hot\text{-}air}$ produced by the SAH-TFCP is always higher than that produced by the SAH-FP.
- The thermal efficiency of the two systems varies significantly with solar irradiance. The thermal efficiency of the SAH-TFCP is higher than that of the SAH-FP. The percent improvement in thermal efficiency is 44–85%, depending on the airflow rate. A higher improvement potential is found when increasing the airflow rate.

- The moisture content wet basis (MC_w) of the SAH-TFCP is lower than that of the SAH-FP with the same drying time. This indicates that the turmeric slices are dried faster via the SAH-TFCP. A lower MC_w with the same drying time is made possible by increasing the mass flow rate.
- The drying efficiency of the SAH-TFCP is much higher than that of the SAH-FP throughout the test period. The improvement potential is 32–44%. An increase in the airflow rate yields a higher improvement potential.

Hopefully, the test results of the solar air heater for drying applications proposed in this work are useful for researchers who are interested in developing high-performance SAHs for drying applications. The test results show that installing the tree-like fractal cylindrical pin is an efficient way to improve the thermal and drying performance of the SAH. This is evidence that solar drying applications can be made more sustainable with lower energy usage. Since the thermal and drying performance of the SAH was acceptable under hot and humid conditions (Thailand's climatic conditions), the improvement technique proposed in this work is expected to further increase thermal and drying efficiency in low-humidity conditions.

Author Contributions: Conceptualization, C.P., T.T., and K.S.; methodology, C.P., T.T., and K.S.; software, C.P.; validation, C.P., T.T., and K.S.; formal analysis, C.P., T.T., and K.S.; investigation, C.P.; data curation, C.P., T.T., and K.S.; writing—original draft preparation, C.P., T.T., and K.S.; writing—review and editing, C.P., T.T., and K.S.; supervision, T.T., and K.S.; funding acquisition, K.S. All authors have read and agreed to the published version of the manuscript.

Funding: This research budget was allocated by the National Science, Research and Innovation Fund (NSRF) and King Mongkut's University of Technology North Bangkok (Project No. KMUTNB-FF-67-B-41).

Data Availability Statement: The original contributions presented in this study are included in the article. Further inquiries can be directed to the corresponding author.

Acknowledgments: The authors would like to thank Pornjit Pratumsuwan at department of Department of Teacher Training in Mechanical Engineering, King Mongkut's University of Technology North Bangkok for his generous guidance and valuable comments.

Conflicts of Interest: The authors declare no conflicts of interest.

Abbreviations

SAH	Solar air heater
TFCP	Tree-like fractal cylindrical pins
FP	Flat plate
IP	Improvement potential
Nomenclature	
A	Cross-sectional area of the duct, m^2
A_{absorb}	Surface area of collector plate, m^2
C_p	Specific heat of air at constant pressure, kJ/(kg K)
D_h	Hydraulic diameter of the air passage, m
D_k	Diameter of the *k*th branching level
DR	Drying rate
e	Roughness height, mm
e/D	Height ratio
H	Height of air channel, m
h	Heat transfer coefficient, W m^{-2} K^{-1}
I	Solar irradiance, W/m^2
k	Thermal conductivity of air, W m^{-2} K^{-1}

L	Duct length, m
L_k	Length of the *k*th branching level
MC	Moisture content
M_t	Total moisture content
m	Mass flow rate of air, kg/s
m_i	Initial product weight, kg
m_d	Weight of the dried product, kg
m_w	Mass of evaporated water, kg
Q_{in}	The solar irradiance falling on absorber plate, W
Q_{out}	Useful heat gain, W
T_{amb}	Ambient air temperature
$T_{hot\text{-}air}$	Air temperature of solar air heater outlet
V	Velocity of air through the duct
W	Duct width, m
Greek letters	
η_{th}	Thermal efficiency, %
η_d	Drying efficiency, %
ρ	Air density, kg/m^3
Δ	Diameter exponent

References

1. Panda, S.; Panda, B.; Kumar, R.; Jena, C. Energy & Exergy Analysis of a PV Panel with Passive cooling mechanism. *Suranaree J. Sci. Technol.* **2023**, *30*, 6.
2. Matsunaga, J.; Kikuta, K.; Hirakawa, H.; Mizuno, K.; Tajima, M.; Hayashi, M.; Fukushima, A. An Assessment of Heating Load Reduction by a Solar Air Heater in a Residential Passive Ventilation System. *Energies* **2021**, *14*, 7651. [CrossRef]
3. Sornek, K.; Papis-Fraczek, K. Development and Tests of the Solar Air Heater with Thermal Energy Storage. *Energies* **2022**, *15*, 6583. [CrossRef]
4. Qamar, Z.; Munir, A.; Langrish, T.; Ghafoor, A.; Tahir, M. Experimental and Numerical Simulations of a Solar Air Heater for Maximal Value Addition to Agricultural Products. *Agriculture* **2023**, *13*, 387. [CrossRef]
5. Hao, W.; Liu, S.; Mi, B.; Lai, Y. Mathematical Modeling and Performance Analysis of a New Hybrid Solar Dryer of Lemon Slices for Controlling Drying Temperature. *Energies* **2020**, *13*, 350. [CrossRef]
6. Karami, H.; Kaveh, M.; Golpour, I.; Khalife, E.; Rusinek, R.; Dobrzański, B., Jr.; Gancarz, M. Thermodynamic Evaluation of the Forced Convective Hybrid-Solar Dryer during Drying Process of Rosemary (*Rosmarinus officinalis* L.). *Leaves Energ.* **2021**, *14*, 5835. [CrossRef]
7. Winans, K.; Brodt, S.; Kendall, A. Life cycle assessment of California processing tomato: An evaluation of the effects of evolving practices and technologies over a 10-year (2005–2015) time frame. *Life Cycle Assess.* **2020**, *25*, 538–547. [CrossRef]
8. Cop, M.; Barta, R.B.; Thomas, C.; Hesse, U. Experimental investigation of a heat pump tumble dryer with a zeotropic refrigerant mixture. *Int. J. Refrig.* **2024**, *158*, 190–201. [CrossRef]
9. Hu, Z.; Li, Y.; El-Mesery, H.S.; Yin, D.; Qin, H.; Ge, F. Design of new heat pump dryer system: A case study in drying characteristics of kelp knots. *Case Stud. Therm. Eng.* **2022**, *32*, 101912. [CrossRef]
10. Afzal, A.; Iqbal, T.; Ikram, K.; Anjum, M.N.; Umair, M.; Azam, M.; Akram, S.; Hussain, F.; Zaman, M.A.; Ali, A.; et al. Development of a hybrid mix-mode solar air heater for product drying. *Heliyon* **2023**, *9*, e14144. [CrossRef]
11. Karwa, R. Thermo-Hydraulic performance of solar air heater with finned absorber plate forming multiple rectangular air flow passages in parallel under laminar flow conditions. *Appl. Therm. Eng.* **2023**, *221*, 119673. [CrossRef]
12. Singh, V.P.; Jain, S.; Karn, A.; Kumar, A.; Dwivedi, G.; Meena, C.S.; Dutt, N.; Ghosh, A. Recent Developments and Advancements in Solar Air Heaters: A Detailed Review. *Sustainability* **2022**, *14*, 12149. [CrossRef]
13. Jongpluempiti, J.; Pannucharoenwong, N.; Benjapiyaporn, C.; Vengsungnle, P. Design and Construction of the Flat Plate Solar Air Heater for Spray Dryer. *Energy Procedia* **2017**, *138*, 288–293. [CrossRef]
14. He, Y.; Yu, H.; Duan, G.; Wang, Y.; Yang, Q.; Feng, L.; Zhang, J. Thermal performance and experimental analysis of stainless steel flat plate solar collector with full-flow channels. *Heliyon* **2024**, *10*, e28255. [CrossRef] [PubMed]
15. Boulemtafes-Boukadoum, A.; Abid, C.; Benzaoui, A. 3D Numerical study of the effect of aspect ratio on mixed convection air flow in upward solar air heater. *Int. J. Heat Fluid. Flow* **2020**, *84*, 108570. [CrossRef]
16. Lingayat, A.B.; Chandramohan, V.P.; Raju, V.R.K.; Meda, V. A review on indirect type solar dryers for agricultural crops—Dryer setup, its performance, energy storage and important highlights. *Appl. Energy* **2020**, *258*, 114005. [CrossRef]

17. Baissi, M.T.; Brima, A.; Aoues, K.; Khanniche, R.; Moummi, N. Thermal behavior in a solar air heater channel roughened with delta-shaped vortex generators. *Appl. Therm. Eng.* **2020**, *165*, 113563. [CrossRef]
18. Saravanan, S.; Murugan, M.; Reddy, M.S.; Ranjit, P.S.; Elumalai, P.V. Thermo-hydraulic performance of a solar air heater with staggered C-shape finned absorber plate. *Int. J. Therm. Sci.* **2021**, *168*, 107068. [CrossRef]
19. Ravi, R.K.; Saini, R.P. Nusselt number and friction factor correlations for forced convective type counter flow solar air heater having discrete multi V shaped and staggered rib roughness on both sides of the absorber plate. *Appl. Therm. Eng.* **2018**, *129*, 735–746. [CrossRef]
20. Alam, T.; Meena, C.S.; Balam, N.B.; Kumar, A.; Cozzolino, R. Thermo-Hydraulic Performance Characteristics and Optimization of Protrusion Rib Roughness in Solar Air Heater. *Energies* **2021**, *14*, 3159. [CrossRef]
21. Yadav, A.S.; Alam, T.; Gupta, G.; Saxena, R.; Gupta, N.K.; Allamraju, K.V.; Kumar, R.; Sharma, N.; Sharma, A.; Pandey, U. A Numerical Investigation of an Artificially Roughened Solar Air Heater. *Energies* **2022**, *15*, 8045. [CrossRef]
22. Khanlari, A.; Guler, H.O.; Tuncer, A.D.; Sirin, C.; Bilge, Y.A.C.; Yilmaz, Y.; Gungor, A. Experimental and numerical study of the effect of integrating plusshaped perforated baffles to solar air collector in drying application. *Renew. Energy* **2020**, *145*, 1677–1692. [CrossRef]
23. Hassan, A.; Nikbahkt, A.; Welsh, Z.; Yarlagadda, P.; Fawzia, S. Experimental and thermodynamic analysis of solar air dryer equipped with V-groove double pass collector: Techno-economic and exergetic measures. *Energy Convers. Manag. X* **2022**, *16*, 100296. [CrossRef]
24. Tefera, E.; Damtie, E.G.; Bezzie, Y.M.; Admass, Z. Experimental testing of mixed-mode double pass solar air heater integrated with aluminium cans for red pepper drying. *Case Stud. Therm. Eng.* **2023**, *50*, 103379. [CrossRef]
25. Mohammed, S.A.; Alawee, W.H.; Chaichan, M.T.; Abdul-Zahra, A.S.; Fayad, M.A.; Ajiuwaya, T.M. Optimized solar food dryer with varied air heater designs. *Case Stud. Therm. Eng.* **2024**, *53*, 103961. [CrossRef]
26. Rani, P.; Tripathy, P.P. CFD coupled heat and mass transfer simulation of pineapple drying process using mixed-mode solar dryers integrated with flat plate and finned collector. *Renew. Energy* **2023**, *217*, 119210. [CrossRef]
27. Akpinae, E.K. Drying of mint leaves in a solar dryer and under open sun: Modelling, performance analyses. *Energy Convers. Manag.* **2010**, *51*, 2407–2418. [CrossRef]
28. Patel, V.; Judal, K.B.; Panchal, H.; Singh, B.; Jomde, A.; Kumar, A.; Patel, A.; Jain, R.; Sadasivuni, K.K. Investigation on drying kinetics analysis of gooseberry slices dried under open sun. *Environ. Chall.* **2023**, *13*, 100778. [CrossRef]
29. Prajapati, C.; Sheorey, T. Exploring the efficacy of natural convection in a cabinet type solar dryer for drying gooseberries: An experimental analysis. *J. Agric. Food Res.* **2023**, *14*, 100684. [CrossRef]
30. Qasem, N.A.A.; Arnous, M.N.; Zubair, S.M. A comprehensive thermal-hydraulic assessment of solar flat-plate air heaters. *Energy Convers. Manag.* **2020**, *215*, 112922. [CrossRef]
31. Jayranaiwachira, N.; Promvong, P.; Thianpong, C.; Skullong, S. Thermal-hydraulic performance of solar receiver duct with inclined punched-ribs and grooves. *Case Stud. Therm. Eng.* **2022**, *39*, 102437. [CrossRef]
32. Altun, A.H.; Gurdal, M.; Berber, A. Effects of sinusoidal turbulator in cylindrical channel on heat transfer and flow characteristics. *Maejo Int. J. Sci. Technol.* **2020**, *14*, 27–42.
33. Said, Z.; Bellos, E.; Ali, H.M.; Rahman, S.; Tzivanidid, C. Nanofluids, turbulators, and novel working fluids for heat transfer processes and energy applications: Current status and prospective. *Appl. Therm. Eng.* **2025**, *258*, 124478. [CrossRef]
34. Sureandhar, G.; Srinivasan, G.; Muthukumar, P.; Senthilmurugan, S. Performance analysis of arc rib fin embedded in a solar air heater. *Therm. Sci. Eng. Prog.* **2021**, *23*, 100891. [CrossRef]
35. Zhang, G.Z.; Sunden, B.; Xie, G. Combined experimental and numerical investigations on heat transfer augmentation in truncated ribbed channels designed by adopting fractal theory. *Int. Commun. Heat Mass Transf.* **2021**, *121*, 105080. [CrossRef]
36. Sureandhar, G.; Srinivasan, G.; Muthukumar, P.; Senthilmurugan, S. Investigation of thermal performance in a solar air heater having variable arc ribbed fin configuration. *Sustain. Energy Technol. Assess.* **2022**, *52*, 102069. [CrossRef]
37. Yu, C.; Wu, S.; Huang, Y.; Yao, F.; Liu, X. Charging performance optimization of a latent heat storage unit with fractal tree-like fins. *J. Energy Storage* **2020**, *30*, 101498. [CrossRef]
38. Zhang, C.; Li, J.; Chen, Y. Improving the energy discharging performance of a latent heat storage (LHS) unit using fractal-tree-shaped fins. *Appl. Energy* **2020**, *259*, 114102. [CrossRef]
39. Zhuang, D.; Yang, Y.; Ding, G.; Du, X.; Hu, Z. Optimization of Microchannel Heat Sink with Rhombus Fractal-like Units for Electronic Chip Cooling. *Int. J. Refrig.* **2020**, *116*, 108–118. [CrossRef]
40. *ASHRAE Standard 93–77*; Method of Testing to Determine the Thermal Performance of Solar Collectors. American Society for Heating, Refrigeration, and Air Conditioning Press: Beijing, China, 2011.
41. Mzad, H.; Bennour, F. On the effect of the transversal aspect ratio on the double-glazed solar air heater performance. *Heat Transf.* **2023**, *52*, 2771–2795. [CrossRef]
42. Osodo, B.; Nyaanga, D.; Kiplagat, J. Simulation of grain quantity, fan and solar collector sizes for an experimental forced convection grain dryer, Agric. *Food Sci.* **2019**, *6*, 98–108.

43. Hassan, H.; Elfadl, S.A.; El-Dosoky, M.F. An experimental investigation of the performance of new design of solar air heater (tubular). *Renew. Energy* **2020**, *151*, 1055–1066. [CrossRef]
44. Abushanab, W.S.; Zaywd, M.E.; Sathyamurthy, R.; Moustafa, E.B.; Elsheikh, A.H. Performance evaluation of a solar air heater with staggered/longitudinal finned absorber plate integrated with aluminium sponge porous medium. *J. Build. Eng.* **2023**, *73*, 106841. [CrossRef]
45. Kline, S.J. Describing uncertainty in single sample experiments. *Mech. Eng.* **1953**, *75*, 3–8.
46. Xia, C.; Fu, J.; Lai, J.; Yao, X.; Chen, Z. Conjugate heat transfer in fractal tree-like channels network heat sink for high-speed motorized spindle cooling. *Appl. Therm. Eng.* **2015**, *90*, 1032–1042. [CrossRef]
47. Prasopsuk, C.; Thongtip, T.; Pratumsuwan, P.; Sutthivirode, K. Experimental study of Thermo-Hydraulic Performance for Solar Air Heater with varied Cylindrical Pins using Tree-like Fractal arrangements. In Proceedings of the 11th International Conference on Technical Education, Bangkok, Thailand, 6–7 June 2024.
48. Promthaisong, P.; Eiamsa-ard, S. Fully developed periodic and thermal performance evaluation of a solar air heater channel with wavy-triangular ribs placed on an absorber plate. *Int. J. Therm. Sci.* **2019**, *140*, 413–428. [CrossRef]
49. Dong, Z.; Lui, P.; Xiao, H.; Liu, Z.; Liu, W. A study on heat transfer enhancement for solar air heaters with ripple surface. *Renew. Energy* **2021**, *172*, 477–487. [CrossRef]
50. Sexena, A.; Varun; El-Sebaii, A. A thermodynamic review of solar air heaters. *Renew. Sustain. Energy Rev.* **2015**, *43*, 863–890. [CrossRef]
51. Jia, B.; Lui, F.; Wang, D. Experimental study on the performance of spiral solar air heater. *Sol. Energy* **2019**, *182*, 78–88. [CrossRef]
52. Abubakar, S.; Umaru, S.; Kaisan, M.U.; Umar, U.A.; Ashok, B.; Nanthagopal, K. Development and performance comparison of mixed-mode solar crop dryers with and without thermal storage. *Renew. Energy* **2018**, *128*, 285–298. [CrossRef]
53. Ramirez, C.; Palacio, M.; Carmona, M. Reduced Model and Comparative Analysis of the Thermal Performance of Indirect Solar Dryer with and without PCM. *Energies* **2020**, *13*, 5508. [CrossRef]
54. Matias Gonçalves, L.; Mendoza-Martinez, C.; Alves Rocha, E.P.; Coutinho de Paula, E.; Cardoso, M. Solar Drying of Sludge from a Steel-Wire-Drawing Industry. *Energies* **2023**, *16*, 6314. [CrossRef]
55. Barik, A.K.; Mohanty, A.; Senapati, J.R.; Awad, M.M. Constructal design of different ribs for thermo-fluid performance enhancement of a solar air heater (SAH). *Int. J. Therm. Sci.* **2021**, *160*, 106655. [CrossRef]
56. An, B.-H.; Choi, K.-H.; Choi, H.-U. Heat Transfer Augmentation and Friction Factor Due to the Arrangement of Rectangular Turbulators in a Finned Air Channel of a Solar Air Heater. *Energies* **2023**, *16*, 6891. [CrossRef]

energies

MDPI

Review

Nanoparticle-Enhanced Phase Change Materials (NPCMs) in Solar Thermal Energy Systems: A Review on Synthesis, Performance, and Future Prospects

Wei Lu [1,2], Jay Wang [1,*], Meng Wang [3], Jian Yan [4], Ding Mao [5] and Eric Hu [6]

1 School of Engineering, Computer & Mathematical Sciences, Auckland University of Technology (AUT), Auckland 1010, New Zealand; lw0426@outlook.com
2 Guangdong Provincial Mechanical Engineering Experimental Teaching Center, Guangdong Technology College, Zhaoqing 266041, China
3 Department of Energy and Power Engineering, Tsinghua University, Beijing 100190, China; meng-wang@mail.tsinghua.edu.cn
4 College of Mechanical Engineering, Hunan University of Science and Technology, Xiangtan 411201, China; yanjian1988@hnust.edu.cn
5 Department of Building Environment and Equipment, School of Civil Engineering, Hefei University of Technology, Hefei 230002, China; maoding@hfut.edu.cn
6 School of Electrical and Mechanical Engineering, The University of Adelaide, Adelaide, SA 5000, Australia; eric.hu@adelaide.edu.au
* Correspondence: jay.wang@aut.ac.nz; Tel.: +64-9-921-9001 (ext. 33206)

Abstract: The environmental challenges posed by global warming have significantly increased the global pursuit of renewable and clean energy sources. Among these, solar energy stands out due to its abundance, renewability, low environmental impact, and favorable long-term economic viability. However, its intermittent nature and dependence on weather conditions hinder consistent and efficient utilization. To address these limitations, nanoparticle-enhanced phase change materials (NPCMs) have emerged as a promising solution for enhancing thermal energy storage in solar thermal systems. NPCMs incorporate superior-performance nanoparticles within traditional phase change material matrices, resulting in improved thermal conductivity, energy storage density, and phase change efficiency. This review systematically examines the recent advances in NPCMs for solar energy applications, covering their classification, structural characteristics, advantages, and limitations. It also explores in-depth analytical approaches, including mechanism-oriented analysis, simulation-based modelling, and algorithm-driven optimization, that explain the behavior of NPCMs at micro and macro scales. Furthermore, the techno-economic implications of NPCM integration are evaluated, with particular attention to cost-benefit analysis, policy incentives, and market growth potential, which collectively support broader adoption. Overall, the findings highlight NPCMs as a frontier in materials innovation and enabling technology for achieving low-carbon, environmentally responsible energy solutions, contributing significantly to global sustainable development goals.

Keywords: phase change material; solar energy; nanoparticle-enhanced; NPCM

1. Introduction

Traditional fossil fuel combustion releases substantial carbon dioxide emissions, constituting the dominant source of greenhouse gases. These emissions drive global warming, ecosystem disruptions [1], and extreme weather events [2]. Simultaneously, in accordance with the 2024 United Nations World Population Prospects, the global population is projected to reach 10.1 billion by 2061 and continue its growth trajectory, which directly causes a relentless increase in energy demand [3]. Due to the dual pressures of rapid population growth and surging energy demands, the adoption of abundant, clean, and eco-friendly renewable energy has become imperative.

Solar energy, which possesses notable advantages in terms of its environmental friendliness, abundant supply, and high degree of independence, has been widely used in energy systems such as photovoltaic (PV) power generation [4], solar thermal conversion [5,6], solar cooling [7,8], and solar still [9], thereby offering tremendous potential for sustainable energy systems. In the domain of solar energy utilization, the intermittent and non-stable nature of solar energy poses significant challenges, so high-efficiency heat storage technology has emerged as a pivotal breakthrough for enhancing the utilization efficiency of solar energy [10]. In solar energy systems, solar radiant energy is first converted into thermal energy via heat collection devices, so heat storage technology becomes essential for the spatio-temporal transfer and stable output of this energy [11,12]. Among various heat storage strategies, the latent heat storage technology founded upon phase change material (PCM) has set itself apart due to its distinctive operational mechanism [13,14]. The PCM materials listed in Table 1 were selected for comparison due to their diverse chemical structures (paraffins, salt hydrates, and eutectic mixtures), varied phase change temperatures, and widespread use in thermal storage. This selection enables the systematic evaluation of nanoparticle (NP) dispersion effects on key PCM parameters, such as thermal conductivity and latent heat.

Table 1. Properties of some PCM.

PCM	Density of Solids	Density of Liquids	Specific Heat Capacity	Latent Heat	Thermal Conductivity (TC)	Temp. of Solids	Temp. of Liquids	Melting Temp. Range	Ref.
Unit	kg/m^3	kg/m^3	J/(kg·K)	kJ/kg	W/(m·K)	°C	°C	°C	
RT28	880	770	2000	179	0.2	25.98	29.6	3.63	[15]
RT28HC	-	-	2000	250	0.2	27	28	-	[16]
RT35	860	-	2000	182	0.3	-	-	1.5	[17]
RT42	880	760	2000	165	0.2	-	-	4	[18]
RT50	876	710	2100	160	0.2	-	-	7	[19]
RT55	880	770	2000	170	0.2			6	[20]
RT58	840	-	2100	180	0.2	321	335		[21]
60% $NaNO_3$ + 40% KNO_3	1980	-	1575	140	0.59	495	519	-	[22]
53% $NaNO_3$ + 40%KNO_3 + 7%$NaNO_2$	1800 at 400 °C	-	1500 at 400 °C	-	-	-	-	-	[23]
KNO_3	2109	-	953	95	0.5	-	-	-	[24]
Dodecyl sulfate	-	-	-	200	-	-	-	-	[25]
Lauric acid	940	885	2285	-	0.15	40.48	42.48	2	[26]
HS 29	1681	1530	1510 (S) 2620 (L)	190	0.382 (L) 0.478 (S)	-	-	-	[27]
Coconut Oil	-	-	2350 (S) 3230 (L)	70–100	-	-	-	-	[28]
Eicosane	810	770	1900 (S) 2200 (L)	241	0.4 (S) 0.157 (L)	-	-	-	[29]
$C_{28}H_{58}$	-	806	-	254	-	-	-	-	[30]

The incorporation of high-performance nanoparticles, such as Cu for reducing melting time, Al_2O_3 for accelerating solidification, and graphite for enhancing thermal conductivity (TC), into phase change materials to form nanocomposite phase change materials can significantly improve solar energy conversion efficiency and thermal storage capacity [31,32]. Their core advantages are mainly reflected in the following aspects. The high heat storage density enables the efficient accumulation of thermal energy within confined spaces, thereby maximizing the utilization of available volume [33,34]. The isothermal behavior during the phase change process ensures the thermal stability of the heat storage system, which is crucial for solar-thermal applications that require a consistent heat source [35,36]. Meanwhile, the minimal phase change temperature difference significantly mitigates heat losses during the heat storage cycle, leading to enhanced energy utilization efficiency [37].

Based on the analysis above, NPCM materials have been widely and intensively studied and analyzed in the scientific research community. However, the research gaps lie in the lack of a review paper that summarizes the latest applications of NPCM in high-temperature thermal energy storage systems, particularly in terms of energy performance through mechanism analysis, simulation, and algorithm. Therefore, the general aim of this review article is to systematically introduce various research and analysis methods from the perspective of NPCM models.

This review is primarily divided into six sections. Section 1 covers the latest developments in NPCMs reported in recent literature. Section 2 focuses on introducing various structural characteristics of NPCMs. Section 3 concentrates on the mechanisms and simulation analysis methods of NPCMs, spanning from the material's microscopic structure to its comprehensive applications. Section 4 demonstrates the algorithm-driven optimization of NPCMs, including machine learning methods and uncertainty analysis. Section 5 discusses techno-economic analysis, including cost-benefit analysis, and policy incentives. Finally, the conclusions and future research directions in the field of NPCMs are proposed in Section 6.

2. Nano-Enhanced Phase Change Material (NPCM)

NPCM has high heat storage density for efficient limited space energy storage, isothermal phase change thermal stability, and a minimal phase—change temperature difference that cuts heat storage cycle losses. These unique properties give PCM-based latent heat storage technology extensive application potential across a wide range of fields, including solar thermal power generation, building energy-efficiency improvements, and electronic thermal management. As such, it represents a crucial technological pathway for addressing the intermittent issue of solar energy and facilitating the efficient utilization of renewable energy resources. From a structural perspective, NPCM can be classified based on the configuration of the nanomaterial within the composite: thermophysical NPCM (dispersed NPs), photonic NPCM (NPs with light-responsive surfaces), NPCM fluids (NP suspensions), nano-encapsulated PCM (core-shell nanostructures), and phase-engineered nanomaterials (NPs with designed phase interfaces).

2.1. *Thermophysical NPCM*

Thermophysical NPCM enhances PCM properties by incorporating nanoparticles (NPs) like graphene and metal compounds, which strengthen thermal conductivity [38]. Abdolahimoghadam designed a bio-based NPCM composed of coconut oil, beeswax, and graphene-Cu hybrid NPs. Research shows that the addition of graphene nanosheets (GNP) to the material can significantly increase the melting rate by 5.05% [39]. Gopi selected bismuth oxychloride, CuO, and their mixture (composite) as NPs and stearic acid as the PCM. They found that the latent heats of NPCMs containing 0.5 wt% bismuth oxychloride, CuO, and the composite were 204.6 $kJ\cdot kg^{-1}$, 198.3 $kJ\cdot kg^{-1}$, and 201.7 $kJ\cdot kg^{-1}$, respectively, and the thermal conductivities increased to 0.19 $W/(m\cdot ^{\circ}C)$, 0.22 $W/(m\cdot ^{\circ}C)$, and 0.20 W/(m·°C), respectively, which were 5.9%, 29.4%, and 17.6% higher than those of the pure PCM [40].

2.2. Photonic NPCM

Photonic NPCM enhances photovoltaic-thermal system performance when combined with nanofluids, improving light absorption [41]. Ahmed used 0.6% silicon carbide-enhanced nanofluid and silicon carbide to enhance the PCM in photovoltaic systems, thereby improving the photovoltaic-thermal efficiency. The research results show that the maximum energy power increased from 17.512 W to 26 W [42]. Under the working conditions of 47.94 °C and a flow rate of 5 L/min, Maseer studied a semi-circular absorber-tube-type solar photovoltaic-thermal (PVT) system with combined cooling of nanofluid and NPCM. The research results of this study show that the power-generation efficiency was 12.70%, the heat generation efficiency was 78.99%, and the total efficiency was 91.83% [43]. Máñez has developed a simple yet effective reconfigurable all-dielectric metasurface design that enables full control of the electromagnetic response, including electrical, magnetic, and electromagnetic coupling, by changing the refractive index of the PCM layer [44].

2.3. NPCM Fluids

NPCM fluids are liquid-phase composite materials formed by dispersing NPs, particularly molten salt compounds, within PCM or base fluids, demonstrating superior specific heat capacity. Sathishkumar formed NPCM clusters by using the fatty-acid-based PCM OM05 and a GNP suspension, which led to research results showing that for the Oleic-acid-modified (OM) 05-based NPCM with 0.75 wt% GNP, its TC increased significantly at 30 °C, the total solidification time was shortened by 32%, and as a result, a more efficient cooling and charging rate was achieved (Shown in Figure 1) [45]. Baiju devised a conical-fin latent heat storage system utilizing HS-89 hydrated salt as the matrix, incorporating 0.5 wt% graphene nanoplatelets (GNP) and 0.5 wt% nano-alumina (nAl_2O_3), achieving a 56% reduction in charging time [46]. Wu added 5 wt% of OM and nanoAl (3:1) to binary alkanes, which caused TC to increase to 0.31 W/(m·K) and the melting time to decrease by 38.3% [47].

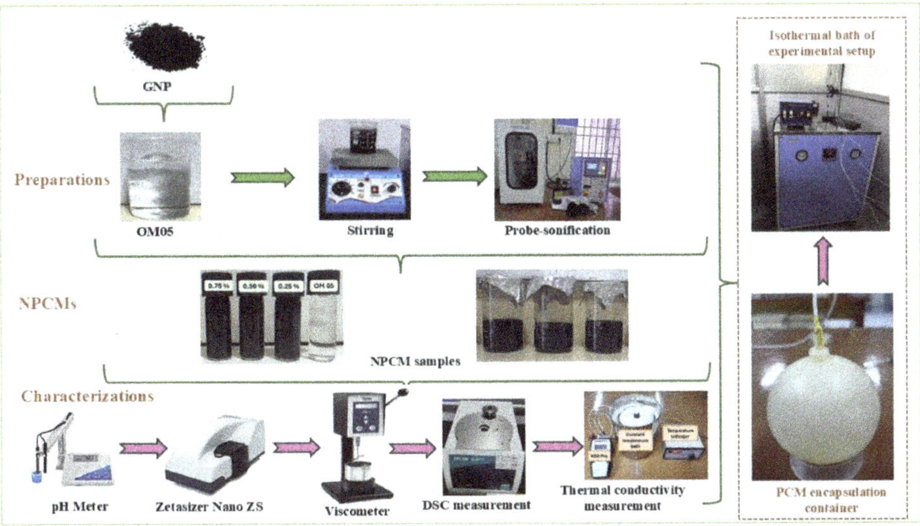

Figure 1. The preparation process of the OM05-based NPCM. (Reproduced with permission from [45], Elsevier, 2025).

2.4. Encapsulated NPCM

Encapsulated NPCM is a composite PCM that combines NPs or nanomaterials with PCM, which can effectively prevent liquid leakage during equipment operation. Shijina added the developed GNE-PCM to an ethylene-glycol-water mixture to prepare a graphene nano-encapsulated composite material, and the research showed that TC, specific heat capacity, and cooling efficiency of the material were all improved [48]. To study the effect of GNP on the thermal properties of NPCM for encapsulated heat sinks, Sivashankar added different concentrations of GNP to the OM35 base to prepare NPCM. The research results showed that 0.5 vol% of GNP led to a significant increase in both TC and thermal diffusivity of NPCM [49]. Sutheesh encapsulated tetradecan-1-ol with polymethyl methacrylate to form NPCM for thermal regulation of Lithium-ion Battery (LIB). The research results showed that for LIB with added NPCM, heat generation was reduced, and both the convective heat transfer coefficient and the discharge rate were improved [50].

2.5. Phase-Engineered Nanomaterials

Phase-Engineered Nanomaterials refer to the optimization of the physical and chemical properties of materials themselves by designing the crystal-phase structure of nanomaterials [51,52]. Liu used palladium-selenide as a model system, which was then transformed into $Pd_{17}Se_{15}$ and Pd_4Se phases by thermally regulating the component ratio. Through precise control of the electrode thickness and spacing, different phase-configuration arrangements could be obtained, which enabled multiple in-situ functions and extended to 29 elemental combinations between metals and chalcogens, as reported in reference [53]. Zang developed an amorphous-crystalline CuxS–Ag_2S composite with a significant temperature gradient, which achieved a synergistic enhancement of optimal thermoelectric therapy and multi-enzyme activity, as indicated in reference [54]. Yang developed a bismuth-doped ruthenium oxide single-atom alloy oxide catalyst, which achieved a current density of 1.0 A cm^{-2} at only 1.59 V under industrial conditions, according to reference [55]. Some NPCM performance improvements are listed in Table 2.

Table 2. Properties of some NPCM.

Reference	Materials Set-Up	Analysis of Performance Improvement
Abdolahimoghadam et al. [39]	Graphene-Cu NPs are added to the bio-nano-based materials of coconut oil and beeswax.	The melting time of the four-tube system with 2 wt% Graphene-Cu hybrid NPs added is shortened by 80.25%, and the solidification time is shortened by 74.28%.
Kannan et al. [40]	CuO NPs are added to stearic acid.	The enhanced TC of NPCM with 0.5 wt% CuO NPs is 29.4% higher than that of pure PCM, and the congruent melting rate is increased by 31%.
Ahmed et al. [42]	SiC NPs are added to water to prepare nanofluids, and SiC NPs are added to paraffin to prepare NPCM.	When nanofluid with 0.6 vol% and NPCM with 1 vol% are added to the collector, the thermal efficiency improves by 86.2%, the electrical energy increases by 11.24%, and the maximum photovoltaic power output grows from 17.512 W to 26 W.
Maseer et al. [43]	0.1% of Al_2O_3 NPs are introduced into water to prepare nanofluids. Subsequently, 0.65% of paraffin is introduced into the nanofluids to prepare NPCM.	The PVT system with NPCM coolant achieves a maximum power of 73.25 W, with a thermal efficiency of 70.5%, an electrical efficiency of 16.9% and a maximum overall efficiency of 84.23%.
Máñez-Espina et al. [44]	Ge, Ge2Sb2Te5, CaF_2	Energy coupling efficiency, energy efficiency, and unidirectional absorption asymmetry can all be improved.
Sathishkumar et al. [45]	0.75% of GNP is added to OM to prepare NPCM.	TC of the cold-heat energy storage system has been significantly increased by 54%, and the total freezing time has been significantly reduced by 32%.
Baiju et al. [46]	Hydrated salt is prepared with 0.5 wt% GNP and 0.5 wt% nano-Al.	The charging time of the latent heat storage device that uses conical fins and NPCM in combination is shortened by 56%, and the annual life-cycle cost is 162.3 euros per year.
Wu et al. [47]	Binary alkane containing 5 wt% modified nano-Al_2O_3.	TC of NPCM with 5 wt% modified NPs is 0.31 W/m·K, and the melting time is reduced by 38.3%.

Table 2. *Cont.*

Reference	Materials Set-Up	Analysis of Performance Improvement
Shijina et al. [48]	Graphene flakes are affixed to polystyrene spherical balls encapsulated with a composite PCM of 80% paraffin and 20% octadecane, and then these are dispersed in an ethylene-glycol-water mixture to obtain nanofluid.	The subcooling degree of the coolant is reduced by 50%, and TC and viscosity of the nanofluid increase by 14.7% and 19%, respectively, and the heat capacity is increased by 56.7%.
Sivashankar et al. [49]	Organic PCM is prepared with 0.5 vol% GNP and OM35.	TC of the solid phase has increased by 93%, the conductivity of the liquid phase has increased by 78%, the thermal diffusivity of the solid phase has increased by 119%, the thermal diffusivity of the liquid phase has increased by 124%, the starting temperature has decreased to 31.21 °C, the peak temperature has decreased to 37.46 °C, the viscosity has increased by 76%, and the latent heat has increased by 19.82%.
Sutheesh et al. [50]	Tetradecan-1-ol serves as the PCM, while deionized water acts as the base fluid. The shell is constituted by polymethyl methacrylate, and MgO functions as the filling material.	The lowest Reynolds number during the highest cell discharge optimizes the phase transition of PCM. When the Reynolds number is 121.09 and the discharge rate is 5C, the convective HTC increases by 4.53 times and 6.15 times for 1% and 4% NPCM, respectively.
Liu et al. [53]	Pd–Se	Through regulating the thickness and the distance between electrodes on the apparatus, we bring about emerging phases. These phases display superconducting orders or can be manipulated to produce an ultralow resistance contact for 2D semiconductor transistors. Additionally, the phase engineering carried out on-device enables in-situ and customized synthesis of outstanding electrocatalysts, which has the ability to surpass the traditional modulation limit of composition proportion.
Zang et al. [54]	Cu_xS-Ag	Abundant active sites are furnished for improving the therapeutic activity by the dangling bonds on the amorphous Cu_xS surface. Moreover, through non-radiative relaxation, the amorphous state boosts the photothermal effect.
Yang et al. [55]	Bi-RuO_2	The Bi-RuO_2 single-atom alloy oxide catalyst demonstrates a relatively low overpotential of 192 mV. It also shows excellent stability lasting over 650 h at a current density of 10 mA cm^{-2}. These properties make possible a practical PEMWE which requires merely 1.59 V to achieve a current density of 1.0 A cm^{-2} under industrial conditions.

3. Analytical Methods for NPCMs

In this section, we will concisely discuss the key methods pivotal for NPCM performance analysis. These mainly include Mechanism-oriented Analysis, which delves into fundamental interactions. Then there is Simulation-based Analysis, using computational models to accurately predict properties. Finally, Algorithm-driven optimization leverages algorithms to efficiently find optimal configurations for NPCM. Each method significantly contributes to a comprehensive understanding and enhances NPCM performance.

3.1. Mechanism-Oriented Analysis

Mechanism-Oriented Analysis is a crucial approach in understanding the behavior of NPCM in high-temperature thermal energy storage systems. This method delves deep into the fundamental mechanisms underlying the material's performance, aiming to establish a clear connection between its microscopic structure and macroscopic properties. At the microscopic level, NPCM consists of a base PCM embedded with NPs. These NPs can significantly alter the properties of PCMs. For example, they may enhance heat transfer by increasing the effective TC [56]. The mechanism-oriented analysis aims to explore how the size, shape, and distribution of these NPs interact with the molecular structure of the PCM, studying the interactions between NPs and the PCM which can affect the atomic morphology, pore structure, and chemical bond interactions within the PCM, thereby influencing its heat storage and heat release capabilities [57,58]. By understanding these mechanisms, researchers can predict and optimize the performance of relevant NPCM. Commonly used mechanism-analysis methods for studying NPCM include molecular dynamics (MD) simulation, density functional theory (DFT), and stochastic simulation approaches, for which a review and analysis of the most recent relevant literature will be conducted hereinafter.

3.1.1. Molecular Dynamics (MD) Simulation

MD simulation has become an important tool for studying NPCMs with excellent performance. In research, MD enables researchers to delve deep into the complex atomic and molecular-level behaviors underlying the macroscopic properties of materials [59]. As shown in Figure 2, the corresponding atomic arrangement characteristics can be observed. For composite materials that are typically composed of matrix PCMs embedded with NPs, MD can accurately model the interactions between PCM molecules and NPs, which are of crucial importance as they significantly influence the thermal, mechanical, and phase change characteristics of the composite materials [60]. Additionally, MD assists in understanding the phase change processes in NPCM, which involve the complex phenomenon of PCMs transitioning from the solid to the liquid state or vice versa [61]. With the help of MD, researchers can track the energy changes, atomic displacements, and intermolecular forces during these phase transitions. The analysis of these principles is vital for optimizing the design of NPCM-based thermal energy storage systems, as it enables the optimization and regulation of phase change temperature, latent heat capacity, and other key parameters. In conclusion, MD plays a key role in uncovering the change mechanisms of NPCM and promoting the development of advanced thermal energy storage solutions. The following is an overview of the research on the application of the latest MD analysis method to the analysis of NPCM.

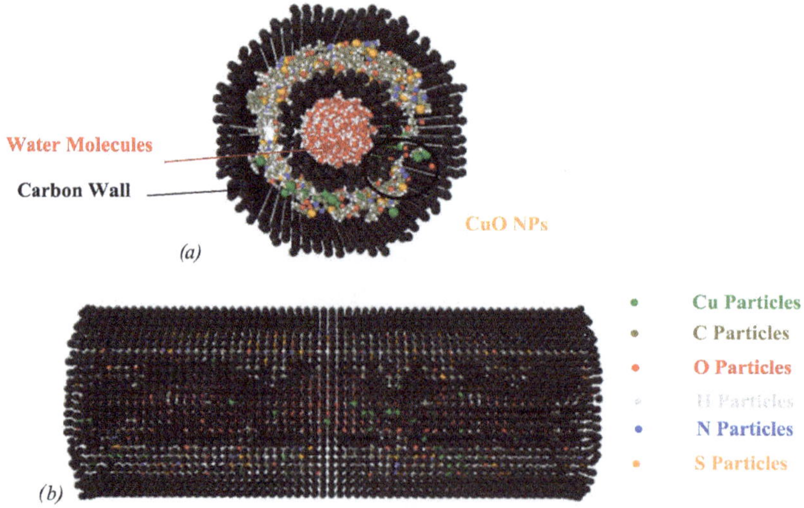

Figure 2. Schematic diagram of the atomic arrangement of SA-paraffin of CuO NPs. (Reproduced with permission from [60], Elsevier, 2025). (**a**) front view; (**b**) side view.

Cu NPs are frequently added to PCMs with the aim of enhancing the latter's thermal properties. Cu NPs are frequently added to PCMs due to their balanced cost, thermal conductivity, and chemical stability. Karimipour added Cu NPs to paraffin and analyzed the trends of TC and energy-efficiency transformation of paraffin-Cu-NFs placed between parallel plates under the influence of different panel temperatures [59]. Liu also used the same materials in his study, where it was found that when the concentration of Cu NPs increased from 1% to 7%, the maximum density decreased by 1.62%, the maximum velocity increased to 0.00493 Å/fs, the maximum temperature increased by 3.9%, the heat transfer flux increased by 2.1%, the conductivity increased to 0.68 W/m·K, the charging time decreased by approximately 1.42%, and the discharging time decreased by approximately 0.14%, yet as the concentration of Cu NPs continued to increase to 10%, these parameters rebounded [62]. Singh's research also confirmed that different concentrations of Cu NPs had an impact on both TC and heat flux, but the turning-point concentration in the research results was 3%, which differed to some extent from previous studies, a difference caused by factors such as the setup of the simulation model and the force-field analysis method [63]. Cao carried out molecular-dynamics simulation and analysis using two types of PCM (hydrocarbon-water composite and paraffin) and three wall-metals (Pt, Cu, and Al), and found that the thermal performance of the combination of paraffin and Cu NPs was superior to that of other combinations [64]. Huang added Cu NPs to Sodium sulfate decahydrate and analyzed the heat and mass transfer process and phonon process [65].

CuO is also a high-quality NP. Gao analyzed the changing trends of the maximum velocity, temperature, TC, and heat flux of the SA/PCM/CuO nanostructure within a cylindrical pipe [66]. Yang considered the influence of the concentration of Cuo NPs on the thermal-performance parameters of the composite material, including density, maximum velocity, maximum temperature, heat flux, TC, charging time, and discharging time [67]. Ru discovered that when the atomic radius increased from 4 Å to 10 Å, the maximum density decreased to 0.0825 atoms/Å^3, the maximum velocity decreased, the maximum temperature dropped, TC decreased by 1.57 W/m·K, the charging time shortened by 6.09 ns, and the discharging time decreased by 8.28 ns [68]. An analysis was also conducted to evaluate the influence of the radius of NPs on these parameters. Notably, the charging time increased to 6.28 ns, which was contrary to the previous research results, so the study on the influence mechanism of the atomic radius on the charging time of the material still warrants further in-depth analysis [69]. Yu analyzed the influence of CuO NPs on the density, TC, and melting enthalpy of the lauric-acid PCM (as shown in Figure 3) [70]. Fang added SiO_2 NPs to the fundamental molten salt (a mixture of Na_2CO_3 and K_2CO_3) and analyzed the thermal conductivity and shear viscosity using MD [71]. The main materials and research contents of the above-mentioned study are listed in Table 3.

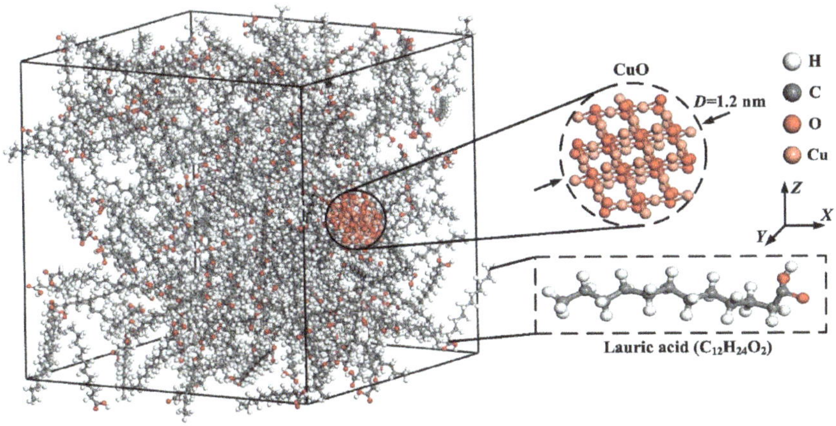

Figure 3. Schematic diagram of Lauric acid-CuO model (Reproduced with permission from [70], Elsevier, 2024).

Table 3. The main materials and research contents about MD.

Reference	Materials Set-Up	Analysis of Performance Improvement
Karimipour et al. [59]	Cu/paraffin	TC and energy-efficiency transformation
Liu et al. [62]	Cu/paraffin	7% Cu NPs: density, velocity, temperature, heat transfer flux, TC, charging time, discharging time
Singh et al. [63]	Cu/paraffin	3% Cu NPs: density, velocity, temperature, heat transfer flux, TC, charging time, discharging time
Cao et al. [64]	hydrocarbon-water composite, and paraffin/platinum, Cu, and Al)	The combination of paraffin and Cu NPs exhibits the highest cost-performance ratio in terms of thermal performance
Huang et al. [65]	Cu, Sodium sulfate decahydrate	TC, radius, diffusion coefficient, phonon density of states, phonon resonance, Debye temperature
Gao et al. [66]	Silica aerogel (SA)/PCM/CuO	velocity, temperature, TC, and heat transfer flux
Yang et al. [67]	SA/paraffin/CuO	The concentration of Cuo NPs: density, maximum velocity, maximum temperature, heat transfer flux, TC, charging time, and discharging time
Ru et al. [68]	SA/paraffin/CuO	The radius of Cuo NPs: density, velocity, temperature, TC, charging time, discharging time
An et al. [69]	SA/paraffin/CuO	The charging time increased to 6.28 ns
Yu et al. [70]	CuO/lauric-acid	density, TC, and melting enthalpy
Fang et al. [71]	a mixture of Na_2CO_3 and K_2CO_3, SiO_2	Thermal conductivity, shear viscosity, test factors, temperature

3.1.2. Density Functional Theory

In the research of NPCMs, DFT is an important quantum-mechanical method. At its core, it takes the electron density as the fundamental variable, positing that the ground-state energy of a system is a functional of the electron density [72]. Solving the relevant equations to determine the electron density significantly simplifies the computational complexity of multi-electron systems [73]. In the study of NPCM, DFT can precisely analyze the interactions between the base PCM and NPs at the atomic scale [74]. This helps in understanding how these interactions alter the heat transfer capabilities and other properties of NPCM. Moreover, DFT can delve deep into its phase change processes, facilitating the optimization of its applications in thermal energy storage and other aspects. Overall, DFT provides crucial support for unlocking the potential of NPCM in various applications.

As shown in Figure 4, Liang employed DFT to reveal enhanced van der Waals interactions and elevated non-covalent energy in acetylated lignin/fatty acid systems, achieving 40.8% particle yield, 57.2% fatty acid loading, and 70 °C phase transition temperature [75]. Zhang used DFT to demonstrate that graphitized carbon coating transforms non-uniform electric fields around Ni_0 NPs into spherical uniformity under solar irradiation [76]. Yao applied DFT to confirm 8.9 nm mesopore concentration in NPCM skeletons, aligning with scanning electron microscope (SEM) data [77]. Tan leveraged DFT-based quantum chemistry to prove modified lotus root starch enhances hydrogen bonding in sodium acetate trihydrate–sodium thiosulfate pentahydrate clusters, boosting water retention and thermal storage [78]. Yuan utilized DFT to characterize MIL-101(Fe) pore structures, identifying dominant 1.19 nm micropores and 15.17 nm mesopores [79]. Additional DFT studies include Wang's pore distribution analysis of a Zeolitic Imidazolate Framework-8 based biocompatible porous material of micron-size [80], Zeng's calculation of higher formation energy for C-substituted Sb_7Se_3 superlattices [81], and Xin's electronic density of states examination of Design and synthesize a composite PCM of $Na_2HPO_4{\cdot}12H_2O$ (DHPD)—vermiculite nanosheets (VMTNS) systems [82]. The main materials and research contents of the above-mentioned study are listed in Table 4.

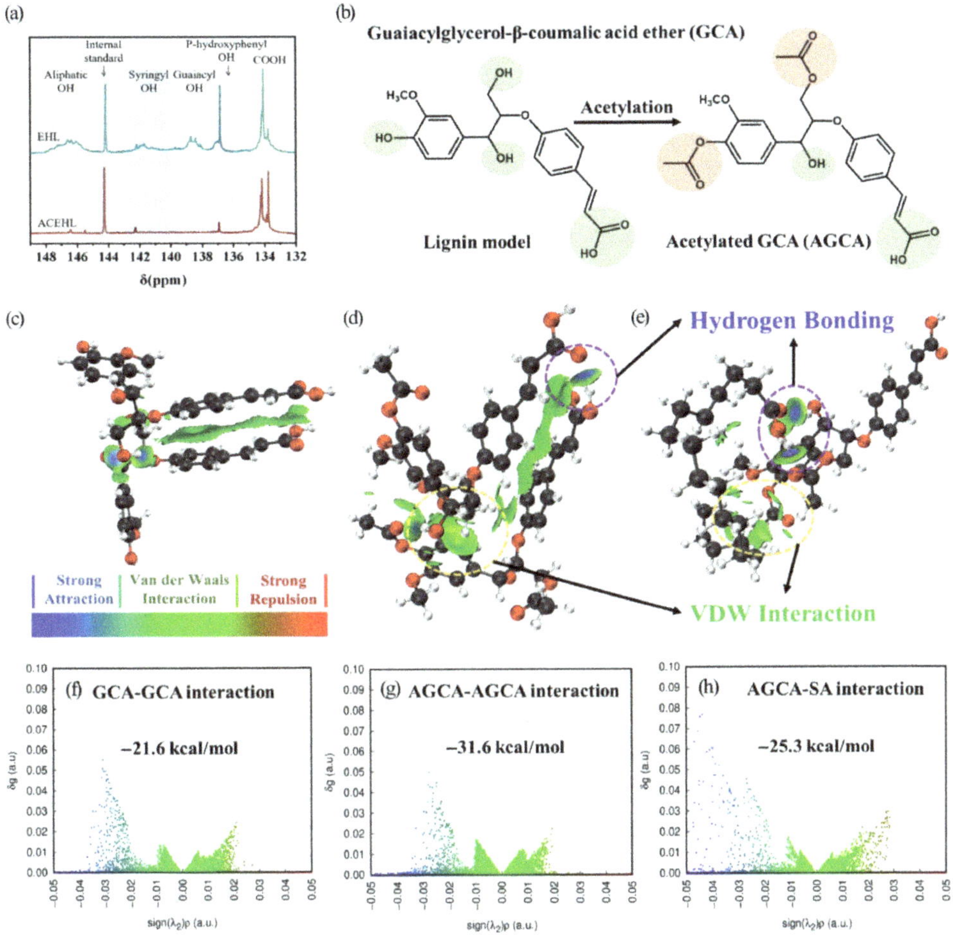

Figure 4. Analysis diagram of the formation mechanism of lignin-stearic acid capsules (Reproduced with permission from [75], Elsevier, 2025). (**a**) ^{31}P NMR spectra of EHL and ACEHL; (**b**) acetylation of GCA; (**c**–**e**) Isosurface and (**f**–**h**) scatter graph of independent gradient model based on Hirshfeld partitioning (IGMH) of the unacetylated lignin (GCA) dimer (**c**,**f**), acetylated lignin (AGCA) dimer (**d**,**g**), AGCA combined with stearic acid (**e**,**h**).

Table 4. The main materials and research contents about DFT.

References	Materials Set-Up/References	Research Methods	Research Points
Liang et al. [75]	Lignin fatty-acid capsules were prepared using an anti-solvent method after lignin was modified with a 6:4 ethanol-ethyl acetate mixture.	Characterization: Transmission electron microscope (TEM), Dynamic light scattering, Fourier transform infrared spectroscopy (FTIR), Nuclear magnetic resonance, X-ray photoelectron spectroscopy (XPS), Differential scanning calorimetry (DSC), Thermogravimetric analysis (TGA), DFT. Model: MD, Independent gradient model based on Hirshfeld Partitioning.	The solubility matching problem between lignin and stearic acid. The interaction relationship between lignin and stearic acid. A self-assembly mechanism for acetylated lignin and stearic acid. The heat storage performance, supercooling degree, and thermal stability of the modified lignin-stearic acid nanocapsules

Table 4. *Cont.*

References	Materials Set-Up/References	Research Methods	Research Points
Zhang et al. [76]	Octadecanol (ODA) is encapsulated into the hierarchically porous carbon structure infiltrated with Ni@graphite (Ni@C/C).	Characterization: SEM, nitrogen adsorption-desorption isotherm, TEM, Energy-Dispersive Spectroscop, X–ray diffraction, raman spectroscopy XPS, FTIR, TGA, DSC, UV-Vis-NIR spectroscopy, DFT. Models: Analytical method for phase change kinetics, MD, Zhishen Mo method, mean square displacement.	The crystallization characteristics of the PCM and Ni@C/C. The interaction energy, radius of gyration, kinetic energy, and diffusion coefficient between the PCM and Ni@C/C. The crystallization time and relative crystallinity of the pristine ODA and ODA/Ni@C/C. The non-isothermal charging and discharging performance, Localized Surface Plasmon Resonance effect, photothermal conversion ability, heat release characteristics, and extreme cold-heat management capacity of ODA/Ni@C/C-4.
Yao et al. [77]	The radio-frequency polymer framework (in-situ polymerized in the presence of a sodium hydroxide catalyst) and the phase change component (resorcinol (R) and furfural (F) are dissolved in molten phase change component).	Characterization: FTIR, SEM, Nitrogen adsorption-desorption isotherm, Brunauer–Emmett–Teller (BET), DSC, TC Test, Heat conductivity coefficient test, Leakage rate analysis, DFT, Flexural strength test, Temperature control test.	The adsorption effect between the RF framework and phase change component. Analyzed the mesoporous nanostructure characteristics of RF-PF. The mass retention rate, shape stability, TC, and contact thermal resistance of the RF- composite PCM specimens. The temperature and temperature difference of the RF- composite PCM plates
Tan et al. [78]	The composite PCM is prepared by using high-density carbon-nanofiber encapsulation of modified broad-leaved root starch enhanced with carbon nanofibers and graphene nanosheets and choosing sodium acetate trihydrate and sodium thiosulfate pentahydrate.	Characterization: DSC, SEM, X–ray diffractometer (XRD), TC analysis, Viscosity analysis, DFT.	The pore structure characteristics, interactions, solidification characteristics, crystallization characteristics, TC, melting enthalpy, and molecular structure features of the composite material.
Yuan et al. [79]	Tetradecanol is confined in the micro-mesopores of the MIL-101(Fe) nanoporous supporting matrix.	Characterization: BET, DFT, Porosity analysis, SEM, XRD, FTIR, DSC, TGA, Thermal cycling stability, Phase change loading rate test.	The interaction between MIL-101(Fe) and tetradecanol. The pore characteristics, adsorption characteristics, and crystallization characteristics of the composite PCM. The influence of different concentrations of tetradecanol on the phase change enthalpy transformation process of the composite material, weight loss rate and weight loss temperature, cycle stability, melting process characteristics, hydrothermal tolerance, and material loss degree.
Wang et al. [80]	Porous Zeolitic Imidazolate Framework-8 Microcrystal/Doxorubicin Hydrochloride @ Tetraol-Polydopamine Composite Material	Characterization: FTIR, TEM, XRD, SEM, Thermal infrared imaging, BET, DFT, Dynamic light scattering, Zeta potential analysis, Nanodrop one spectrophotometer	Drug release behaviors at different pH values and temperatures. The morphological characteristics, pore size characteristics, drug-loading capacity, and photothermal conversion ability of the composite material.
Zeng et al. [81]	Prepare C/Sb_7Se_3 thin film using Sb_7Se_3 and C materials.	Characterization: Magnetron sputtering instrument, XRD, Raman spectroscopy, near-infrared spectrophotometer, XPS, X–ray reflectometry, Atomic force microscope, Ion thinning process, Dual-beam focused ion beam process, TEM, Model: Utilize a signal system constructed from an arbitrary waveform generator, a digital source meter, and a micro-control probe observatory. Employ the Vienna ab initio simulation package, the projector-augmented wave potential, the generalized gradient approximation of Perdew–Burke–Ernzerhof, and DFT-D2.	Amorphous resistance characteristics, crystalline resistance characteristics, crystallization activation energy, crystallization process, grain size, interactions, and wide bandgap of the thin film

Table 4. *Cont.*

References	Materials Set-Up/References	Research Methods	Research Points
Xin et al. [82]	DHPD-VMTNS with both surface and confined space using $Na_2HPO_4 \cdot 12H_2O$.	Characterization: X–ray fluorescence, field emission scanning electron microscope, TEM, Atomic force microscopy, Zeta potential, XRD, FTIR, DSC, Constant-temperature water-bath method, TGA, DFT. Model: MD, optimization method of CASTEP software package, universal force field, analysis of electrostatic force and van der Waals force, Gamma analysis based on ultrasoft pseudopotential.	The interaction and diffusion mechanism between VMT and DHPD. The adsorption characteristics, crystallinity, compatibility, morphological features, phase change and heat storage properties of DHPD in VMTNS.

3.1.3. Characterization Method

Characterization methods, which involve using experimental equipment to collect relevant signals and conduct principled analyses of specific material properties, are employed to better observe the microstructure and performance characteristics of NPCMs when applied to their analysis. This provides crucial support for developing new high—heat and high—stability NPCMs. Multiple experimental approaches are often combined with modeling techniques for analyzing a single NPCM material. Common characterization techniques include SEM, TEM, XRD, FTIR, DSC, TGA, and Temperature history methods (T-history methods). SEM is commonly employed to observe the microstructure and offers a high-resolution surface view [83,84]. TEM is also utilized for microstructure observation, but it enables internal-structure analysis at the micro-scale [85,86]. XRD is utilized to determine the crystal structure of NPCM components [87,88]. FTIR identifies functional groups by detecting chemical bond vibrations [89,90]. DSC was employed to characterize its unique thermal properties, including thermal hysteresis, phase transition temperature, latent heat, and thermal stability [91,92]. TGA is employed to investigate its phase change characteristics [93,94]. The T-history method is used to analyze the temperature curves of NPCM and further derive the thermophysical properties of the materials [95,96]. The main materials and research contents of the above-mentioned study are listed in Table 5.

Table 5. The main materials and research contents about characterization methods.

Research Method	Research Content	Ref.
SEM	Observe the surface morphology and microstructure of NPCM and assess its particle size and distribution.	[83,84]
TEM	Analyze the internal microstructure of NPCM and examine its crystalline structure and defects.	[85,86]
XRD	Identify the crystalline composition and crystallinity of NPCM and investigate its phase transition.	[87,88]
FTIR	Detect functional groups and chemical bonds in NPCM and investigate its chemical structure and interactions.	[89,90]
DSC	Analyze the thermal properties of NPCM, including thermal hysteresis, phase transition temperature, latent heat, and thermal stability.	[91,92]
TGA	Assess the thermal stability and decomposition temperature of NPCM and investigate its thermal decomposition process and composition.	[93,94]
T-history methods	Analyze the temperature curves of NPCM and research the thermophysical properties of the materials.	[95,96]

3.2. Simulation-Based Analysis

Simulation-based analysis plays a pivotal role in the study of NPCM within high-temperature thermal energy storage systems. This approach utilizes computational models to replicate and analyze the complex behaviors of NPCM, offering insights that are often difficult to obtain through experimental means alone. In Simulation-Based Analysis, various physical phenomena such as heat transfer, phase transitions [97], and fluid flow within the NPCM are numerically modeled. For heat transfer, models consider the enhanced TC imparted by the NPs in NPCM. They simulate how heat diffuses through the composite material, taking into account the interactions between the base PCM and the NPs. When it comes to phase transitions, simulations can precisely capture the temperature-dependent behavior of NPCM. By inputting the thermo-physical properties of the components, these models can predict the onset and duration of phase changes, which are crucial for understanding the energy storage and release processes. Moreover, in systems where NPCM might be used in flowing configurations, fluid-flow simulations help analyze how the presence of NPs affects the flow characteristics. This information is valuable for designing efficient heat exchange systems. Overall, Simulation-Based Analysis provides a cost-effective and time-efficient way to explore different scenarios, optimize NPCM-based systems, and guide experimental research. Commonly used simulation-based analyses for studying NPCM include finite element analysis (FEA), computational fluid dynamics (CFD), stochastic simulation approaches, and multiphysics coupling models, for which a review and analysis of the most recent relevant literature will be conducted hereinafter.

3.2.1. Finite Element Analysis

FEA discretizes complex structures into finite elements to build simulation models, predicting the temperature field, stress distribution, and energy transfer efficiency of materials during phase transitions [98,99]. FEA can quantitatively analyze the impact of NP dispersibility on the thermal performance parameters of PCMs, thereby optimizing the design of porous encapsulation to reduce the leakage rate. Meanwhile, it simulates the regulatory mechanism of nanoscale interfacial thermal resistance on phase—change kinetics at the micro—scale, providing theoretical support for the efficient thermal management of NPCMs in scenarios such as electronic heat dissipation and intelligent buildings. The following is an overview of the research on the application of the latest FEA analysis method to the analysis of NPCM.

Gürsoy investigated the thermal performance of RT58-based composites incorporating spherical, square, and elliptical dimpled fins with varied Al_2O_3 NP concentrations [100]. Hekmat explored heat transfer enhancement in a spiral circular thermal storage system (as shown in Figure 5) using RT-50 with differently dosed NPs [101]. Dora developed a hierarchical composite by blending nano-silica/coconut fiber with decanol-ethanol-treated exfoliated vermiculite PCM, subsequently integrating it into foam concrete for building thermal analysis [102]. Yang engineered graded-porosity foam Cu/paraffin NPCMs to optimize structural configurations based on porosity uniformity, thermal response, and energy storage dynamics [103]. Lastly, Zhao quantified Herringbone Graphite Nanofiber agglomeration effects in paraffin via spherical clustering and linear percolation models, revealing enhanced TC in linear networks under high interfacial conditions [104]. Sun used artificially cultivated diatom microcrystals to modulate carbilose nanofibers to fabricate porous frameworks to fabricate PCMs with stable shapes, and modeled and studied the influence of pore properties on leakage resistance (as shown in Figure 6) [105]. Çiçek established a symmetric radiator model to analyze the effects of different NP concentrations and heat flux on thermal performance [56]. Saleh modeled the flexible structure of a square cavity using the Arbitrary Lagrangian–Eulerian method and analyzed the effects of temperature, Stefan number, and other factors on the system's overall performance [106]. Egami added CuO nano-powders to PCM in a rectangular container with H-shaped fins, analyzing the effects of nano-powders' concentration and shape factor on freezing performance [107]. The main materials and research contents of the above-mentioned study are listed in Table 6.

Table 6. The main materials and research contents about FEA approaches.

Reference	Materials Set-Up	Simulation	Analysis Method	Key Parameters
Gürsoy et al. [100]	RT58 paraffin, Al-based metal foam, nanoscale particles of Al_2O_3	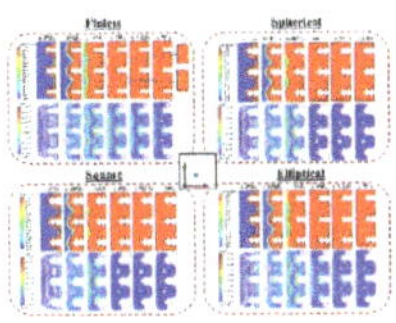	Spherical, square and oval rectangular containers with integrated structural walls; Based on the finite volume method, the Brinkman–Forchheimer extended Darcy model and the local thermal equilibrium model, the enthalpy-porosity method.	Melting time, temperature difference, Nusselt (Nu), energy change, energy storage parameters.
Hekmat et al. [101]	RT-50, Al_2O_3 NPs	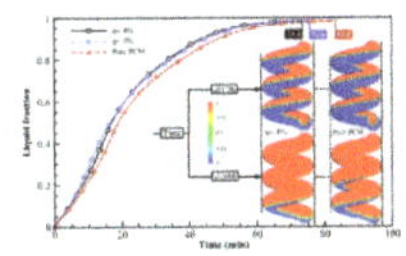	Three-dimensional spiral circular structure; Enthalpy-porosity model and Boussinesq approximate.	Melting time, solidification time, volume fraction NPs, feed port temperature, Reynolds number.
Dora et al. [102]	Nano silica/coir, decanol-ethanol treated stripped vermiculite, concrete	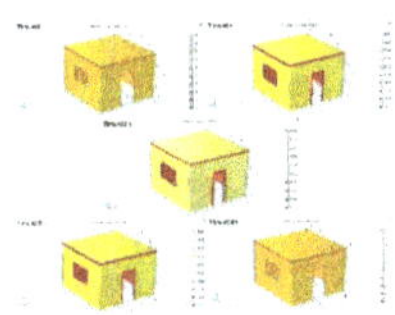	3D heat transfer building model; A general heat conduction equation for structures with no internal heat source.	TC, compressive strength, melting and freezing parameters, latent heat capacity, mass loss, compatibility and high heat resistance.
Yang et al. [103]	Cu foam/paraffin	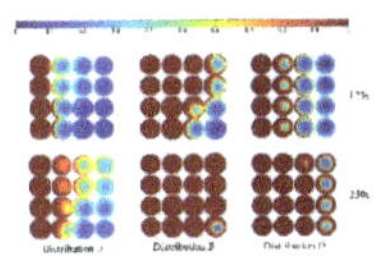	Tetrahedral models of elements with different porosities, continuity equations, energy equations, heat transfer mechanisms.	Porosity uniformity, temperature difference, heat storage capacity.
Zhao et al. [104]	Cu foam, carbon fiber reinforced plastic composite and octadecane	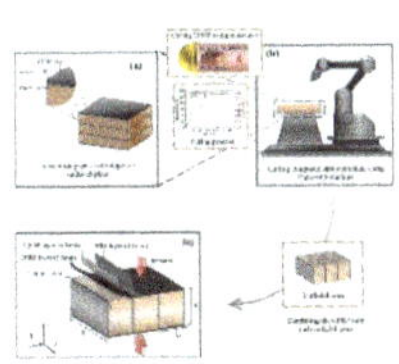	A multifunctional thermal management composite sandwich structure: 3D failure model, universal testing machine, 2D transient numerical model.	Specific strength, specific energy, composite plate thickness, thermal rate, TC.
Sun et al. [105]	Artificially cultured diatom microlenses regulate carbocellulose nanofibers	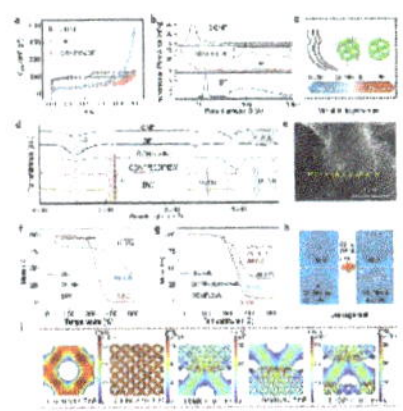	Multi-level pore model, SEM, XPS, Al Kα radiation, Raman spectroscopy, N2 adsorption and desorption isotherms, BET, Barrett–Joyner–Halenda, FTIR, TGA	Shrinkage, specific surface area, enthalpy of phase-change, TC.

Table 6. *Cont.*

Reference	Materials Set-Up	Simulation	Analysis Method	Key Parameters
Çiçek et al. [56]	Fe_3O_4, MgO, ZnO, and xGNP, added to RT-35HC with volume fractions of 0.02, 0.04, and 0.06. Al6061 as a heat sink and fin material	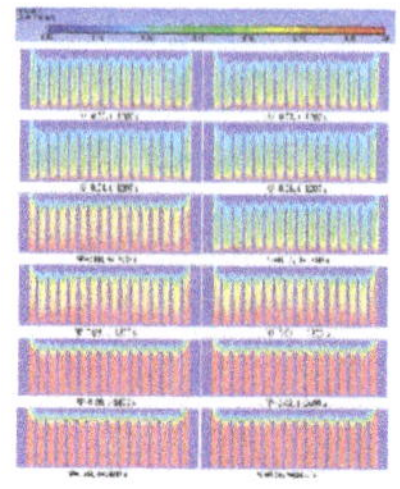	A symmetrical 3D NPCM heat sink (114 × 114 × 25 mm) model was constructed using Ansys fluent to analyze the effect of the input heat flux generated by the heater cap (100 × 100 × 1 mm) with a grid and a duration of 900,000 and 1 s. Continuity momentum equation, energy equation, thermal conductivity equation, viscosity equation.	NP type, vessel temperature profile, melt profile, melting time and volume fraction, reference temperature, thermal conductivity, heat flux.
Saleh et al. [106]	Nano-capsules are structured with polyurethane as the outer shell and nonadecane as the inner core.	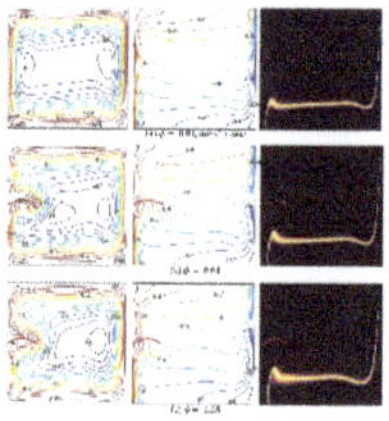	The square cavity filled with nano—encapsulated PCM and a base fluid has a fin of variable length at the midpoint of its left side. The Arbitrary Lagrangian–Eulerian method, Boussinesq approximation, Laplace equation, and Winslow formula.	Rayleigh number, Stefan number, fusion temperature, Capsule PCM concentration, Melting temperature, Young's modulus.
Egami et al. [107]	CuO nano-powders, PCM	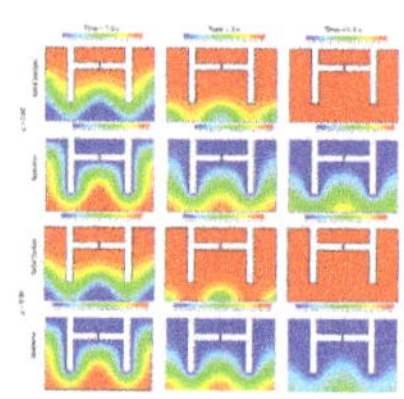	A rectangular container with H-shaped fins. Galerkin Method.	Concentration and shape factor of nano-powders

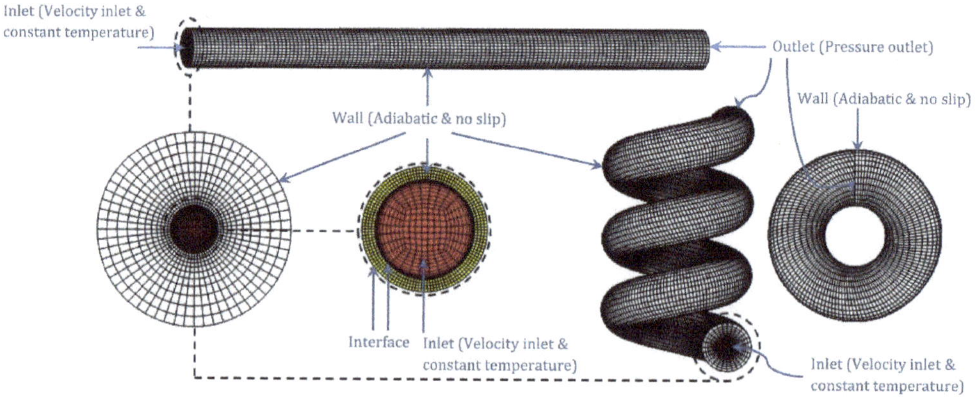

Figure 5. Schematic diagram of the three-dimensional spiral circular structure (Reproduced with permission from [101], Elsevier, 2025).

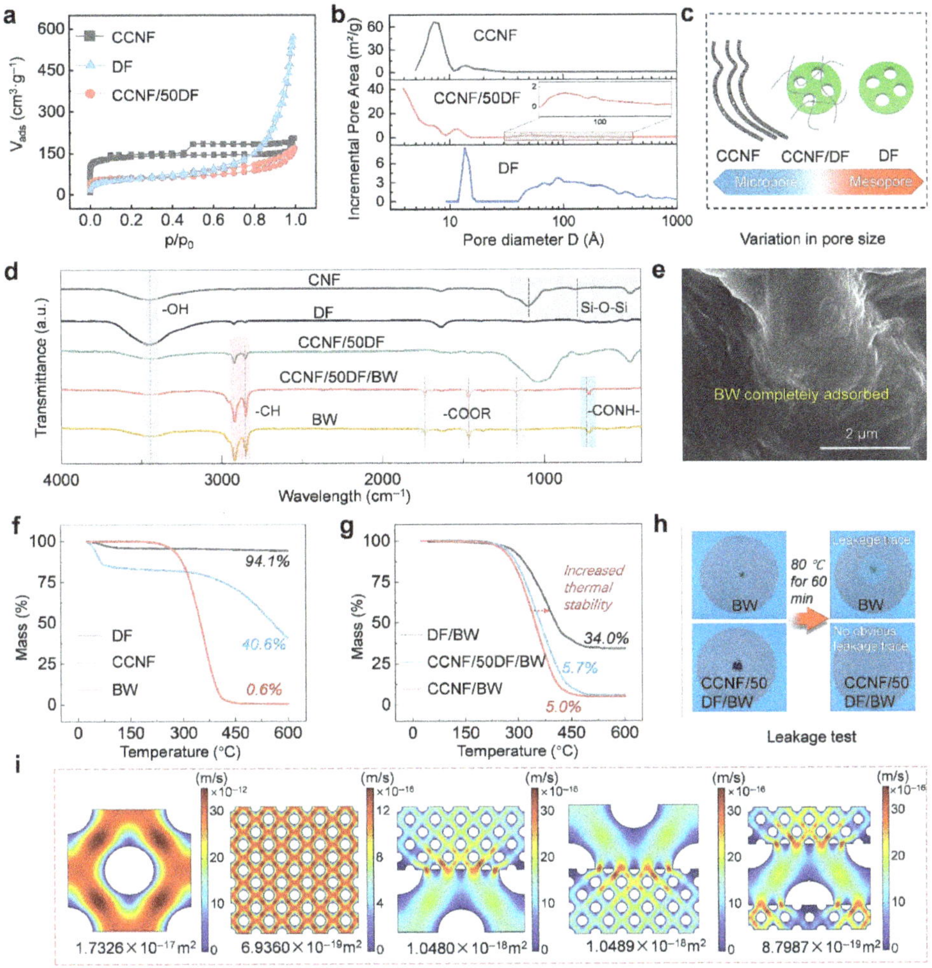

Figure 6. Simulation analysis of carbilose nanofibers and the impact on leaks (Reproduced with permission from [105], Elsevier, 2025). (**a**) N_2 isothermal adsorption–desorption curves; (**b**) pore size distribution of CCNF, DF, and CCNF/50DF; (**c**) pore size regulation; (**d**) FTIR spectra of the CCNF composite porous framework and BW composites; (**e**) SEM image of CCNF/50DF/BW. TGA curves of (**f**) CCNF, BW and DF as well as (**g**) the prepared ss-PCMs. (**h**) Schematic diagram of liquid flow in the porous framework, along with (**i**) simplified 2D models constructed.

3.2.2. Computational Fluid Dynamics

CFD plays a key role in the study of NPCM applications as a powerful tool for simulating fluid flow and heat transfer phenomena [108,109]. NPCMs use the phase transformation process to absorb or release a large amount of latent heat, but their thermal properties are highly dependent on the nanoscale structure of the material itself, its dispersion stability, and its interaction with the surrounding fluid. By simulating complex multiphase flow systems containing NPCM particles or capsules, CFD can reveal micro/macroscale flow, heat transfer, and phase transition coupling mechanisms to guide the design of NPCMs and evaluate their thermal management performance in energy storage systems. The following is an overview of the research on the application of the latest CFD analysis method to the analysis of NPCM.

Olivares-Robles designed a ring thermoelectric cooler composed of semiconductors, PCM (OM32 and Polyethylene glycol 1500), base fluid (water), and Al_2O_3 NPs, and analyzed the effects of NP concentration and temperature on the thermal performance of the cooler [110]. Benyahia developed a simulation model of a finned triangular inner tube and paraffin PCM enhanced with Cu NPs to analyze the effects of different NP concentrations, tube eccentricity, and inclination on the thermal properties of materials [111]. Bouzennada established a simulation model of NPCM fluids contained in different wave side wall cavities to analyze the influence mechanism of different wall structures, cylindrical radii, and rotational speed (as shown in Figure 7), and NPs concentration on thermal performance [112]. Glolami built Cu fins and Al heat sinks, analyzed the phase transformation process of $Na_2SO_4 \cdot 10H_2O$, paraffin, and eicosane, and analyzed the efficiency of photovoltaic panels [113]. Nandi used CuO NPs, lauric acid, RT35HC, and P58 materials to analyze the affected mechanism of improved solar irradiation by designing a rectangular geometry with different parallel structure shapes [114]. Shaik used CuO NPs, RT82, to analyze the effects of different liquid fractions and shapes on melting time and melting rate by designing a modified fin (as shown in Figure 8) with different vertical structure shapes [115]. Waqas used multi-walled carbon nanotube (MWCNT), TiO_2, and molten-salt PCM materials to design different rotationally symmetrical fin structures in hollow cylinders (as shown in Figure 9) to analyze the thermal performance of tubular heat exchangers [116]. Mahdavi designed PV laminates using Cu NPs, porous Cu media, and RT35-HC paraffin, and analyzed the effects of water pipe placement, NP concentration, and porous media on the thermal and electrical energy of the PV panels [117]. Alam analyzed the melting performance of latent heat thermal energy storage units made of different materials and shaped PCM [118]. Kiyak analyzed the effect mechanism of curvature and different concentrations of Cu NPs on the melting and solidification processes of thermal energy storage systems [119]. The main materials and research contents of the above-mentioned study are listed in Table 7.

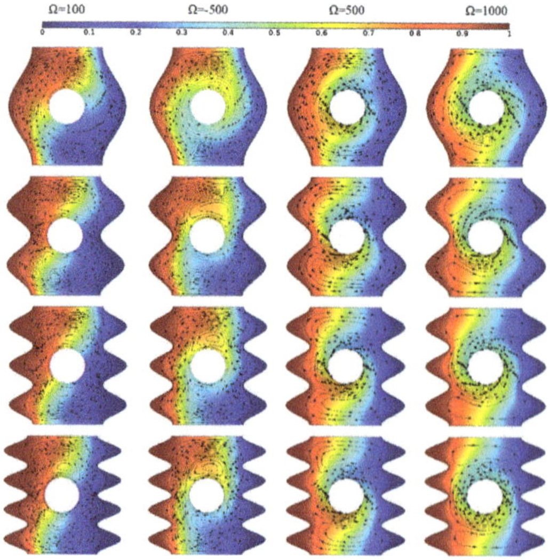

Figure 7. Simulation results under different wall structures and speeds (Reproduced with permission from [112], Elsevier, 2024).

Table 7. The main materials and research contents about CFD approaches.

Reference	Materials Set-Up	Simulation	Analysis Method	Key Parameters
Olivares-Robles et al. [110]	OM32 and Polyethylene glycol 1500, Al_2O_3 NPs	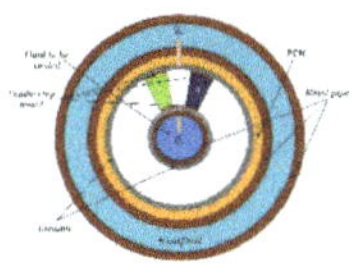	The annular thermoelectric cooler model consists of p-type and n-type semiconductor components made of bismuth telluride (Bi_2Te_3), with the components insulated by a ceramic plate and cooled using PCM. Finite difference method discrete transient equations	Coefficient of performance, TC, Reynolds number, Prandtl number, Nu, NP concentration
Benyahia et al. [111]	Paraffin, Cu NP	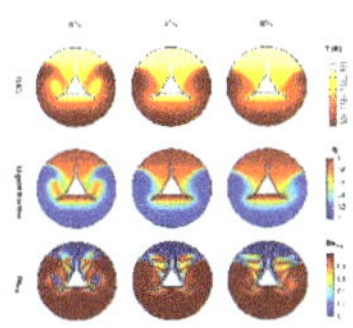	The cylindrical shell of the inner hollow triangular tube was established; the two-dimensional axisymmetric formula was used to simulate the thermal and flow behavior in the latent heat thermal energy storage (LHTES) element, the numerical solution was obtained by the generalized α time integral method, the second-order Lagrangian finite element was spatially discretized with a time step of 1 s, Boussinesq approximate, the Carman–Kozeny relation kinetic energy equation, and the Carman–Kozeny equation	The concentration of Cu NPs, TC, melting time, inclination angles, Nu, eccentricity, porosity
Bouzennada et al. [112]	Polyurethane/Nonadecane dispersed in water	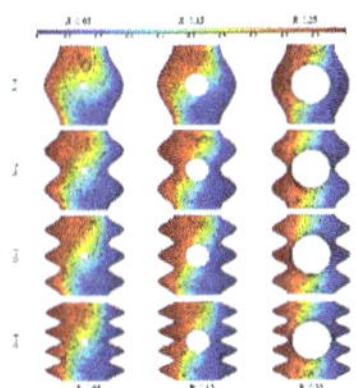	A 2D enclosed body filled with NPCM fluid; Galerkin weighted residual finite element method, Galkin weighted residual finite element method	Volume fraction, cylindrical radius, rotational speed, melting temperature, flow structure, heat transfer rate, Stefan number, Nu
Gholami et al. [113]	$Na_2SO_4{\cdot}10H_2O$ was mixed with Al_2O_3 and MgO NPs, paraffin and eicosane	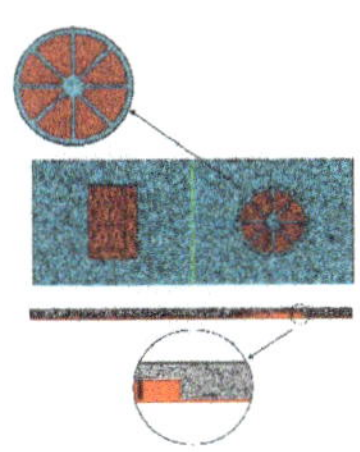	A photovoltaic panel and PCM cavity structure were established; A first-order upwind approach for the discretization of the momentum equation and a second-order approach for the energy equation, and a second-order scheme for the energy equation	the concentration of NPs, efficiency and temperature
Nandi et al. [114]	CuO NPs, lauric acid, RT-35HC and P-58	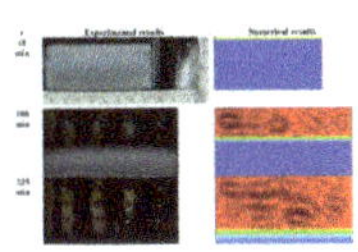	a rectangular geometry characterized by specific dimensions	The concentration of NPs, fin structure, energy storage rate, melting rate
Shaik et al. [115]	CuO NPs, RT82	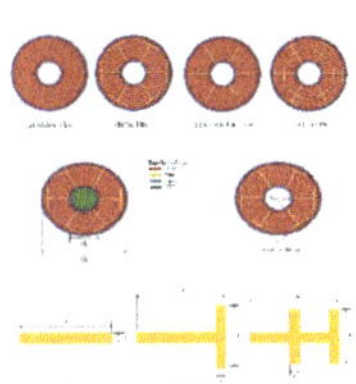	PISO method, PRESTO, second-order upwind methods, second-order upwind algorithms	NP concentration, shape, melting time, melting speed

Table 7. *Cont.*

Reference	Materials Set-Up	Simulation	Analysis Method	Key Parameters
Waqas et al. [116]	MWCNT, TiO_2, molten-salt PCM	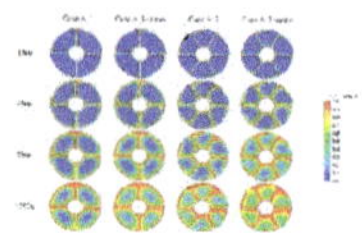	Boussinesq approximation, Semi-implicit method for pressure linked equations (SIMPLE), PRESTO system	Fin shape, compactness, melting time
Mahdavi et al. [117]	Cu NPs, porous Cu media, and RT35-HC paraffin	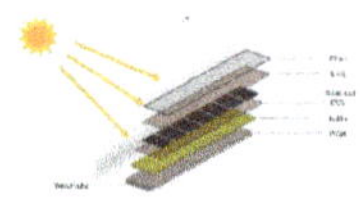	Five-layer PV panels; discrete ordinate method; Boussinesq approximation	Electrical efficiency, economic analysis, pipe placement, NP concentration
Alam et al. [118]	n-Octadecane, Capric acid, MWCNTs, Cu metal foam	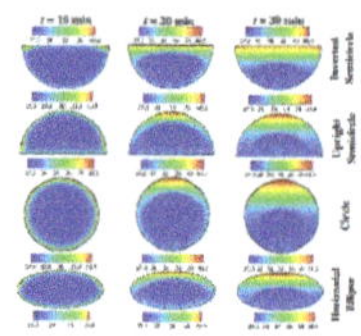	A two-dimensional computational model of square geometry, finite volume method	Enclosure geometries, Prandtl numbers, Grashof number, Stefan number, NPs concentration, melting rate
Kiyak et al. [119]	CuO NPs, RT19HC	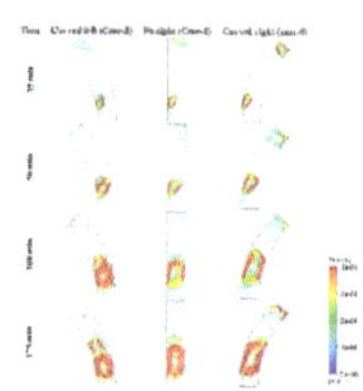	a curved geometry and a straight geometry; the Navier–Stokes equations, enthalpy-porosity technique. Boussinesq approximation, finite volume method, SIMPLE, the implicit Euler method	Nu, concentration of NPs, geometric curvature, phase change time, melting time, discharge efficiency

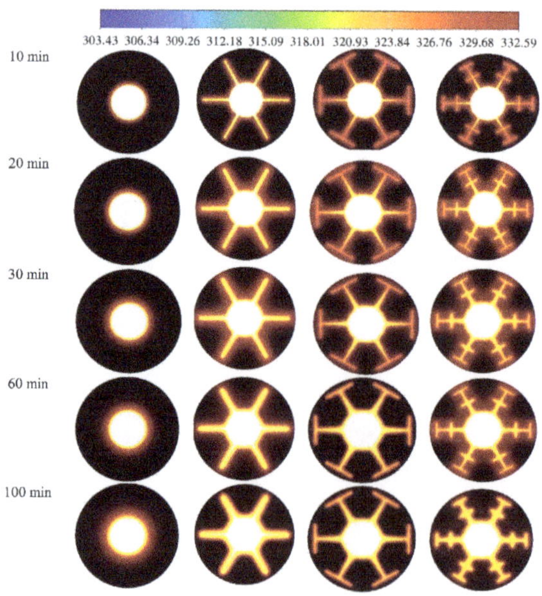

Figure 8. Simulation results of modified fin with different parallel structure shapes (Reproduced with permission from [115], Elsevier, 2025).

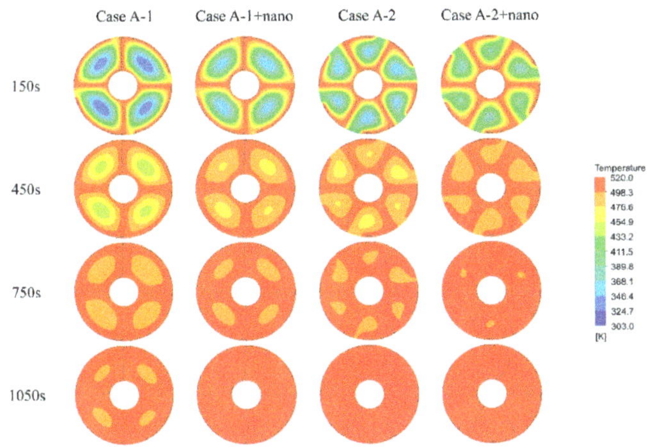

Figure 9. Simulation results of different rotationally symmetrical fin structures (Reproduced with permission from [116], Elsevier, 2025).

3.2.3. System Performance Simulation

System performance simulation provides core support for design optimization by building a highly integrated system simulation coupling model to efficiently simulate the overall dynamic response and comprehensive performance of complex systems in an experimental or simulation environment. With its high thermal conductivity, latent heat of phase change, and fast response, NPCMs are often used in the design of various photothermal conversion system materials, which enhance the performance characteristics of the system in thermal management or electrical efficiency management, accelerate the research and development of photothermal systems, and improve the reliability of the system. The following is an overview of the research on the application of the latest system performance simulation analysis method to the analysis of NPCM.

Amin added graphene to beeswax to make distillation tubes and designed an experimental system for analyzing freshwater production performance by combining K-type thermocouples, insolation intensity meters, graduated cylinders, and data collectors [120]. Singh added different concentrations of Al_2O_3 particles to OM37 to make NPCMs for the analysis of the thermal performance of solar ponds [121]. Emara integrated the design of a freshwater system by adding different concentrations of TiO_2 and Al_2O_3 nanofluids to paraffin with a melting temperature of 62 °C [122]. Al-Karboly integrated a twisted band absorber, a twisted absorption tube, a 1% SiC NP-reinforced paraffin preparation vessel, and a silicon carbide nanofluid, comprehensively designed a photovoltaic system (Shown in Figure 10), and analyzed its thermal efficiency [123]. Kamrava used ZnO, CuO, and Al_2O_3 in paraffin wax as NPCM to further construct the floor heating system and analyzed the thermal performance characteristics of the floor temperature and heat flux of the floor heating system [124]. Gür added B_4C to the RT35HC to design the heat sink, further constructed an ultra-low energy consumption home heat sink simulation system, and analyzed the thermal performance factors such as room temperature and water temperature [125]. Sathishkumar used GNP and strong nitric acid to prepare NPCM materials to design the energy storage tank system, and analyzed the thermal performance parameters such as stored energy and solidification time [126]. Bassam used nanofluids (SiC and water preparation) and NPCM (SiC and paraffin preparation) to design micro-finned tubes, and the comprehensive design of the photovoltaic system showed a 77.5% increase in thermal efficiency and an increase in output power to 14.5 W [127,128]. Kibria used NPCM (paraffin wax with Al_2O_3 and ZnO NPs) to design a photovoltaic thermal system, and the analysis results showed that the surface temperature of the system was reduced by 5.4 °C, the power was increased by 2.75%, the electrical efficiency was increased by 2.71%, the maximum heat output power was increased by 17.4%, and the overall efficiency was increased by 15.82% [129]. Bharathiraja designed a collector system incorporating MWCNTs and SiO and discovered that this design led to a higher output temperature and a longer duration for hot water production [130]. The main materials and research contents of the above-mentioned study are listed in Table 8.

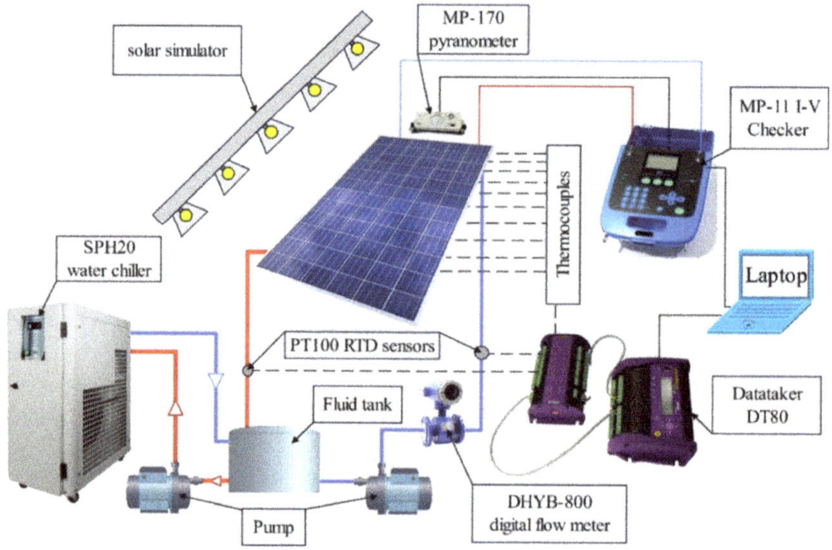

Figure 10. Photovoltaic system (Reproduced with permission from [123], Elsevier, 2025).

Table 8. The main materials and research contents about system performance simulation approaches.

Reference	Materials	System	Devices	Key Parameters
Amin et al. [120]	Graphene, beeswax	Freshwater production	Distillation tubes, K type thermocouples, insolation intensity meters, graduated cylinder connections, data acquisition equipment	Ambient temperature, solar radiation, thermal conductivity, temperature, freshwater productivity, thermal efficiency, energy efficiency
Singh et al. [121]	Al_2O_3, OM37	Solar ponds	Data collectors, thermocouples, solar pond installations	Latent calorific value, thermal conductivity, thermal power, thermal efficiency
Emara et al. [122]	TiO_2, Al_2O_3, paraffin	Distilled water	Radiators, water tanks and their heaters, liquid supply tanks, centrifugal pumps, fans, flow meters, thermocouples, temperature sensors, temperature indicators, solid state relays	Flow rate, rotational speed, heat transfer coefficient, NP concentration
Al-Karboly et al. [123]	SiC, PCM	PV	Solar simulator, pyranometer, I-V checker, Laptop, datataker, digital flow meter, pump, fluid tank, water chiller, RTD sensors, Themocouples	Electrical efficiency, thermal efficiency, overall efficiency, nanofluid concentration
Kamrava et al. [124]	ZnO, CuO, Al_2O_3, paraffin	Underfloor heating	A structured framework for integrating NPCMs	NP concentration, floor temperature, heat flux, thermal conductivity, melting time
Gür et al. [125]	B4C, RT35HC	radiator	PVT collectors, heat sinks, PCM layers, Cu tubes	Water temperature, room temperature, heat retention

Table 8. *Cont.*

Reference	Materials	System	Devices	Key Parameters
Sathishkumar et al. [126]	GNP, strong nitric acid	energy storage tank	A chiller unit, energy storage tank, circulating pump, flow meter and energy meter	Flow rate, desolidification time, charging time, melting time, heat transfer rate, temperature, charging power, voltage drop, latent heat energy, accumulated energy, specific energy consumption
Bassam et al. [127,128]	SiC, paraffin, Polyvinylpyrrolidone	PV	Fluid tank, pump, flow meter, data logger, photovoltaic panel, pyranometer, I–V curve checker, computer cooling unit, heat exchanger, PCM expansion tank	Electrical efficiency, thermal efficiency, temperature
Kibria et al. [129]	Al_2O_3, ZnO, paraffin,	PV	PV, hot water storage, PVT/PCM, Structure, PVT/HNPCM, water tap, overhead tank, pipes	Thermal conductivity, latent heat capacity, temperature, electricity efficiency
Bharathiraja et al. [130]	MWCNTs, SiO	Collector	Cup type anemometer, data logger, solar intensity meter, laptop, storage tank, copper tube embedded PCM, thermocouple, absorber plate	Outlet water temperature, temperature, hot water time

3.2.4. Magnetic Field Coupling Models

The magnetic field changes the Lorentz and Kelvin forces in the convective heat transfer process through the coupling of the changes in magnetic induction intensity with the conductivity and magnetism of different NPs. This regulates the melting and solidification behaviors in the phase transformation process of the NPCM, further altering the relevant thermodynamic characteristic parameters of the NPCM. Additionally, the coupling effect can also impact the dynamic behavior between NPs and PCM, resulting in current changes that adjust the system's charging and discharging processes, thereby optimizing the system's power characteristic parameters [131]. The following is an overview of the research on the application of the latest magnetic field coupling models analysis method to the analysis of NPCM.

Izadi added different concentrations of Al_2O_3 nanoadditives to paraffin wax to construct a hexagonal heat storage unit and analyzed the convective melting mechanism of the tilt angle and magnetic field strength [132]. Hassan designed a U-shaped baffle cavity for the nano-encapsulated PCMs and analyzed the mechanism of the natural convection effect of the inclined magnetic field on the structure [133]. Younis designed a cavity structure to analyze the mechanism by which magnetic field strength affects the thermal properties of nano-encapsulated PCMs [134]. Selimefendigil designed a triangular-shaped structure containing Ag-MgO hybrid nanofluids to analyze the convective influence mechanism in a heterogeneous magnetic field [135]; designed a multi-jet impingement system to study the numerical effects of Ag-MgO nanofluids on the thermal properties of PCMs under different magnetic field strengths and clarity [136]; studied the phase transition and convective heat transfer mechanisms of different types of NPs and encapsulated PCMs under magnetic fields [137]. Nguyen designed an external magnetic field for the analysis of the thermal performance effects of NPCM capsules incorporated into roof structures [138]. Lu designed a stable rotating magnetic field applied to paraffin wax doped with Fe_3O_4 to analyze the effect of the magnetic field on the melting mechanism [139]. Li added Al_2O_3 and Fe_3O_4 to paraffin wax to prepare NPCMs, and analyzed the melting and energy storage mechanisms inside the materials using ultrasonic and magnetic fields (Shown in Figure 11) [140]. Zhuang added Al_2O_3 and Fe_3O_4 to paraffin wax to prepare NPCMs, and analyzed the heat transfer mechanism of NPs to the materials by thermochromic liquid crystal combined with a magnetic field [141]. Farahani analyzed the effects of fin factors and Fe_3O_4 EMPs concentration on the thermal performance of energy storage systems under the application of magnetic fields [142]. The main materials and research contents of the above-mentioned study are listed in Table 9.

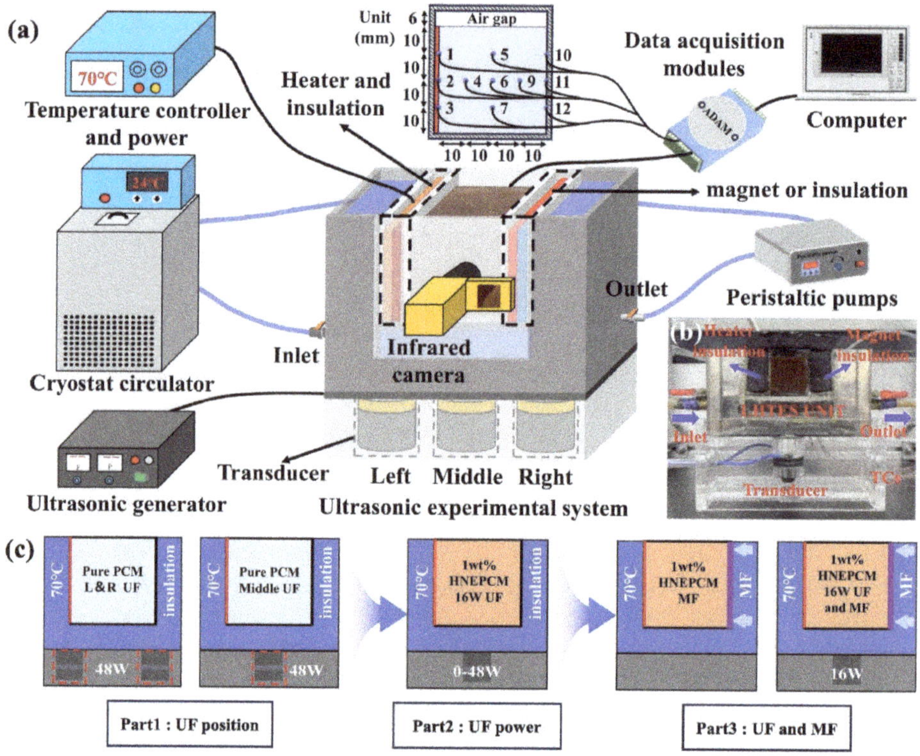

Figure 11. Ultrasonic experimental system diagram (Reproduced with permission from [140], Elsevier, 2024). (**a**) Schematic diagram; (**b**) experimental setup of the ultrasonic and magnetic fields experimental system; (**c**) Boundary conditions under different cases.

Table 9. The main materials and research contents about the magnetic field.

Reference	Materials	Physical Model	Initial and Boundary Conditions	Governing Equations	Approach	Key Parameters
Izadi et al. [132]	Al_2O_3, paraffin	Ahexagonal heat storage unit	$\theta(\hat{x},\hat{y},0)=0$, $\hat{u}(\hat{x},\hat{y},0)=\hat{v}(\hat{x},\hat{y},0)=0$; $\theta(\hat{x},\hat{y},\tau)=0$, $\hat{u}(\hat{x},\hat{y},\tau)=\hat{v}(\hat{x},\hat{y},\tau)=0$; $\frac{\partial\theta(\hat{x},\hat{y},\tau)}{\partial N}=0$, $\hat{u}(\hat{x},\hat{y},\tau)=\hat{v}(\hat{x},\hat{y},\tau)=0.$	Continuity, momentum in two coordinate directions and energy equations	The enthalpy-porosity method, the finite element method	The tilted angle and intensity of the magnetic field, the Hartmann number, Nu, the nanoadditives concentration
Hassan et al. [133]	Shell-core structured particles suspended in water	A U-shaped baffled cavity	$t=0, u=x=0$; $\theta_b=1$; $\theta_t=0$; $\frac{\partial\theta}{\partial n}=0.$	Continuity, momentum conservation with additional source terms, and energy balance	The Arbitrary Lagrangian–Eulerian approach, the deformation gradient tensor and the second Piola–Kirchhoff stress tensor	Rayleigh number, Stefan number, fusion temperature, NP volume fraction, oscillation amplitude, Young's modulus,
Younis et al. [134]	A graphene core encapsulated by a polyurethane shell	A prism cover drive chamber	$\frac{\partial U}{\partial X}+\frac{\partial V}{\partial Y}=0.$	Continuity, momentum conservation, and energy balance	Galerkin method, Boussinesq approximation	Hartmann number, the variable time Reynolds number, Darcy number, Darcy number, the angle of deflection

Table 9. *Cont.*

Reference	Materials	Physical Model	Initial and Boundary Conditions	Governing Equations	Approach	Key Parameters
Selimefendigil et al. [135]	Water-Ag-MgO hybrid NF	A triangular shape	-	Continuity, momentum conservation, and energy balance	Kozeny–Carmen model, method, the finite element method, Streamline Upwind Petrov–Galerkin approach, Biconjugate Gradient Stabilized finite element method, Lagrange	Port size, Reynolds number, Reynolds number.
Selimefendigil et al. [136]	Ag-MgO, paraffin	Multi jet impingement system	-	Energy equation, conservation equation.	FEs of various orders, SUPG, BICGStab. Kozeny–Carmen method	Nu, Reynolds number, magnetic field strength and inclination
Selimefendigil et al. [137]	Spherical Al_2O_3, cylindrical MWCNT, platelet GNP	Trapezoidal vented cavity	$u = u_g, T = T_g, T = T_h, \frac{\partial T}{\partial n} = 0.$	Continuity, momentum conservation, and energy balance	Kozeny–Carman model, Galerkin weighted residual finite element method	Reynolds number, Hartmann number, inclination parameter, non-uniform amplitude, NP loading.
Nguyen et al. [138]	Al_2O_3, paraffin	A closed fined rectangular chamber	$\theta = 0, U = V = 0 \forall F_0, X, Y \exists F_0 = 0$; $\theta = 1, U = V = 0 \forall F_0, X, Y \exists 0 \leq X \leq 1, Y = 0$; $\theta = 1, U = V = 0 \forall F_0, X, Y \exists 0 \leq X \leq 1, Y = 1$; $\frac{\partial \theta}{\partial x} = 0, U = V = 0 \forall F_0, X, Y \exists X = 0, 0 \leq Y \leq 1$; $\frac{\partial \theta}{\partial x} = 0, U = V = 0 \forall F_0, X, Y \exists X = 1, 0 \leq Y \leq 1.$	Continuity, momentum conservation, and energy balance	The effective thermal conductivity of ferromagnetic-PCM, Boussinesq approximation	Capsule location, Hartmann number,
Lu et al. [139]	Fe_3O_4, paraffin	a glass beaker filled with PCM	-	-	uncertainty analysis	NP mass fraction, magnetic field rotational speed
Li et al. [140]	Al_2O_3, Fe_3O_4 acetylacetone	Ultrasonic experimental system	-	-	SEM, EDS, uncertainty analysis	Nu, filling height, melting time, melting rate
Zhuang et al. [141]	Al_2O_3, paraffin	A set of visualized experimental systems based on thermochromic liquid crystal (TLC)	-	-	TLC, GetData software Version 2.26, uncertainty analysis	Nu, filling height, melting time, melting rate
Farahani et al. [142]	RT82, Fe_3O_4.	Annular structure with strip fins	-	Continuity, momentum conservation, and energy balance	SIMPLE algorithm, center difference and quick method	Uniform magnetic field pattern, fin effect, melting time, NPs concentration.

4. Algorithm-Driven Optimization

Algorithm-driven optimization has become an important research strategy for enhancing the performance of NPCMs in high-temperature thermal storage systems [143]. This strategy utilizes the powerful computational capability of algorithms to finely optimize parameters such as the composition and structural design of NPCM, in pursuit of maximizing thermal efficiency and thermal stability. In the NPCM field, algorithms play a critical role in determining the ideal mass ratio of the base PCMs to NPs. They filter optimal solutions from numerous combinations of NP characteristics and PCM properties based on preset material performance benchmarks, which may encompass multiple dimensions ranging from maximizing thermal conductivity and energy storage capacity to enhancing thermal and electrical efficiency [144]. Furthermore, algorithm-driven optimization extends to the manufacturing processes of NPCM. During the synthesis of composite materials, algorithms take into account parameters such as mixing intensity, temperature regulation, and pressure control. By optimizing these factors, it is possible to enhance the performance of NPCM while also reducing costs, making NPCM more attractive for large-scale applications in high-temperature thermal storage and thus promoting the development of more efficient and sustainable energy solutions [145].

4.1. Machine Learning Methods

Machine learning methods serve as a system analysis and prediction paradigm based on the inherent laws of massive data. By deeply mining the hidden correlations within complex systems, these methods break through the simplification dependence of traditional physical models on complex boundary conditions and multi-field coupling mechanisms. In the thermal performance study of NPCMs, this approach demonstrates significant advantages: through high-dimensional data analysis of the synergistic effects of multiple parameters, it can accurately capture the distribution of NPs, the evolution of phase change interfaces, and the nonlinear response patterns of flow within porous media, revealing cross-scale coupling mechanisms that are difficult to quantify with traditional theoretical models. This data-centric research paradigm not only optimizes the thermal management performance prediction of nano PCM in complex geometric structures but also provides a new pathway for the intelligent design of industrial-grade phase change energy storage systems. The following is an overview of the research on the application of the latest machine learning methods to the analysis of NPCM.

Shakibi uses CFD simulation to design key factors, construct an Artificial neural network (ANN) model (Shown in Figure 12), and integrate long short-term memory (LSTM), grey wolf optimizer (GWO), Particle swarm optimization (PSO) and other methods to analyze and solve the intelligent model of the optimal thermal performance parameters [146]. Aly explored the heat and mass transfer properties of NPCM in a wave porous cavity using the incompressible smoothed particle hydrodynamics method and the extreme gradient boosting (XGBoost) machine learning model, Darcy number, Rayleigh number, Fusion temperature, Soret number, Dufour number, Hot zone length, Hot zone Key parameters such as height reveal the influencing characteristics of internal thermal performance [147]. Abdolahimoghamad established the multilayer perceptron (MLP)-ANN model for estimating the optimal melting and solidification time of Gr-Cu hybrid NPs and latent heat, thermal conductivity of bio-based NPCMs [148,149]. Gao established an automatic correlation determination model to predict liquid-state thermal conductivity, k-nearest neighbor and polynomial regression composite models to predict the thermal conductivity of NPCMs, and minimum absolute shrinkage and choice operators to predict freezing-state latent heat [150]. Ghasemi established an ANN model and input heat flux, Reynolds number, and concentration to predict performance evaluation criterion and average Nu [151]. Kiaghadi used four models: linear regression, lasso regression, polynomial regression, and auto-regressive integrated moving average to verify the correlation between time and liquid fraction, and confirmed that the data analysis method can be effectively applied to the prediction of phase transition behavior [152]. The auto-regressive integrated moving average algorithm was used to predict parameters such as solar unit time, PV temperature, melting fraction, NPCM temperature, electrical efficiency, and other parameters based on historical data [153]. Yang established an ANN model with Reynolds number and porosity of nickel foam as inputs, which could effectively predict the maximum temperature of LIB [154]. Xu established an ANN model with shear rate and mass fraction as inputs, which could effectively predict the dynamic viscosity, density, and shear stress of the LIB of the NPCM [155]. Salari established a Harris Hawks optimizer model optimized by a random forest algorithm to predict printed circuit board (PCB) temperature [156]. Khadem established a GWO-support vector regression with a hybrid composite model (Shown in Figure 13), and predicted the chipset temperature parameters through the melting point, thermal conductivity, specific heat capacity, and enthalpy parameters of the system [157]. Fini developed a composite model of an ANN-multiobjective genetic algorithm for predicting the Nu and pressure drop of microencapsulated PCM fluid [158]. The main materials and research contents of the above-mentioned study are listed in Table 10.

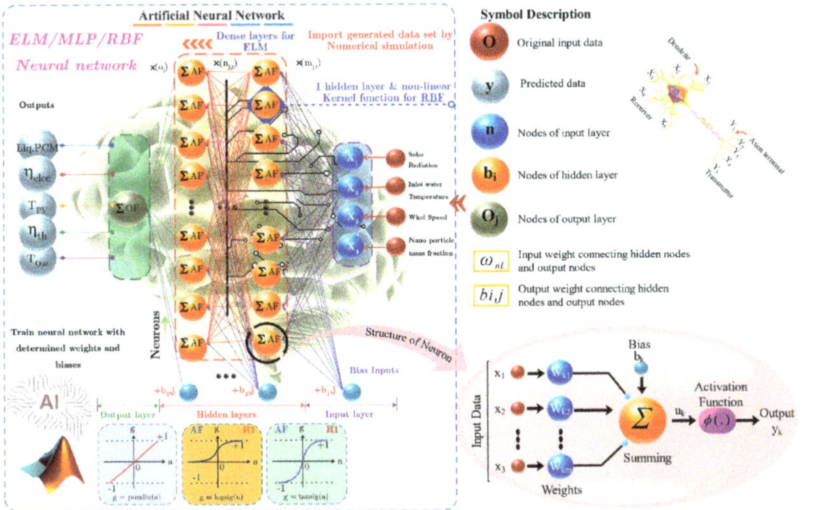

Figure 12. Schematic diagram of machine learning model (Reproduced with permission from [146], Elsevier, 2023).

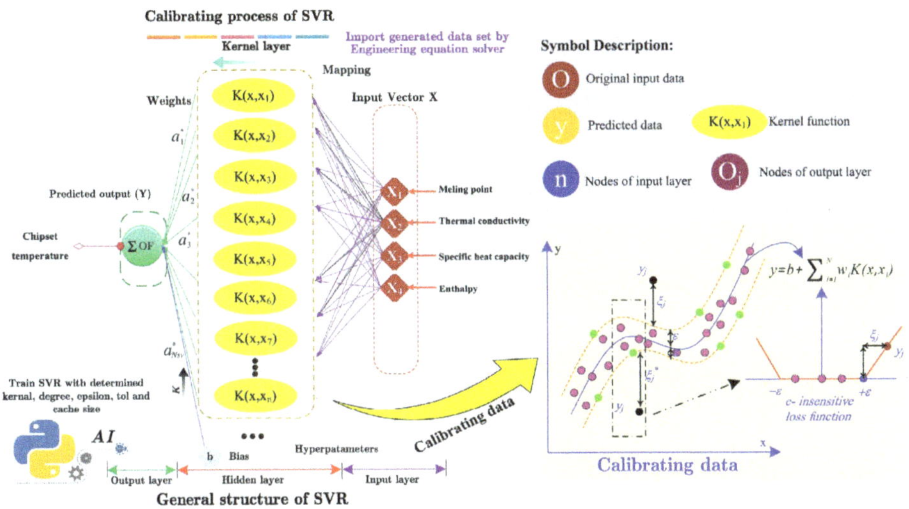

Figure 13. Support vector regression with hybrid model (Reproduced with permission from [157], Elsevier, 2023).

Table 10. The main materials and research contents about the machine learning methods.

Reference	Materials	Model	A pproach	Input	Output	Mean Absolute Percentage Error
Shakibi et al. [146]	MWCNT/PCM	ANN	CFD, LSTM, Extreme Learning Machine, Radial basis function, MLP, GWO, bat algorithm (BA), optimization based on biogeography (BBO), PSO, TOPSIS, LINMAP, and Shannon entropy	Surface temperature, coolant outlet temperature, electrical efficiency, thermal performance, melted PCM	Surface temperature, coolant outlet temperature, electrical efficiency, thermal performance, melted PCM	PSO: 0.09714; BBO: 0.07890; GWO: 0.07800; BA: 0.07837

Table 10. *Cont.*

Reference	Materials	Model	A pproach	Input	Output	Mean Absolute Percentage Error
Aly et al. [147]	PMMA, n-octadecane, water	XGBoost	the least squares boosting method	Nu, Sherwood number	Nu, Sherwood number (SH)	Nu: 5×10^{-7}; SH: 2.1×10^{-7}
Abdolahimoghadam et al. [148]	Beeswax, coconut oil, Gr-Cu hybrid NPs	MLP-ANN	Tansig, radbas	Temperatures, concentrations, thermal conductivity, liquid or solid state of PCM, latent heat of melting and solidification processes	Latent heat, thermal conductivity	TC: 1.1253×10^{-4} Latent heat: 4.7×10^{-4}
Abdolahimoghadam et al. [149]	Beeswax, coconut oil, Gr-Cu hybrid NPs	MLP-ANN	Trainlm, trainbr, tansig, radbas, logsig	Concentration, density, average temperature, specific heat, viscosity, latent heat, thermal conductivity	The melting's and solidification's liquid fraction and time,	Solidification: 6.1019×10^{-7}, 0.0458; Melting: 8.05×10^{-5}, 0.00935
Gao et al. [150]	MWCNT, GNPs, nano-graphite (NG), myristic acid (MA)	Automatic correlation determination; k nearest neighbors and polynomial regression; Minimum absolute shrink and selection operators	-	Solid-state thermal conductivity; thermal conductivity of pure myristic acid, temperature; Melting-state latent heat	Liquid-state thermal conductivity; thermal conductivity of NPCMs; Freezing-state latent heat	GNPs/MA: 1.762×10^{-4}; MWCNTs/MA: 5.45×10^{-7}; NG/MA: 5.035×10^{-6}
Ghasemi et al. [151]	Al, microencapsulated PCM-37	ANN	Levenberg-Marquardt, sigmoid, backpropagation algorithm, Regularization techniques	Heat flux, Reynolds number, concentration	Performance evaluation criterion, average Nu	-
Kiaghadi et al. [152]	RT44HC, Silver	Auto-regressive integrated moving average	Linear regression, lasso regression, polynomial regression	Time	Liquid fraction	0.0093
Kiaghadi et al. [153]	RT28HC, Cu	Auto-regressive integrated moving average	Linear regression, lasso regression, polynomial regression	Time, temperature of PV, melting fraction, temperature of NPCM, electrical efficiency	Time, temperature of PV, melting fraction, temperature of NPCM, electrical efficiency	-
Yang et al. [154]	$LiFeO_4$, n-nonadecane	ANN	ReLU	Reynolds number, porosity of nickel foam	Maximum temperature	-
Xu et al. [155]	sodium lauryl sulfate, n-eicosane	ANN	Gradient descent algorithm	Shear rate and mass fraction	Dynamic viscosity, density, shear stress	0.00066
Salari et al. [156]	MWCNT, RT-35 HC, RT-47	Harris Hawks optimizer	Random forest	-	The temperature of PCB	-
Khadem et al. [157]	RT-35 HC, aluminum foam	Support vector regression with hybrid	GWO	Meling point, thermal conductivity, specific heat capacity, enthalpy	Chipset temperature	-
Fini et al. [158]	n-Octadecane, Polymethyl methacrylate	ANN-multiobjective genetic algorithm	NSGA-II, LINMAP, TOPIS, Shannon Entropy	Reynolds number, volume fraction	Nu, pressure drop	-

Monte Carlo methods, which often conduct research by leveraging the Bayesian approach optimized by Markov chain Monte Carlo, play a crucial role. Regarding data processing, the distribution of independent variables, such as normal distribution or log-normal distribution, is set according to different scenarios, and the corresponding dependent variables are determined [159]. Multiple models are established, and a large amount of input—output data is utilized, with the training set and the test set divided in a specific proportion, and metrics such as mean squared error, root mean squared error, and coefficient of determination are employed to analyze the prediction accuracy of the models [160]. Some studies also use the mean value, the standard deviation of the number of iterations, and the standard deviation for precise data analysis. In the research on the dispersion of highly conductive nanomaterials in D-mannitol to enhance heat transfer in the LHTES system, parameters related to the thermophysical properties of NPCM and the inlet conditions of the heat-conducting fluid are set as independent variables following a normal distribution, with the Fourier number (Fo) as the dependent variable. Based on

153 CFD data, four models are established by dividing the training set and the test set into an 8:2 ratio. The curves of the Fo values predicted and the actual Fo values by the four models are shown in Figure 14. Eventually, a model that can best predict the Fo of the actual dataset is constructed based on relevant metrics [161]. In the study of graphite nanosheets reinforcing dodecanol ($C_{12}H_{26}O$), the melting rate and the surface-average Nu are set as independent variables with a normal or log-normal distribution, and the Stefan number, Grashof number, and Fo are set as dependent variables. Based on 573 melting-rate datasets and 516 surface-average Nu-number datasets, five models are established by dividing the training set and the test set in a 7:3 ratio. Subsequently, the models that can best predict the dataset values and the Nu values are analyzed, fully demonstrating the application of the Monte Carlo method in practical research [162]. The main materials and research contents of the above-mentioned study are listed in Table 11.

Table 11. The main materials and research contents about the Monte Carlo Method.

Ref.	Materials Set-Up	Analysis Method
Alazwari et al. [161]	Highly conductive nanomaterials dispersed in D-mannitol can be utilized to enhance heat transfer in LHTES systems	The Bayesian method optimized by Markov chain Monte Carlo is implemented. Independent variables (normally distributed): TC, density, specific heat capacity of the thermophysical properties of NPCM, as well as Reynolds number and inlet temperature of the inlet conditions of the heat conducting fluid. Dependent variable: Fo. Data analysis: Four models are established with 153 input-output CFD data. The ratio of the training set to the test set is 8:2. Mean Squared Error, Root Mean Squared Error, and Coefficient of Determination (R^2) are used to analyze the accuracy of the model's predicted values. A model that can best predict the Fo of the actual dataset is established to predict the dimensionless melting time.
Goodarzi et al. [162]	Graphite nanosheets reinforcing dodecanol ($C_{12}H_{26}O$)	The Bayesian method optimized by Markov Chain Monte Carlo is implemented. Independent variables (normally distributed or log-normally distributed): Melting rate and surface-average Nu. Dependent variables: The Stefan number, which is the ratio of sensible heat to latent heat and represents the driving force for melting. The Grashof number indicates the intensity of natural convection of liquid PCM during the melting process. Finally, the Fo represents the dimensionless time in transient problems. The model that can best predict the dataset values was analyzed. Another model that can best predict the Nu values was analyzed. Data analysis: Five models are established, using 573 melting-rate datasets and 516 surface-average Nu-number datasets. The ratio of the training set to the test set is 7:3. The mean, the standard deviation of the number of iterations, and the standard deviation are used for precise data analysis. The coefficient of determination, root-mean-squared error, and mean-squared error are used to evaluate the accuracy of the predicted values of the proposed models.

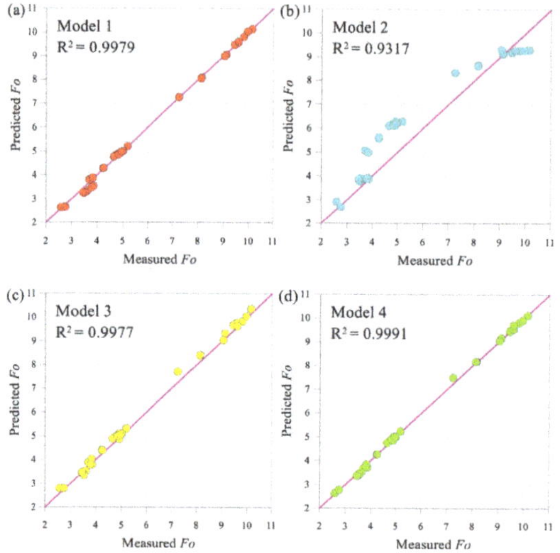

Figure 14. The curves of the Fo values by the four models (Reproduced with permission from [161], Elsevier, 2022). (**a**) Model 1; (**b**) Model 2; (**c**) Model 3; (**d**) Model 4.

4.2. Uncertainty Quantification Techniques

The uncertainty quantification techniques are uncertain to comprehensively assess the impact of uncertainty factors in systems on outcomes [163]. By meticulously identifying and precisely quantifying various sources of uncertainty, including measurement errors, model assumptions, and parameter variability, they provide strong assurance for the accuracy and reliability of models, enabling effective improvements on this basis [164]. In many fields, uncertainty quantification techniques play a crucial role, particularly in the research and application of complex systems based on NPCM materials. They assist researchers in gaining a deeper understanding of the intrinsic uncertainty mechanisms within the system, thereby allowing for a full consideration of uncertainty factors in model construction, optimization, and decision-making processes, ensuring the scientific and rational nature of the results. The following is an overview of the research on the application of the latest uncertainty quantification techniques to the analysis of NPCM.

Al-Aasam calculated that the error value of the PVT system caused by temperature, mass glow rate, solar irradiance, voltage, and current is 3.83% [165]. Daneshazarian determined that the error value of the ground source heat pump system resulting from temperature outlet, temperature inlet, and flow rate is 0.5% [166]. Gür analyzed the error value of the electronics system attributed to the temperatures of the PCM and the temperatures of the empty system, which is 1.5% [167]. Venkatesh conducted an analysis of the error in the heat exchanger system using parameters of radiation, temperature, wind velocity, and flow rate [168]. Ganeshkumar examined the error values of parameters such as mass flow rate, Nu, Convective heat transfer coefficient (CHTC), heat gain, solar energy, and energy efficiency for the solar air heater [169]. Amin adopted the Holman method to calculate the error values caused by solar radiation, temperature, and productivity in the freshwater production system, which were ±1.55%, ±1.2%, and ±1.58%, respectively [120]. Singh analyzed that the total error of the thermal performance system of solar ponds using NPCM was 2.59% [121]. Al-Karboly integrated the design of the photovoltaic system and analyzed that the total error caused by sensors or instruments was less than 4% [123]. Bassam utilized SiC and water to prepare nanofluids, and SiC and paraffin to prepare NPCMs, comprehensively designed photovoltaic systems, and analyzed the error values caused by temperature, radiation, voltage, and flow measurements [127,128]. The main materials and research contents of the above-mentioned study are listed in Table 12.

Table 12. The main materials and research contents about uncertainty quantification techniques.

Reference	Materials	System	Parameters	Total Uncertainty (%)
Al-Aasam et al. [165]	SiC, paraffin	PVT	Temperature, mass glow rate, solar irradiance, voltage and current	3.83
Daneshazarian et al. [166]	Ag, Cu, Al_2O_3, CuO, TiO_2, GNP, MWCNT, paraffin	Ground source heat pump	Temperature outlet, temperature inlet, flow rate	0.5
Gür et al. [167]	Y_2O_3, myristic acid	Electronics	The temperatures of PCM, the temperatures of empty system	1.5
Venkatesh et al. [168]	SiC, Ag, paraffin	Heat exchanger	Radiation, temperature, wind velocity, flow rate	-
GaneshKumar et al. [169]	Al_2O_3, paraffin	Solar air heater	Mass flow rate; Nu; CHTC; heat gain; solar energy; energy efficiency	±2.5; ±3.9; ±4.1; ±4.32; ±5.45; ±4.21
Amin et al. [120]	Graphene, beeswax	Freshwater production	Intensity of solar radiation, temperature, productivity	±2; ±0.1; ±5
Singh et al. [121]	Al_2O_3, OM-37	Solar ponds	Magnetic stirrer, weight of NP and PCM, thermal analysis in DSC, hot disk thermal constants analyzer, solar intensity, wind speed, temperature	2.59
Al-Karboly et al. [123]	SiC	PV	temperature, mass flow rate, solar irradiance, voltage and current.	3.82
Bassam et al. [127,128]	SiC, paraffin	PV	Temperature, radiation, voltage, flow	±0.1; ±1.5; ±1.0; ±0.5

5. Techno-Economic Analysis

The techno-economic analysis within the framework of NPCM is a comprehensive evaluation that combines considerations of both technological and economic feasibility [170]. This analysis delves deeply into the specifics of NPCM-related technologies, exploring everything from fundamental scientific principles to practical applications, and accurately assesses the costs and potential benefits associated with research and development, production, and maintenance. It not only predicts short-term financial impacts but also evaluates long-term economic sustainability. By integrating technological and economic dimensions, techno-economic analysis provides stakeholders with a holistic view, assisting them in making informed investment decisions in NPCM projects and ensuring the efficient allocation of resources in this emerging field [171].

5.1. Cost-Benefit Analysis

Cost-benefit analysis is a systematic decision-making tool, the core logic of which lies in quantifying and comparing all expected costs and benefits of an NPCM product over its entire lifecycle, while considering the time value of the product [172,173]. Ultimately, it calculates key indicators to evaluate the economic feasibility of the product as well as the relative advantages of different proposals [174]. For NPCM products, cost-benefit analysis effectively avoids resource waste by preemptively identifying projects where expected benefits will not cover costs, thus preventing the allocation of valuable funds to inherently unprofitable innovations. Furthermore, in the context of limited resources, it optimizes resource allocation by horizontally comparing the net benefit priorities of various NPCM proposals, ensuring investments are directed towards the avenues with the highest economic potential. This transforms innovation risks into measurable investment propositions, maximizing the likelihood of creating long-term value [175]. The following is an overview of the research on the application of the latest cost-benefit analysis method to the analysis of NPCM.

Sharshir analyzed that the hemispherical solar still device with metal organic framework is the most cost-effective and efficient, with the highest unit production cost and life cycle cost for producing fresh water [176]. Rasaiah studied multiple nanofluid improved double slope solar still devices and found that CuO nanofluid has the highest cost-effectiveness in terms of economy and environmental protection [177]. Kumar's study found that the configuration of aluminum foam and PCM with microchannels helps maintain efficient flow conditions but has higher maintenance costs [178]. Gürbüz designed a collector embedded with S NPs and analyzed it from multiple perspectives, such as efficiency, energy, and cost [179]. Mohtasim studied the energy recovery period and economic and environmental factors, such as CO_2 reduction of solar still systems and PVT systems with different structures [180]. Singh combined multiple NPCMs, the configuration of the device, and the analysis of daily freshwater production and investment payback period parameters under various conditions were carried out [181]. Anika comprehensively analyzed the performance, economy, environment, and other factors of the distiller constructed with mixed NPCMs and proposed improvement suggestions [182]. Rad designed a PVT system using nanofluid as a coolant and found that the unit production cost was 0.134 $/kWh [183]. Essa placed PCM and Ag NPs under the absorber to design the best-performing solar still, and the research results showed that the use of the device increased freshwater productivity by 245% [184]. Paul regarded different NPs as important factors affecting the thermal and environmental economic performance of the solar distiller [185]. Elamy combined Ag NPs for comprehensive solar still research; the results showed that the productivity of the slipped material increased by 233%, and both the environmental and economic cost-effectiveness were improved [186]. Attia's work demonstrated that incorporating graphene materials into the design of solar stills led to a remarkable 574.1% enhancement in energy efficiency [187]. El-Gazar combined the characteristics of TiO_2 and SiO_2 to design a solar still system and found that the production cost of distilled water was lower in summer than in winter [188]. The main materials and research contents of the above-mentioned study are listed in Table 13.

Table 13. The main materials and research contents about cost-benefit analysis.

System	Product	Unit Production Cost	Life Cycle Cost	Payback Period	CO_2 Reduction	Ref.
Solar still	Distilled water	0.0115 $/L	130 $	-	1.57 ton/year	[176]
Solar still	Distilled water	0.09 $/L	219 $	496 days	16.80 ton/payback period	[177]
Battery	Flow condition	-	49–70 $	-	-	[178]
Collector	Electrical energy	0.0138 €/kWh	3.6624 €	0.058 year	-	[179]
Solar still/PVT	Distilled water; electrical energy	0.01495 $/L	-	7.263 year; 0.26 year	3322.12 kg (emissions), 564.82 kg (mitigation); 6.98 ton/payback period	[180]
Solar still	Distilled water	0.012 $/L	-	129 days	-	[181]
Solar still	Distilled water	0.0192 $/L	-	7.26 year	322.12 kg (emissions); 565 kg (mitigation)	[182]
PVT	Electrical energy	0.134 $/kWh	-	-	-	[183]
Solar still	Distilled water	0.011 $/L	-	-	-	[184]
Solar still	Distilled water	-	-	-	448 kg/m^2 year	[185]
Solar still	Distilled water	0.01 $/L		420 days	28.8 ton/year	[186]
Solar still	Distilled water	0.00713 $/L	65.85$	-	-	[187]
Solar still	Distilled water	0.0156 $/L (in summer), 0.0264 $/L (in winter)	-	-	16.38 ton/year (in summer), 19.47 ton/year (in winter)	[188]

After delving into a series of cases where the latest cost-benefit analysis method was applied to NPCM analysis, the remarkable practical advantages and potential of this method are evident. These cases span multiple industries like freshwater, photovoltaic, and electric power, and involve complex NPCM scenarios. They offer decision-makers clear, intuitive, and actionable references. This approach effectively overcomes the blind spots and limitations of conventional analysis. By applying it, businesses can optimize resource allocation in NPCM implementation. This, in turn, enhances economic efficiency while advancing sustainable development goals.

5.2. Policy Incentives and Market Growth Potential

Policy incentives are constructing a multi-level environmental governance system, driving the global green transition. Under the UN framework, conference of the parties (COP) 29 established the climate finance target, mandating developed countries to provide $300 billion annually in public funds by 2035 to leverage $1.3 trillion in support for developing countries' infrastructure development [189]; COP30 proposes establishing a carbon tariff revenue-sharing mechanism and a deforestation accountability mechanism to ensure biofuel and mineral development do not harm the ecological rights of developing countries [190]. At the international institutional level, the International Energy Agency (IEA) demonstrated through its Strategy for an Affordable and Fair Clean Energy Transition that renewable energy costs are lower than fossil fuels [191], and its 2023 Energy Efficiency Report requires doubling the global annual energy efficiency improvement rate to 4% by 2030 [192]; the tripling renewable energy capacity target (to 11.2 TW by 2030) jointly advanced by the IEA and International Renewable Energy Agency (IRENA) has shown tangible results that China contributed 64% of the 585 GW global new installations in 2024 [193]. The EU established a 40% minimum domestic clean energy production capacity target by 2030 through its Net Zero Industry Act [194], complemented by the Batteries and Waste Batteries Regulation mandating full lifecycle carbon footprint disclosure (implementing a battery passport by 2027) [195], combined with the Carbon Border Adjustment Mechanism imposing comprehensive tariffs on six high-carbon import sectors [196]. Supported by the Critical Raw Materials Act aiming to increase strategic raw material processing to 40% and recycling rates to 25% [197], and augmented by the EU Wind Power Action Plan streamlining permitting processes (targeting 120 GW for North Sea wind power by 2030) [198], this forms a tripartite "technological autonomy-supply chain decarbonization-resource circularity" system. The US revised the Inflation Reduction Act, raising clean electricity tax credits to 35% (with domestic manufacturing requirements) and removing battery mineral restrictions [199], while New York State enacted the Climate Change Superfund Act, imposing a $75 billion levy on fossil fuel companies [200]. Developing countries are advancing concurrently: China's Energy Law established a renewable energy consumption guarantee mechanism and initiated a unified energy market [201], while its Draft Ecological Environment Code consolidated over 30 laws to create a dedicated "Green and Low-Carbon" chapter [202]; India launched a national carbon market covering power and petrochemical sectors, paired with a 280 GW solar target [203]; Vietnam's Revised Electricity Law introduced direct green power purchase agreements exempting them from bidding [204]; Brazil invested R$21 billion in its National Hydrogen Program [205]; the UAE's National Biofuel Policy established production standards promoting waste cooking oil conversion, targeting a 20% increase in biodiesel consumption by 2050 [206]. These policies, relying on mandatory fund allocation, technical standards, market mechanism restructuring, and ecological rights protection, transform global environmental commitments into enforceable binding rules, highlighting the critical role of policy-driven approaches in environmental governance. The main research contents of the above-mentioned study are listed in Table 14.

Table 14. The main research content about policy.

Policy/Legislation Name	Country/Region	Effective/Revised Date	Core Content	Ref.
New Collective Quantified Goal on Climate Finance	UN (COP29)	Nov 2024	Requires developed countries to provide $300 billion in public funding annually by 2035, mobilizing $1.3 trillion annually to support developing countries.	[189]
-	UN (COP30)	-	A carbon tariff revenue-sharing and deforestation accountability mechanism; Biofuel production and mineral extraction do not sacrifice the ecological rights of developing countries.	[190]
Strategies for Affordable and Fair Clean Energy Transitions	IEA	May 2024	Argues clean energy costs less than fossil fuels; proposes energy efficiency subsidies, promotion of efficient appliances, and clean transport.	[191]
2023 Energy Efficiency Report	IEA	Nov 2023	Requires doubling the global annual energy efficiency improvement rate to 4% by 2030.	[192]
Renewable Energy Capacity Tripling Goal	IEA/IRENA	2023–2025 (Progress)	Global renewable energy capacity to reach 11.2 TW by 2030; 585 GW added in 2024 (China accounted for 64%).	[193]
Net Zero Industry Act	EU	2025 (Implementation)	Local net-zero tech production capacity ≥40% by 2030; limits dependence on any single third country to >50%.	[194]
Batteries and Waste Batteries Regulation	EU	Feb 2025	Mandates disclosure of full-lifecycle carbon footprint for batteries; implements "battery passport" from 2027.	[195]
Carbon Border Adjustment Mechanism	EU	1 January 2025	Full carbon tax implementation on imports across six sectors (steel, Al, etc.).	[196]
Critical Raw Materials Act	EU	2023 (Adopted)	By 2030: 10% domestic extraction, 40% domestic processing, and 25% domestic recycling capacity for strategic raw materials.	[197]
European Wind Power Action Plan	EU	Oct 2023	Simplifies wind permitting; targets 120 GW North Sea offshore wind capacity by 2030.	[198]
Inflation Reduction Act Amendments	USA	2025 (Revision)	Raises clean electricity tax credit to 35% (requires domestic manufacturing); removes battery mineral "Foreign Entity of Concern" restrictions.	[199]
Climate Change Superfund Act	USA	2025	New York State to levy $75 billion over 25 years on fossil fuel companies.	[200]
Energy Law of the People's Republic of China	China	1 January 2025	Establishes renewable energy consumption guarantee; defines financial energy attributes; builds a unified energy market.	[201]
Ecological and Environmental Code (Draft)	China	Apr 2025 (Review)	Consolidates over 30 environmental laws; first-ever "Green and Low-Carbon Development" chapter.	[202]
National Carbon Market	India	Apr 2025 (Launch)	Covers power and petrochemical sectors; coupled with 280 GW solar PV target.	[203]

Table 14. *Cont.*

Policy/Legislation Name	Country/Region	Effective/Revised Date	Core Content	Ref.
Revised Electricity Law	Vietnam	1 February 2025	Introduces Direct Power Purchase Agreements, allowing green projects to bypass bidding.	[204]
National Hydrogen Program	Brazil	2023–2025	Invests BRL 21 billion; aims to establish a low-carbon hydrogen industrial chain by 2030.	[205]
National Biofuels Policy	UAE	Apr 2024 (Approved)	Sets biofuel production standards and trading rules; targets 20% increase in biodiesel consumption by 2050; promotes waste cooking oil conversion to biofuel.	[206]

6. Conclusions and Future Work

This review comprehensively analyzes NPCM's high thermal performance research methods and economic benefits, delves into its melting mechanisms, and optimal design configurations. Starting with the fundamental types of NPCM, it offers in-depth explanations of these mechanisms and designs, highlighting their advantages in thermal energy management. The review also incorporates cost-benefit analyses of application systems and the latest international environmental policies to fully assess NPCM's energy-saving and eco-friendly attributes. These findings demonstrate NPCM's great potential in materials innovation and low-carbon environmental protection, strongly supporting sustainable development.

6.1. Key Findings and Contributions

(1) NPCMs improve thermal conductivity, reduce supercooling, and accelerate melting/solidification rates.

(2) Various NPCM structures, such as thermophysical, photonic, fluid-based, encapsulated, and phase-engineered, enable targeted applications across different solar thermal systems.

(3) Mechanism-level analysis using MD and DFT provides critical insights into nanoparticle-PCM interactions and nano-scale heat transfer behavior.

(4) Techno-economic analyses confirm that NPCMs offer high cost-effectiveness, improved energy efficiency, and extended lifecycle benefits for solar energy systems.

(5) The identical Cu NPs and paraffin-based NPCM structure exhibit analogous thermal property inflection points at significantly disparate Cu NP concentrations. Consequently, it is imperative to establish a rigorous simulation model to analyze the correlation between nanoparticle concentration and the thermal performance characteristics of NPCM.

(6) The unit production cost of distilled water for solar distillers using NPCM can be as low as \$0.00713/L, and other models of solar distillers also demonstrate good, low-carbon, and environmentally friendly economic performance.

(7) The combined use of mechanistic analysis methods and simulation methods is more conducive to studying the phase change mechanism of NPCM during the melting process.

(8) The combination of a finite element analysis model and experimental data is more advantageous for the analysis of the melting phase transition mechanism of NPCM under the action of a magnetic field.

(9) The application of algorithm optimization and analysis methods, particularly integrated algorithmic models based on artificial neural networks (ANNs), can be

harnessed to more effectively project the material configurations of devices or systems that achieve high thermal efficiency performance.
(10) The incorporation of graphene materials significantly enhances the performance characteristics of PCMs. For instance, in solar still applications, it improves energy efficiency by 574.1%. Therefore, graphene and other high-performance materials should be prioritized in future NPCM research.
(11) Increasing policy incentives, clean energy mandates, and investments in renewable infrastructure are accelerating the commercial potential of NPCM-based solutions.

6.2. Remaining Challenges and Future Opportunities

Although NPCM exhibits strong thermal performance in melt heat transfer, the design and selection of its nano materials or structures face challenges in practical applications. As different researchers have discussed NPCM from various research perspectives, integrating innovative research ideas into the design of NPCM application systems has become a mainstream research direction. Research on NPCM holds extensive promise, and future leading research methods are likely to be associated with it.

(1) The accuracy of ideal model simulation methods, which is related to mesh size, models, quantities, and solution methods, necessitates further research to design high-precision physical simulation models and reduce interference from inaccuracies in real-world models.
(2) In algorithm optimization analysis, which demands extensive data collection with accuracy being critical, the impact of varying parameter choices in different algorithms on model outcomes necessitates further exploration into constructing high-quality datasets and developing efficient, accurate algorithmic models.
(3) During the experiment, the inaccuracy caused by measuring devices is inevitable, highlighting the necessity for more precise system accuracy control and making it necessary to conduct uncertainty analysis on key steps.
(4) In actual solar energy system applications, unforeseen circumstances often arise during operation, so future research could integrate simulations of additional weather factors to develop a more realistic and production-oriented simulation model.
(5) An intelligent optimization model for solar-powered freshwater production systems should be established, since the study of nanomaterials within a single system is limited in scope, failing to take into account the substitutability of materials and the acceptability of economic costs.
(6) Although significant progress has been made in the research of thermodynamic melting associated with magnetic field convection across various fields, it is necessary to further explore the economic benefits of these findings and address the insufficient studies on the impact of eddy current characteristics in magnetic fields on the thermal properties of materials.

Author Contributions: Conceptualization, J.W. and W.L.; methodology, W.L.; software, M.W.; validation, M.W., J.Y. and D.M.; formal analysis, W.L.; investigation, W.L. and M.W.; resources, J.Y.; data curation, W.L.; writing—original draft preparation, W.L. and J.W.; writing—review and editing, E.H. and J.W.; visualization, D.M.; supervision, J.W.; project administration, J.W.; funding acquisition, J.W. All authors have read and agreed to the published version of the manuscript.

Funding: This research received no external funding.

Data Availability Statement: No new data were created or analyzed in this study.

Conflicts of Interest: The authors declare no conflict of interest.

Abbreviations

ANN	Artificial neural network
BA	Bat algorithm
BBO	Optimization based on biogeography
BET	Brunauer–Emmett–Teller
CFD	Computational fluid dynamic
CHTC	Convective heat transfer coefficient
COP	Conference of the parties
DFT	Density functional theory
DHPD	$Na_2HPO_4 \cdot 12H_2O$
DSC	Differential scanning calorimetry
F	Furfural
FEA	Finite element analysis
Fo	Fourier number
FTIR	Fourier transform infrared spectroscopy
GNP	Graphene nanosheets
GWO	Grey wolf optimization
IEA	International energy agency
IRENA	International renewable energy agency
L	Liquid
LHTES	Latent heat thermal energy storage
LIB	Lithium-ion battery
LSTM	Long short-term memory
MA	Myristic acid
MD	Molecular dynamics
MLP	Multilayer perceptron
MWCNT	Multi-walled carbon nanotube
NG	Nano-graphite
Ni@C/C	The hierarchically porous carbon structure infiltrated with Ni@graphite
NP	Nanoparticle
NPCM	Nanoparticle-enhanced phase change material
Nu	Nusselt
ODA	Octadecanol
OM	Oleic-acid-modified
PCB	Printed circuit board
PCM	Phase change material
PSO	Particle swarm optimization
PV	Photovoltaic
PVT	Photovoltaic-thermal
R	Resorcinol
RBF	Radial basis function
S	Solid
SA	Silica aerogel
SEM	Scanning electron microscope
SIMPLE	Semi-implicit method for pressure linked equations
SH	Sherwood number
TC	Thermal conductivity
TEM	Transmission electron microscope
TGA	Thermogravimetric analysis
TLC	Thermochromic liquid crystal
VMTNS	Vermiculite nanosheets

XGBoost	Extreme gradient boosting
XPS	X–ray photoelectron spectroscopy
XRD	X–ray diffractometer

References

1. Wang, J.; Gullo, P.; Ramezani, H. Review on the trend of ultra-low-GWP working fluids for small-capacity vapour-compression systems. *Sustain. Energy Technol. Assess.* **2024**, *66*, 103803. [CrossRef]
2. Alizadeh, O. A review of ENSO teleconnections at present and under future global warming. *WIREs Clim. Change* **2024**, *15*, e861. [CrossRef]
3. Rodés-Guirao, H.R.a.L. Peak Global Population and Other Key Findings from the 2024 UN World Population Prospects. Available online: https://ourworldindata.org/un-population-2024-revision?weekend-reading-link-040824/ (accessed on 1 May 2025).
4. Ngagoum Ndalloka, Z.; Vijayakumar Nair, H.; Alpert, S.; Schmid, C. Solar photovoltaic recycling strategies. *Sol. Energy* **2024**, *270*, 112379. [CrossRef]
5. Yu, J.; Cheng, H.; Wang, Y.; He, C.; Zhou, B.; Liu, C.; Feng, Y. Multiple shearing-induced high alignment in polyethylene/graphene films for enhancing thermal conductivity and solar-thermal conversion performance. *Chem. Eng. J.* **2024**, *480*, 148062. [CrossRef]
6. Wang, J.; Hu, E.; Blazewicz, A.; Ezzat, A.W. Investigation on the long-term performance of solar thermal powered adsorption refrigeration system based on hourly accumulated daily cycles. *Heat Mass Transf.* **2021**, *57*, 361–375. [CrossRef]
7. Ali, F.H.; Al-Amir, Q.R.; Hamzah, H.K.; Alahmer, A. Unveiling the potential of solar cooling technologies for sustainable energy and environmental solutions. *Energy Convers. Manag.* **2024**, *321*, 119034. [CrossRef]
8. Wang, J.; Hu, E.; Blazewicz, A.; Ezzat, A.W. Simulation of accumulated performance of a solar thermal powered adsorption refrigeration system with daily climate conditions. *Energy* **2018**, *165*, 487–498. [CrossRef]
9. Kasaeian, A.; Nazari, N.S.; Masoumi, A.; Shabestari, S.T.; Jadidi, M.; Fereidooni, L.; Bidhendi, M.E. A review on phase change materials in different types of solar stills. *J. Energy Storage* **2024**, *99*, 113430. [CrossRef]
10. Wang, B.; Liu, Y.; Wang, D.; Song, C.; Fu, Z.; Zhang, C. A review of the photothermal-photovoltaic energy supply system for building in solar energy enrichment zones. *Renew. Sustain. Energy Rev.* **2024**, *191*, 114100. [CrossRef]
11. Pintaldi, S.; Perfumo, C.; Sethuvenkatraman, S.; White, S.; Rosengarten, G. A review of thermal energy storage technologies and control approaches for solar cooling. *Renew. Sustain. Energy Rev.* **2015**, *41*, 975–995. [CrossRef]
12. Suresh, C.; Saini, R.P. Review on solar thermal energy storage technologies and their geometrical configurations. *Int. J. Energy Res.* **2020**, *44*, 4163–4195. [CrossRef]
13. Carnie, J.-T.; Hardalupas, Y.; Sergis, A. Decarbonising building heating and cooling: Designing a novel, inter-seasonal latent heat storage system. *Renew. Sustain. Energy Rev.* **2024**, *189*, 113897. [CrossRef]
14. Zhang, X.; Yu, C.; Zhang, C. Advances in latent heat storage technology for electronic cooling. *Renew. Sustain. Energy Rev.* **2025**, *215*, 115614. [CrossRef]
15. Ou, S.; Wang, P.; Yang, S.; Su, H. Thermal characteristics and energy-saving research of P-PCM FinTwall in hot summer and cold winter regions. *Energy* **2025**, *329*, 136670. [CrossRef]
16. Karaağaç, M.O. Experimental investigation of a novel PCM-enhanced underfloor heating system for energy-efficient sustainable buildings. *Appl. Therm. Eng.* **2025**, *275*, 126891. [CrossRef]
17. Chen, X.; Su, Y.; Zhang, Y.; Shen, J.; Xu, X.; Wang, X.; Zhou, F. Performance of thermal management system based on PCM/forked liquid-cold plate for 18650 cylindrical battery. *J. Energy Storage* **2024**, *91*, 112071. [CrossRef]
18. Kurşun, B.; Balta, M.; Karabulut, K. Exploring the impact of inner and middle channel geometries on the melting rate of PCM-metal foam composition in a triplex tube heat exchanger. *Therm. Sci. Eng. Prog.* **2024**, *51*, 102621. [CrossRef]
19. Palaniappan, M.; El-Shafay, A.S.; Shanmugan, S. Improving heat retention properties of steeped M-shape basin solar distillers utilizing paraffin RT50-enhanced silver nanoparticles and Manihot esculenta extracts. *Desalination* **2024**, *586*, 117836. [CrossRef]
20. Gürbüz, H.; Demirtürk, S.; Akçay, H.; Topalcı, Ü. Experimental investigation on electrical power and thermal energy storage performance of a solar hybrid PV/T-PCM energy conversion system. *J. Build. Eng.* **2023**, *69*, 106271. [CrossRef]
21. Manikandan, S.; Muthuvairavan, G.; Samykano, M.; Natarajan, S.K. Numerical simulation of various PCM container configurations for solar dryer application. *J. Energy Storage* **2024**, *86*, 111294. [CrossRef]
22. Ji, C.; Liu, D.; Waqas, H.; Sun, S.-L.; Jalili, B.; Abdelmohimen, M.A.H. Melting performance of Molten Salt with nanoparticles in a novel triple tube with leaf-shaped fins. *Case Stud. Therm. Eng.* **2024**, *61*, 105063. [CrossRef]
23. Ma, T.; Li, Z.; Lv, K.; Chang, D.; Hu, W.; Zou, Y. Design and performance analysis of deep peak shaving scheme for thermal power units based on high-temperature molten salt heat storage system. *Energy* **2024**, *288*, 129557. [CrossRef]
24. Reddy, B.D.; Rahul, S.V.S.; Harish, R. Impact of fin number and nanoparticle size on molten salt NanoPCM melting in finned annular space. *J. Energy Storage* **2023**, *72*, 108705. [CrossRef]
25. Gui, X.; Wang, S.; Ding, L. Simulation study of a novel phase change cooling garment for electricians in a high-temperature environment. *Int. J. Refrig.* **2024**, *168*, 79–88. [CrossRef]

26. Anter, A.G.; Sultan, A.A.; Hegazi, A.A.; El Bouz, M.A. Thermal performance and energy saving using phase change materials (PCM) integrated in building walls. *J. Energy Storage* **2023**, *67*, 107568. [CrossRef]
27. Madurai Elavarasan, R.; Nadarajah, M.; Pugazhendhi, R.; Gangatharan, S. An experimental investigation on coalescing the potentiality of PCM, fins and water to achieve sturdy cooling effect on PV panels. *Appl. Energy* **2024**, *356*, 122371. [CrossRef]
28. Zhan, H.; Mahyuddin, N.; Sulaiman, R.; Khayatian, F. Phase change material (PCM) integrations into buildings in hot climates with simulation access for energy performance and thermal comfort: A review. *Constr. Build. Mater.* **2023**, *397*, 132312. [CrossRef]
29. Gado, M.G. Thermal management and heat transfer enhancement of electronic devices using integrative phase change material (PCM) and triply periodic minimal surface (TPMS) heat sinks. *Appl. Therm. Eng.* **2025**, *258*, 124504. [CrossRef]
30. Madurai Elavarasan, R.; Pugazhendhi, R.; Shafiq, S.; Gangatharan, S.; Nadarajah, M.; Shafiullah, G.M. Efficiency enhancement of PV panels with passive thermal management using PCM: An exhaustive review on materials, designs and effective techniques. *Appl. Energy* **2025**, *382*, 125217. [CrossRef]
31. Anya, B.; Mohammadpourfard, M.; Akkurt, G.G.; Mohammadi-Ivatloo, B. Exploring geothermal energy based systems: Review from basics to smart systems. *Renew. Sustain. Energy Rev.* **2025**, *210*, 115185. [CrossRef]
32. Suo, Y.; Tang, C.; Jia, Q.; Zhao, W. Influence of PCM configuration and optimization of PCM proportion on the thermal management of a prismatic battery with a combined PCM and air cooling structure. *J. Energy Storage* **2024**, *80*, 110340. [CrossRef]
33. Al-Amayreh, M.I.; Alahmer, A. Efficiency enhancement in direct thermal energy storage systems using dual phase change materials and nanoparticle additives. *Case Stud. Therm. Eng.* **2024**, *59*, 104577. [CrossRef]
34. Murali, G.; Ramani, P.; Murugan, M.; Elumalai, P.V.; Ranjan Goud, N.U.; Prabhakar, S. Improved solar still productivity using PCM and nano-PCM composites integerated energy storage. *Sci. Rep.* **2024**, *14*, 15609. [CrossRef]
35. Kazaz, O.; Abu-Nada, E. Thermal performance of nano-architected phase change energetic materials for a next-generation solar harvesting system. *Energy Convers. Manag.* **2025**, *327*, 119541. [CrossRef]
36. Ulker, N.; Bulut, H.; Demircan, G. Comparative study on the effect of surface-modified nanoparticles on PCM for solar energy applications. *J. Therm. Anal. Calorim.* **2024**, *149*, 12053–12070. [CrossRef]
37. Bestas, S.; Aktas, I.S.; Bayrak, F. A bibliometric and performance evaluation of nano-PCM-integrated photovoltaic panels: Energy, exergy, environmental and sustainability perspectives. *Renew. Energy* **2024**, *226*, 120383. [CrossRef]
38. Goel, V.; Dwivedi, A.; Kumar, R.; Kumar, R.; Pandey, A.K.; Chopra, K.; Tyagi, V.V. PCM-assisted energy storage systems for solar-thermal applications: Review of the associated problems and their mitigation strategies. *J. Energy Storage* **2023**, *69*, 107912. [CrossRef]
39. Abdolahimoghadam, M.; Rahimi, M. The effect of the number of tubes on the charging and discharging performances of a novel bio-nPCM within a vertical multi-tube TES system. *Energy* **2025**, *319*, 135010. [CrossRef]
40. Dhivagar, R.; Singh, P.K.; Nandhakumar, E. Synergistic effect of stearic acid/bismuth oxychloride/cupric oxide for thermal storage applications: Preparation, stability, rheological and thermophysical analysis. *Phys. Scr.* **2024**, *99*, 065039.
41. Azeez, H.L.; Ibrahim, A.; Ahmed, B.O.; Dol, S.S.; Al-Waeli, A.H.A.; Jaber, M. Experimental evaluation of a novel photovoltaic thermal System: Energy, economic, and exergy-based sustainability analysis. *Case Stud. Therm. Eng.* **2025**, *71*, 106167. [CrossRef]
42. Ahmed, B.O.; Ibrahim, A.; Azeez, H.L.; Dol, S.S.; Al-Waeli, A.H.A.; Jaber, M. Energy and exergy analysis of a newly designed photovoltaic thermal system featuring ribs, petal array, and coiled twisted tapes: Experimental analysis. *Case Stud. Therm. Eng.* **2024**, *63*, 105388. [CrossRef]
43. Maseer, M.M.; Ismail, F.B.; Kazem, H.A.; Hachim, D.M.; Al-Gburi, K.A.H.; Chaichan, M.T. Performance enhancement of photovoltaic/thermal collector semicircle absorber tubes using nanofluid and NPCM. *Renew. Energy* **2024**, *233*, 121152. [CrossRef]
44. Máñez-Espina, L.M.; Díaz-Rubio, A. Control of the local and nonlocal electromagnetic response in all-dielectric reconfigurable metasurfaces. *Phys. Rev. Appl.* **2024**, *22*, 054022. [CrossRef]
45. Sathishkumar, A.; Sundaram, P.; Prabakaran, R.; Kim, S.C. Graphene based fatty acid phase change material as an alternative to water: An experimental investigation on thermal energy storage characteristics. *Colloids Surf. A Physicochem. Eng. Asp.* **2025**, *715*, 136589. [CrossRef]
46. Baiju, V.; Priya, K.L.; Shafi, K.A.; Shahid, P.A.M.; Lamba, R.; Singh, H.; Ichiyanagi, M.; Suzuki, T.; Shajan, S. Enhanced latent heat storage for solar thermal applications using tapered fins and hybrid nano-additives. *J. Energy Storage* **2025**, *118*, 116308. [CrossRef]
47. Wu, Z.; Shao, Y.; Li, M.; Zou, M.; Wu, K.; Chen, W.; Huang, X.; Ma, C.; Ran, J. Stability and discharging performance improvements of modified nano particle enhanced binary alkanes capsule for cold latent heat storage tank. *J. Energy Storage* **2025**, *117*, 116169. [CrossRef]
48. Shijina, S.S.; Akbar, S.; Sajith, V. Graphene functionalized nano-encapsulated composite phase change material based nanofluid for battery cooling: An experimental investigation. *Appl. Therm. Eng.* **2025**, *259*, 124893. [CrossRef]
49. Sivashankar, M.; Selvam, C.; Suresh, S. Experimental study on the performance of low concentrated solar photovoltaic system with nano-enhanced phase change material encapsulated heat sink. *Appl. Therm. Eng.* **2025**, *262*, 125254. [CrossRef]

50. Sutheesh, P.M.; Thomas, R.; Bandaru, R. Battery thermal management systems by using DI water and PCF with Nano-encapsulated Tetradecan-1-ol for Lithium-ion batteries. *Int. Commun. Heat Mass Transf.* **2025**, *165*, 108992. [CrossRef]
51. Han, Y.; Wang, L.; Cao, K.; Zhou, J.; Zhu, Y.; Hou, Y.; Lu, Y. In Situ TEM Characterization and Modulation for Phase Engineering of Nanomaterials. *Chem. Rev.* **2023**, *123*, 14119–14184. [CrossRef]
52. He, Q.; Sheng, B.; Zhu, K.; Zhou, Y.; Qiao, S.; Wang, Z.; Song, L. Phase Engineering and Synchrotron-Based Study on Two-Dimensional Energy Nanomaterials. *Chem. Rev.* **2023**, *123*, 10750–10807. [CrossRef]
53. Liu, X.; Shan, J.; Cao, T.; Zhu, L.; Ma, J.; Wang, G.; Shi, Z.; Yang, Q.; Ma, M.; Liu, Z.; et al. On-device phase engineering. *Nat. Mater.* **2024**, *23*, 1363–1369. [CrossRef]
54. Zang, P.; Yu, C.; Zhang, R.; Yang, D.; Gai, S.; Liu, B.; Shen, R.; Yang, P.; Lin, J. Phase Engineered CuxS–Ag2S with Photothermoelectric Activity for Enhanced Multienzyme Activity and Dynamic Therapy. *Adv. Mater.* **2024**, *36*, 2400416. [CrossRef] [PubMed]
55. Yang, Z.; Ding, Y.; Chen, W.; Luo, S.; Cao, D.; Long, X.; Xie, L.; Zhou, X.; Cai, X.; Liu, K.; et al. Phase-Engineered Bi-RuO_2 Single-Atom Alloy Oxide Boosting Oxygen Evolution Electrocatalysis in Proton Exchange Membrane Water Electrolyzer. *Adv. Mater.* **2025**, *37*, 2417777. [CrossRef] [PubMed]
56. Çiçek, B. A numerical comparison of the thermal performances of nano-PCM heat sinks with Fe_3O_4, MgO, ZnO and xGNP nanoparticles: Key role of increased thermal conductivity. *Therm. Sci. Eng. Prog.* **2025**, *63*, 103712. [CrossRef]
57. Praveenkumar, N.; Lingadurai, K.; Ramkumar, T.; Bharathiraja, R. Characterization and thermophysical property evaluation of ceramic and metal oxide-based hybrid nanoparticles enhanced paraffin PCM. *J. Therm. Anal. Calorim.* **2025**, *150*, 3319–3334. [CrossRef]
58. Zhai, Z.; Zhu, J.; Cao, S.; Zhang, Y.; Van der Bruggen, B. Supramolecular Engineering of Metal–Organic Membranes with Metal-Support Interactions for Nanoconfined Synergistic Catalysis. *Small Methods* **2025**, 2500409. [CrossRef]
59. Karimipour, A.; Asiri, S.A.; Alfawaz, K.M.; Alogla, A.F.; Abu-Hamdeh, N.H.; Viet, P.H.H. The use of molecular dynamics method to evaluate the thermo-physical properties of Cu nanoparticles dispersed in Paraffin wax PCM. *J. Taiwan Inst. Chem. Eng.* **2024**, *165*, 105785. [CrossRef]
60. Geng, L.; Ali, A.B.M.; Babadoust, S.; Kumar, A.; Abdullaeva, B.; Hussein, R.A.; Salahshour, S.; Esmaeili, S. Modeling the influence of external heat flux on thermal characteristics of the silica aerogel/paraffin in a cylindrical atomic duct. *Case Stud. Therm. Eng.* **2025**, *65*, 105633. [CrossRef]
61. Mim, M.; Habib, K.; Farabi, S.N.; Zaed, M.A.; Saidur, R. Investigation of the thermophysical properties of PCMs with novel ionic liquid assisted nanocomposite for sustainable thermal energy storage application. *Case Stud. Therm. Eng.* **2025**, *70*, 106117. [CrossRef]
62. Liu, Y.; Basem, A.; Al-zahy, Y.M.A.; Singh, N.S.S.; Al-Bahrani, M.; Abduvalieva, D.; Salahshour, S.; Esmaeili, S. Molecular dynamics simulation of thermal behavior of paraffin/Cu nanoparticle PCM in a non-connected rotating ribbed tube. *Int. Commun. Heat Mass Transf.* **2025**, *165*, 109058. [CrossRef]
63. Sawaran Singh, N.S.; Hassan, W.H.; Thiab, R.F.; Al-zahy, Y.M.A.; Salahshour, S.; Hekmatifar, M. Effect of copper nanoparticle volume fraction on flow in a 3D lid-driven cavity with phase change materials using molecular dynamics simulation. *Case Stud. Chem. Environ. Eng.* **2025**, *11*, 101181. [CrossRef]
64. Cao, C.; Ali, A.B.M.; Hussein, Z.A.; Sawaran Singh, N.S.; Abdullaeva, B.; Zubair, A.; Salahshour, S.; Baghaei, S. Analysis of different phase change materials (PCMs) and wall material in a nano-circular space thermal energy storage (TES) system: A molecular dynamics approach. *Int. J. Therm. Sci.* **2025**, *215*, 109954. [CrossRef]
65. Huang, W.; Zhou, G. Molecular dynamics simulation of heat, mass and phonon transport performances of nanoparticle enhanced $Na_2SO_4{\cdot}10H_2O$. *J. Mol. Liq.* **2024**, *397*, 124157. [CrossRef]
66. Gao, Y.; Basem, A.; Mohammad Sajadi, S.; Jasim, D.J.; Nasajpour-Esfahani, N.; Salahshour, S.; Esmaeili, S.; Baghaei, S. The effect of initial pressure on the thermal behavior of the silica aerogel/PCM/CuO nanostructure inside a cylindrical duct using molecular dynamics simulation. *Case Stud. Therm. Eng.* **2024**, *54*, 104064. [CrossRef]
67. Yang, J.; Ali, A.B.M.; Al-zahy, Y.M.A.; Sawaran Singh, N.S.; Al-Bahrani, M.; Orlova, T.; Rahimi, M.; Salahshour, S.; Esmaeili, S. The effect of copper oxide nanoparticles on the thermal behavior of silica aerogel/paraffin as a phase change material in a cylindrical channel with molecular dynamics simulation. *Prog. Nucl. Energy* **2025**, *181*, 105645. [CrossRef]
68. Ru, Y.; Ali, A.B.M.; Qader, K.H.; Singh, N.S.S.; Jhala, R.; Soliyeva, M.; Salahshour, S.; Esmaeili, S. Investigating the effect of copper oxide nanoparticles radius on thermal behavior of silica aerogel/paraffin nanostructure using molecular dynamics simulation. *Int. Commun. Heat Mass Transf.* **2025**, *161*, 108547. [CrossRef]
69. An, Q.; Bagheritabar, M.; Basem, A.; Ghabra, A.A.; Li, Y.; Tang, M.; Sabri, L.S.; Sabetvand, R. The effect of size of copper oxide nanoparticles on the thermal behavior of silica aerogel/paraffin nanostructure in a duct using molecular dynamics simulation. *Case Stud. Therm. Eng.* **2024**, *60*, 104666. [CrossRef]
70. Yu, Y.; Liu, Y.; Chai, H.; Tian, H.; Wu, X.; Zhao, C. Nanoscale insight into the thermal properties of lauric acid and CuO based phase change material used for thermal energy storage. *J. Mol. Liq.* **2024**, *393*, 123630. [CrossRef]

71. Fang, M.; Cao, L.; Cao, Y. Optimization of thermophysical properties of molten salt-based date-occupied composites based on orthogonal tests. *J. Energy Storage* **2024**, *102*, 114183. [CrossRef]
72. Badreldin, A.; Bouhali, O.; Abdel-Wahab, A. Complimentary Computational Cues for Water Electrocatalysis: A DFT and ML Perspective. *Adv. Funct. Mater.* **2024**, *34*, 2312425. [CrossRef]
73. Rajabi, A.; Grotjahn, R.; Rappoport, D.; Furche, F. A DFT perspective on organometallic lanthanide chemistry. *Dalton Trans.* **2024**, *53*, 410–417. [CrossRef] [PubMed]
74. Chen, Z.; Zhang, J.; Deng, S.; Hou, M.; Zhang, X.; Jiang, Z.; Lai, N.-C. Morphology-controlled synthesis of Cu_2O encapsulated phase change materials: Photothermal conversion and storage performance in visible light regime. *Chem. Eng. J.* **2023**, *454*, 140089. [CrossRef]
75. Liang, S.; Wang, X.; Chen, L.; Qiu, X. Utilizing heterogeneity of lignin to diminish supercooling of phase change material nano-capsules with high latent heat. *J. Colloid Interface Sci.* **2025**, *683*, 833–840. [CrossRef]
76. Zhang, G.; Ren, S.; Dong, C.; Dong, W.; Du, Q.; Chen, X.; Li, A. Ni@graphite carbon synergistic reinforcement sites penetrated hierarchical porous carbon boosting PCMs encapsulation and solar-thermal energy storage. *Compos. Sci. Technol.* **2025**, *268*, 111228. [CrossRef]
77. Yao, Z.; Xie, J.; Fu, T.; Luo, Y.; Yang, X. One-pot preparation of phase change material employing nano-scaled resorcinol-furfural frameworks. *Chem. Eng. J.* **2024**, *484*, 149553. [CrossRef]
78. Tan, X.; Gu, X.; Nian, H.; Wang, X.; Li, J.; Wang, Q.; Wang, Z.; Zhao, Y.; Zhou, Y. Enhanced thermal performance of $NaCH_3COO{\cdot}3H_2O$-$Na_2S_2O_3{\cdot}5H_2O$ eutectic based composite phase change materials with hybrid dimensional carbon nanomaterials and modified lotus root starch. *J. Energy Storage* **2024**, *95*, 112418. [CrossRef]
79. Yuan, K.; Zhang, A.; Li, S.; Zhou, T.; Wang, Q.; Cao, W. Nanopore constrained thermal characteristics investigation of MIL-101 (Fe)/tetradecanol composite phase change materials. *Surf. Interfaces* **2025**, *60*, 106013. [CrossRef]
80. Wang, K.; Xue, Y.; Lin, J.; Yu, J.; Wang, Q.; Xie, Z.; Yu, H.; Qiu, X. ZIF-8 based microspheres with ordered mesopores and dual responsive surfaces. *Surf. Interfaces* **2023**, *38*, 102876. [CrossRef]
81. Zeng, X.; Zhu, X.; Sun, Y.; Hu, Y. Improving the thermal stability and operation speed of Sb_7Se_3 films via carbon nanolayers. *Appl. Surf. Sci.* **2024**, *651*, 159295. [CrossRef]
82. Xin, Y.-X.; Hu, D.; Li, J.-H.; Ge, Y.-X.; Wang, Y.-X. Incorporated surfaces and confined spaces of $Na_2HPO_4{\cdot}12H_2O$–Layered vermiculite nanosheets for phase change materials with enhanced thermal properties. *Appl. Surf. Sci.* **2024**, *675*, 160881. [CrossRef]
83. Kwon, M.S.; Jin, X.; Kim, Y.C.; Hu, J.W. Development of microencapsulated PCM concrete with improved strength and long-term thermal performance using MWCNTs. *Constr. Build. Mater.* **2024**, *442*, 137609. [CrossRef]
84. Yeşilata, B.; Bayram, M.; Ustaoğlu, A.; Kurşuncu, B.; Hekimoğlu, G.; Sarı, A.; Çıkman, İ.Ü.; Gencel, O.; Ozbakkaloglu, T. Effects of Cavity-Shapes in 3D printed PCM encapsulation plates on sustainable thermal energy efficiency in buildings. *Therm. Sci. Eng. Prog.* **2025**, *62*, 103570. [CrossRef]
85. Bonifacio, C.S.; Yu, Y.; Ray, M.L.; Skowronski, M.; Fischione, P. TEM Preparation and Characterization of a GeTe-based Phase Change Memory Device at Partial SET Mode. *Microsc. Microanal.* **2024**, *30*, ozae044-607. [CrossRef]
86. Karimi, Y.; Solaimany Nazar, A.R.; Molla-Abbasi, P. Synthesis, characterization and application of a magnetic nanoencapsulated PCM in enhancing the natural convection in a cube cavity. *Therm. Sci. Eng. Prog.* **2024**, *47*, 102342. [CrossRef]
87. Ennamri, A.; Bencaid, J.; Draoui, K.; Ouarga, A.; Oualid, H.A. Development of a new kaolinite/phase change material (PCM) composite for latent heat thermal energy storage in building applications. *Mater. Sci. Eng. B* **2025**, *319*, 118366. [CrossRef]
88. Hassan, E.A.; Tony, M.A.; Awad, M.M. Thermal Energy Storage Using Hybrid Nanofluid Phase Change Material (PCM) Based on Waste Sludge Incorp Rated ZnO/α-Fe_2O_3. *Nanomaterials* **2024**, *14*, 604. [CrossRef]
89. Jiang, D.; Chen, P.; Liu, J.; Xu, Y. Preparation and performance study of modified diatomite based PCM for construction. *Green Mater.* **2024**, 1–13. [CrossRef]
90. Vignesh, T.; Selvakumar, D.; Madhankumar, S. Drying kinetics, energy, colour and FTIR spectroscopy analysis on indirect solar dryer with paraffin wax and glass pieces as thermal storage material. *Sol. Energy* **2024**, *282*, 112925. [CrossRef]
91. Gao, M.; Zhao, S.; Yang, H.; Wu, X.; Xiao, Y. An Analysis of the Influence of DSC Parameters on the Measurement of the Thermal Properties of Phase-Change Material. *Materials* **2024**, *17*, 5689. [CrossRef]
92. Laasri, I.A.; Charai, M.; Es-sakali, N.; Mghazli, M.O.; Outzourhit, A. Evaluating passive PCM performance in building envelopes for semi-arid climate: Experimental and numerical insights on hysteresis, sub-cooling, and energy savings. *J. Build. Eng.* **2024**, *98*, 111161. [CrossRef]
93. El Majd, A.; Sair, S.; Ousaleh, H.A.; Bouhaj, Y.; Belouaggadia, N.; Younsi, Z.; El Bouari, A. Advancing tent thermoregulation: Integrating shape-stabilized PCM into fabric design. *J. Energy Storage* **2024**, *95*, 112681. [CrossRef]
94. Naqvi, S.N.R.; Haider, M.Z.; Jin, X.; Hu, J.W. Investigation of thermal and mechanical performance of composite PCM mortar incorporated with graphene nanoplatelets. *Constr. Build. Mater.* **2025**, *470*, 140576. [CrossRef]
95. Ayyagari, V.; Shooshtari, A.; Ohadi, M. Glauber's Salt Composites for HVAC Applications: A Study on the Use of the T-History Method with a Modified Data Evaluation Methodology. *Materials* **2025**, *18*, 2998. [CrossRef] [PubMed]

96. Solé, A.; Miró, L.; Barreneche, C.; Martorell, I.; Cabeza, L.F. Review of the T-history method to determine thermophysical properties of phase change materials (PCM). *Renew. Sustain. Energy Rev.* **2013**, *26*, 425–436. [CrossRef]
97. Wang, M.; Wang, J. Flow boiling of nanofluids in microchannel heat exchangers: A critical review. *Therm. Sci. Eng. Prog.* **2024**, *55*, 102930. [CrossRef]
98. Behseresht, S.; Park, Y.H. Additive Manufacturing of Composite Polymers: Thermomechanical FEA and Experimental Study. *Materials* **2024**, *17*, 1912. [CrossRef]
99. Smolnicki, M.; Lesiuk, G.; Duda, S.; de Jesus, A.M.P. A Review on Finite-Element Simulation of Fibre Metal Laminates. *Arch. Comput. Methods Eng.* **2023**, *30*, 749–763. [CrossRef]
100. Gürsoy, E. Melting and energy storage performance enhancement of rectangular cavity with metal foam by nano-PCM and recessed/protruding dimpled fin wall. *J. Energy Storage* **2025**, *119*, 116327. [CrossRef]
101. Hekmat, M.H.; Saharkhiz, S. Impact of helically coiled shell and tube on melting and solidification of PCMs in thermal energy storage systems: A three-dimensional parametric study. *J. Energy Storage* **2025**, *108*, 115172. [CrossRef]
102. Dora, S.; Kuznik, F.; Mini, K.M. A novel PCM-based foam concrete for heat transfer in buildings -Experimental developments and simulation modelling. *J. Energy Storage* **2025**, *105*, 114625. [CrossRef]
103. Yang, B.; Xie, L.; Cao, Y.; Yu, X. The role of porosity gradient distribution on the heat transfer characteristics of copper foam/paraffin composite phase change material. *J. Energy Storage* **2025**, *127*, 117157. [CrossRef]
104. Zhao, W.; Yu, G.; Li, S.; Liu, Z.; Wu, L. Rate capability and Ragone plots for thermal management multifunctional structure designing. *Int. J. Mech. Sci.* **2023**, *252*, 108367. [CrossRef]
105. Sun, H.; Li, M.; Li, T.; Zhao, J.; Du, Q.; Li, B.; Sun, D. Regulation of carbonized cellulose nanofiber pore structure by artificially cultured diatom frustules for leakage-proof phase change composites. *Chem. Eng. J.* **2025**, *504*, 158482. [CrossRef]
106. Saleh, H. Heat transfer enhancement in a nano encapsulated PCM-water filled square cavity with flexible fin. *J. Energy Storage* **2024**, *85*, 111121. [CrossRef]
107. Egami, R.H.; Almalki, A.S.; Alharbi, I.S. Simulation of freezing process of PCM in existence of nano-sized additives within storage tank. *J. Therm. Anal. Calorim.* **2024**, *149*, 4575–4586. [CrossRef]
108. Tominaga, Y. CFD simulations of turbulent flow and dispersion in built environment: A perspective review. *J. Wind Eng. Ind. Aerodyn.* **2024**, *249*, 105741. [CrossRef]
109. Wang, M.; Ju, H.; Wu, J.; Qiu, H.; Liu, K.; Tian, W.; Su, G.H. A review of CFD studies on thermal hydraulic analysis of coolant flow through fuel rod bundles in nuclear reactor. *Prog. Nucl. Energy* **2024**, *171*, 105175. [CrossRef]
110. Olivares-Robles, M.A.; Enciso-Montes de Oca, O.Y.; Vargas-Almeida, A. Optimizing the transient response of an annular thermoelectric cooler: PCM and nanofluid synergy. *J. Energy Storage* **2025**, *115*, 115950. [CrossRef]
111. Benyahia, I.; Al-Ghamdi, M.F.; Abderrahmane, A.; Younis, O.; Laouedj, S.; Guedri, K.; Alahmer, A. Comprehensive thermal analysis of a nano-enhanced PCM in a finned latent heat storage system. *Int. Commun. Heat Mass Transf.* **2025**, *165*, 109106. [CrossRef]
112. Bouzennada, T.; Abderrahmane, A.; Aich, W.; Younis, O.; Ben Ali, N.; Kolsi, L. Heat transfer and fluid flow in nano-encapsulated PCM-filled undulated cavity. *Ain Shams Eng. J.* **2024**, *15*, 102669. [CrossRef]
113. Gholami, A.; Parach, K.; Jafari, B. Numerical analysis of a photovoltaic panel containing hybrid nanoparticles integrated with PCM in the presence of fixed fins. *Case Stud. Therm. Eng.* **2025**, *70*, 106098. [CrossRef]
114. Nandi, A.; Biswas, N. Enhanced melting dynamics of phase change material (PCM) based energy storage system combining modified fin and nanoparticles under solar irradiation. *Int. J. Numer. Methods Heat Fluid Flow* **2025**, *35*, 2104–2143. [CrossRef]
115. Shaik, S.A.; Nigam, P.K.; Gugulothu, S.K. Advanced fin designs for improved thermal management in PCM-based latent heat storage systems. *Appl. Therm. Eng.* **2025**, *272*, 126337. [CrossRef]
116. Waqas, H.; Hussain, M.; Al-Mdallal, Q.M.; Ahammad, N.A.; Elseesy, I.E. The impact of nano-infused phase change materials and blossom-shaped fins on thermal energy storage. *Case Stud. Therm. Eng.* **2025**, *65*, 105623. [CrossRef]
117. Mahdavi, A.; Niloofari, P.; Gorji-Bandpy, M.; Mahmoudi, A. Numerical simulation of a PVT-PCM system by altering fluid tube positioning adding nanoparticles and embedding a porous medium. *Alex. Eng. J.* **2025**, *115*, 30–46. [CrossRef]
118. Alam, M.T.; Garg, V.; Soni, V.; Gupta, A.K. Effect of enclosure design with composite/nano-enhanced/dual phase change material on melting response of latent heat storage systems. *Chem. Eng. Res. Des.* **2025**, *214*, 125–143. [CrossRef]
119. Kiyak, B.; Biswas, N.; Öztop, H.F.; Selimefendigil, F. Geometry curvature influence on melting and solidification performance in nano particle added phase change material to storage energy. *J. Energy Storage* **2025**, *117*, 116172. [CrossRef]
120. Amin, M.; Putra, N.; Rizal, T.A.; Umar, H.; Ginting, S.F.; Amir, F.; Kusuma, M.H. Enhanced tubular solar still performance using nano beeswax/graphene phase change materials and parabolic trough collectors. *J. Energy Storage* **2025**, *124*, 116884. [CrossRef]
121. Singh, A.P.; Tiwari, S.; Sinhmar, H.; Pal, G.; Pandey, A.K. Experimental study on PCM (OM-37) enhanced with nanoparticles for improved heat retention in solar ponds. *J. Energy Storage* **2025**, *123*, 116760. [CrossRef]
122. Emara, K.; Abd-Elgawad, A.M.M.M.; Emara, A. Effect of nanofluids-PCM heat exchanging on engine downsizing and heat transfer enhancement via the heat engine's cooling system: A novel saving tactic. *J. Energy Storage* **2025**, *117*, 115815. [CrossRef]

123. Al-Karboly, A.M.O.; Ibrahim, A.; Fazlizan, A.; Sopian, K.; Al-Aasam, A.B.; Ishak, M.A.A.B.; Al-Waeli, A.H.A.; Elmnifi, M. Experimental evaluation of a photovoltaic thermal collector using twisted tape absorber with nano-enhanced phase change material for thermal storage. *J. Energy Storage* **2025**, *109*, 115122. [CrossRef]
124. Kamrava, M.; Fazilati, M.A.; Toghraie, D. Investigating the use of nano-enhanced phase change material in floor heating system: A numerical approach. *Int. Commun. Heat Mass Transf.* **2025**, *164*, 108962. [CrossRef]
125. Gür, M.; Gürgenç, E.; Coşanay, H.; Öztop, H.F. Solar-assisted radiant heating system with nano-B4C enhanced PCM for nearly zero energy buildings. *Case Stud. Therm. Eng.* **2025**, *65*, 105544. [CrossRef]
126. Sathishkumar, A.; Cheralathan, M. Charging and discharging processes of low capacity nano-PCM based cool thermal energy storage system: An experimental study. *Energy* **2023**, *263*, 125700. [CrossRef]
127. Bassam, A.M.; Sopian, K.; Ibrahim, A.; Al-Aasam, A.B.; Dayer, M. Experimental analysis of photovoltaic thermal collector (PVT) with nano PCM and micro-fins tube counterclockwise twisted tape nanofluid. *Case Stud. Therm. Eng.* **2023**, *45*, 102883. [CrossRef]
128. Bassam, A.M.; Sopian, K.; Ibrahim, A.; Fauzan, M.F.; Al-Aasam, A.B.; Abusaibaa, G.Y. Experimental analysis for the photovoltaic thermal collector (PVT) with nano PCM and micro-fins tube nanofluid. *Case Stud. Therm. Eng.* **2023**, *41*, 102579. [CrossRef]
129. Kibria, M.G.; Mohtasim, M.S.; Paul, U.K.; Das, B.K.; Saidur, R. Impact of hybrid nano PCM (paraffin wax with Al_2O_3 and ZnO nanoparticles) on photovoltaic thermal system: Energy, exergy, exergoeconomic and enviroeconomic analysis. *J. Clean. Prod.* **2024**, *436*, 140577. [CrossRef]
130. Bharathiraja, R.; Ramkumar, T.; Selvakumar, M.; Sasikumar, K. Experimental and numerical analysis of hybrid nano-enhanced phase change material (PCM) based flat plate solar collector. *J. Energy Storage* **2024**, *96*, 112649. [CrossRef]
131. Mousavi, S.M.; Khanmohammadi, F.; Rabienataj Darzi, A.A. Magnetic influence on phase change materials for optimized thermal energy storage: A comprehensive review and prospective insights. *J. Energy Storage* **2024**, *89*, 111625. [CrossRef]
132. Izadi, M.; Sheremet, M.; Hajjar, A.; Galal, A.M.; Mahariq, I.; Jarad, F.; Ben Hamida, M.B. Numerical investigation of magneto-thermal-convection impact on phase change phenomenon of Nano-PCM within a hexagonal shaped thermal energy storage. *Appl. Therm. Eng.* **2023**, *223*, 119984. [CrossRef]
133. Hassan, A.M.; Alomari, M.A.; Alajmi, A.; Sadeq, A.M.; Alqurashi, F.; Flayyih, M.A. Enhanced natural convection in a U-shaped baffled cavity: Synergistic effects of magnetic fields and wall oscillations on Nano-encapsulated PCM. *Int. Commun. Heat Mass Transf.* **2025**, *165*, 109051. [CrossRef]
134. Younis, O.; Laidoudi, H.; Abderrahmane, A.; Belazreg, A.; Qasem, N.A.A.; Homod, R.Z.; Yacine, k.; Rawa, M.; Hassan, A.M. Thermal pattern of nano-encapsulated PCM in a lid-driven cavity with presence of a heated body, magnetic field and limited permeability. *Case Stud. Therm. Eng.* **2023**, *50*, 103469. [CrossRef]
135. Selimefendigil, F.; Oztop, H.F. Effects of location of nano-enhanced PCM-packed bed on the process of heat transfer and phase transition in a nonuniform magnetic field. *Numer. Heat Transf. Part A Appl.* **2024**, 1–21. [CrossRef]
136. Selimefendigil, F.; Öztop, H.F. Multijet impingement heat transfer under the combined effects of encapsulated-PCM and inclined magnetic field during nanoliquid convection. *Int. J. Heat Mass Transf.* **2023**, *203*, 123764. [CrossRef]
137. Selimefendigil, F.; Oztop, H.F. Convection and phase change in a lid-driven trapezoidal vented cavity with encapsulated PCM under non-uniform magnetic field by using ternary nanofluid. *Therm. Sci. Eng. Prog.* **2024**, *48*, 102424. [CrossRef]
138. Nguyen, L.M.Q.; Alshuraiaan, B.; Hajjar, A.; Izadi, M.; El Idi, M.M. Assessment of the thermal resistance of novel roof structures incorporating nano-Phase Change Materials capsules under the influence of an external magnetic field. *J. Build. Eng.* **2023**, *79*, 107851. [CrossRef]
139. Lu, B.; Zhang, Y.; Xiao, J.; Hu, M.; Niu, Y.; Luo, M.; Zhu, J.; Zhang, J. Experimental investigation of the effect of rotating magnetic field on the melting performance enhancement of paraffin/nano-Fe3O4 composite phase change material. *J. Energy Storage* **2024**, *83*, 110751. [CrossRef]
140. Li, H.; Zhuang, Y.; Feng, J.-C. Multi-scale experimental analysis on the coupled effects of ultrasonic field and magnetic field on the melting and energy storage performances for hybrid nano-enhanced phase change materials. *J. Energy Storage* **2024**, *84*, 110801. [CrossRef]
141. Zhuang, Y.; Li, H.; Xu, W.; Huang, S.-M. Experimental study on the melting performance of magnetic NEPCMs embedded in metal foam subjected to a non-uniform magnetic field. *Sol. Energy Mater. Sol. Cells* **2023**, *250*, 112077. [CrossRef]
142. Farahani, S.D.; Farahani, A.D.; Mamoei, A.J.; Yan, W.-M. Enhancement of phase change material melting using nanoparticles and magnetic field in the thermal energy storage system with strip fins. *J. Energy Storage* **2023**, *57*, 106282. [CrossRef]
143. Zhou, Y.; Zhang, W.; Ma, E.; Deringer, V.L. Device-scale atomistic modelling of phase-change memory materials. *Nat. Electron.* **2023**, *6*, 746–754. [CrossRef]
144. Taheri, M.; Pourfayaz, F.; Hemmati, S. A highly accurate model for prediction of thermal conductivity of carbon-based nano-enhanced PCMs using an artificial neural network. *Energy Rep.* **2023**, *10*, 1249–1258. [CrossRef]
145. Sharma, A.; Singh, P.K.; Makki, E.; Giri, J.; Sathish, T. A comprehensive review of critical analysis of biodegradable waste PCM for thermal energy storage systems using machine learning and deep learning to predict dynamic behavior. *Heliyon* **2024**, *10*, e25800. [CrossRef] [PubMed]

146. Shakibi, H.; Shokri, A.; Sobhani, B.; Yari, M. Numerical analysis and optimization of a novel photovoltaic thermal solar unit improved by Nano-PCM as an energy storage media and finned collector. *Renew. Sustain. Energy Rev.* **2023**, *179*, 113230. [CrossRef]
147. Aly, A.M.; Alsedais, N. Heat and mass transfer of NEPCM in a wavy porous cavity with variable hot and high-concentration zones: A study using ISPH and XGBoost models. *Int. Commun. Heat Mass Transf.* **2025**, *160*, 108383. [CrossRef]
148. Abdolahimoghadam, M.; Rahimi, M. New hybrid nano-and bio-based phase change material containing graphene-copper particles hosting beeswax-coconut oil for solar thermal energy storage: Predictive modeling and evaluation using machine learning. *Energy* **2024**, *307*, 132604. [CrossRef]
149. Abdolahimoghadam, M.; Rahimi, M. Experimental, numerical, and machine learning study of vertical thermal energy storage filling with novel hybrid nano-and bio-based phase change material. *J. Energy Storage* **2025**, *106*, 114815. [CrossRef]
150. Gao, Y.; Shigidi, I.M.T.A.; Ali, M.A.; Homod, R.Z.; Safaei, M.R. Thermophysical properties prediction of carbon-based nano-enhanced phase change material's using various machine learning methods. *J. Taiwan Inst. Chem. Eng.* **2023**, *148*, 104662. [CrossRef]
151. Ghasemi, H.; Sheikhzadeh, G.A.; Fattahi, A. Numerical Simulation and ANN Prediction of Nano-Encapsulated PCM Slurry in a Microchannel: A Thermodynamic Analysis. *Arab. J. Sci. Eng.* **2025**. [CrossRef]
152. Kiaghadi, M.; Keshvarinia, M.; Boora, F.M.; Mousavi, S.M. Machine learning applications for predicting liquid fraction in a PV system with NEPCM and fins. *Case Stud. Therm. Eng.* **2024**, *61*, 104819. [CrossRef]
153. Kiaghadi, M.; Sheikholeslami, M.; Alinia, A.M.; Boora, F.M. Predicting the performance of a photovoltaic unit via machine learning methods in the existence of finned thermal storage unit. *J. Energy Storage* **2024**, *90*, 111766. [CrossRef]
154. Yang, Y.; Wang, Z.; Ayed, H.; Alhoee, J. Incorporating nickel foam with nano-encapsulated phase change material and water emulsion for battery thermal management: Coupling CFD and machine learning. *Case Stud. Therm. Eng.* **2024**, *60*, 104672. [CrossRef]
155. Xu, S.; Basem, A.; Al-Asadi, H.A.; Chaturvedi, R.; Daminova, G.; Fouad, Y.; Jasim, D.J.; Alhoee, J. Employing deep learning for predicting the thermal properties of water and nano-encapsulated phase change material. *Int. J. Low-Carbon Technol.* **2024**, *19*, 1453–1459. [CrossRef]
156. Salari, A.; Ahmadi, R.; Vafadaran, M.S.; Shakibi, H.; Sardarabadi, M. Predicting the performance of a heat sink utilized with an energy storage unit using machine learning approach. *J. Energy Storage* **2024**, *83*, 110470. [CrossRef]
157. Khadem, Z.; Salari, A.; Naghdbishi, A.; Shakibi, H.; Sardarabadi, M. Parametric analysis of a PCM-based heat sink for electronic device thermal management. *J. Energy Storage* **2023**, *74*, 109118. [CrossRef]
158. Fini, A.T.; Fattahi, A.; Musavi, S. Machine learning prediction and multiobjective optimization for cooling enhancement of a plate battery using the chaotic water-microencapsulated PCM fluid flows. *J. Taiwan Inst. Chem. Eng.* **2023**, *148*, 104680. [CrossRef]
159. Coletti, K.; Romeo, R.C.; Davis, R.B. Bayesian backcalculation of pavement properties using parallel transitional Markov chain Monte Carlo. *Comput.-Aided Civ. Infrastruct. Eng.* **2024**, *39*, 1911–1927. [CrossRef]
160. Chen, Z.; Mak, S.; Wu, C.F.J. A Hierarchical Expected Improvement Method for Bayesian Optimization. *J. Am. Stat. Assoc.* **2024**, *119*, 1619–1632. [CrossRef]
161. Alazwari, M.A.; Algarni, M.; Safaei, M.R. Effects of various types of nanomaterials on PCM melting process in a thermal energy storage system for solar cooling application using CFD and MCMC methods. *Int. J. Heat Mass Transf.* **2022**, *195*, 123204. [CrossRef]
162. Goodarzi, M.; Elkotb, M.A.; Alanazi, A.K.; Abo-Dief, H.M.; Mansir, I.B.; Tirth, V.; Gamaoun, F. Applying Bayesian Markov chain Monte Carlo (MCMC) modeling to predict the melting behavior of phase change materials. *J. Energy Storage* **2022**, *45*, 103570. [CrossRef]
163. Wang, Z.; Bu, S.; Wen, J.; Huang, C. A comprehensive review on uncertainty modeling methods in modern power systems. *Int. J. Electr. Power Energy Syst.* **2025**, *166*, 110534. [CrossRef]
164. Azarhoosh, Z.; Ilchi Ghazaan, M. A Review of Recent Advances in Surrogate Models for Uncertainty Quantification of High-Dimensional Engineering Applications. *Comput. Methods Appl. Mech. Eng.* **2025**, *433*, 117508. [CrossRef]
165. Al-Aasam, A.B.; Ibrahim, A.; Sopian, K.; Abdulsahib, M.B.; Dayer, M. Nanofluid-based photovoltaic thermal solar collector with nanoparticle-enhanced phase change material (Nano-PCM) and twisted absorber tubes. *Case Stud. Therm. Eng.* **2023**, *49*, 103299. [CrossRef]
166. Daneshazarian, R.; Eslami, R.; Azizi, N.; Zarrin, H.; Berardi, U. Performance evaluation of a novel nano-enhanced phase change material for thermal energy storage applications. *J. Energy Storage* **2023**, *74*, 109376. [CrossRef]
167. Gür, M.; Gürgenç, E.; Coşanay, H.; Öztop, H.F. Novel nano-Y_2O_3/myristic acid nanocomposite PCM for cooling performances of electronic device with various fin designs. *J. Energy Storage* **2024**, *100*, 113646. [CrossRef]
168. Venkatesh, R.; Karthik, K.; Vinayagam, M.; Verma, A.; Chaturvedi, R.; Soudagar, M.E.M.; Bhooshanam, N.N.; Al Obaid, S.; Alharbi, S.A. Enhancement of solar cooking performance by utilizing hybrid SiC/Ag featured nano-PCM heat exchanger integrated with parabolic trough collector. *J. Therm. Anal. Calorim.* **2024**, *149*, 14389–14402. [CrossRef]

169. GaneshKumar, P.; Vigneswaran, V.S.; Murugan, P.; Cheralathan, M.; Velraj, R.; Kim, S.C.; Ramkumar, V. Exploring the thermal performance of a solar air heater with a V-corrugated and shot-blasted absorber plate comprising nano-enhanced phase change materials. *J. Energy Storage* **2024**, *77*, 109955. [CrossRef]
170. Osman, A.I.; Fang, B.; Zhang, Y.; Liu, Y.; Yu, J.; Farghali, M.; Rashwan, A.K.; Chen, Z.; Chen, L.; Ihara, I.; et al. Life cycle assessment and techno-economic analysis of sustainable bioenergy production: A review. *Environ. Chem. Lett.* **2024**, *22*, 1115–1154. [CrossRef]
171. Saez, R.; Boer, D.; Shobo, A.B.; Vallès, M. Techno-economic analysis of residential rooftop photovoltaics in Spain. *Renew. Sustain. Energy Rev.* **2023**, *188*, 113788. [CrossRef]
172. Ma, C.; Van Coile, R.; Gernay, T. Fire protection costs in composite buildings for cost-benefit analysis of fire designs. *J. Constr. Steel Res.* **2024**, *215*, 108517. [CrossRef]
173. Samat, N.; Goh, K.H.; See, K.F. Review of the application of cost–benefit analysis to the development of production systems in aquaculture. *Aquaculture* **2024**, *587*, 740816. [CrossRef]
174. Hekrle, M.; Liberalesso, T.; Macháč, J.; Matos Silva, C. The economic value of green roofs: A case study using different cost–benefit analysis approaches. *J. Clean. Prod.* **2023**, *413*, 137531. [CrossRef]
175. Dong, Z.; Bian, Z.; Jin, W.; Guo, X.; Zhang, Y.; Liu, X.; Wang, C.; Guan, D. An integrated approach to prioritizing ecological restoration of abandoned mine lands based on cost-benefit analysis. *Sci. Total Environ.* **2024**, *924*, 171579. [CrossRef]
176. Sharshir, S.W.; Abdo, M.R.; Yusuf, K.; El-Naggar, A.A.; Joseph, A.; Ismail, M.; Elsaid, A.M.; El-Samadony, M.O.A.; Abdelfatah, M.; El-Shaer, A.; et al. Thermo-economic and environmental evaluation of thin film evaporation using novel metal-organic frameworks to enhance solar desalination performance. *Process Saf. Environ. Prot.* **2025**, *199*, 107223. [CrossRef]
177. Rasaiah, N.; Eugine, R.; Manickam, R. Efficiency, productivity and economic analysis of polystyrene foam–insulated conventional and modified double-slope solar still using nanofluids. *Environ. Sci. Pollut. Res.* **2025**, *32*, 12713–12742. [CrossRef]
178. Kumar, K.; Sarkar, J.; Mondal, S.S. Energy, exergy, and economic evaluations of various cylindrical lithium-ion battery thermal management systems. *Int. Commun. Heat Mass Transf.* **2025**, *165*, 109007. [CrossRef]
179. Gürbüz, E.Y.; Şahinkesen, İ.; Tuncer, A.D.; Güler, O.V.; Keçebaş, A.; Georgiev, A.G. Experimental investigation of a baffled photovoltaic-thermal air collector with SiC nano-embedded thermal paste: A comparative study. *Renew. Energy* **2025**, *244*, 122649. [CrossRef]
180. Mohtasim, M.S.; Kibria, M.G.; Pranto, M.M.H.; Das, B.K. Hybrid PVT integrated pyramid solar still: 11E, sustainability, and sustainable development goals assessment. *Renew. Energy* **2025**, *246*, 122914. [CrossRef]
181. Singh, V.K.; Kumar, D. An experimental investigation and thermo-economic performance analysis of solar desalination system by using nano-enhanced PCM. *Mater. Today Sustain.* **2024**, *27*, 100884. [CrossRef]
182. Anika, U.A.; Kibria, M.G.; Kanka, S.D.; Mohtasim, M.S.; Paul, U.K.; Das, B.K. Exergy, exergo-economic, environmental and sustainability analysis of pyramid solar still integrated hybrid nano-PCM, black sand, and sponge. *Sol. Energy* **2024**, *274*, 112559. [CrossRef]
183. Vaziri Rad, M.A.; Rousta, M.; Khanalizadeh, A.; Kouravand, A.; Akbari, M.; Mousavi, S.; Razi Astaraei, F.; Sadeghitabar, E.; Yan, W.-M.; Kasaeian, A. Critical analysis of enhanced photovoltaic thermal systems: A comparative experimental study of PCM, TEG, and nanofluid applications. *Energy Convers. Manag.* **2024**, *314*, 118712. [CrossRef]
184. Essa, F.A.; Alawee, W.H.; Abdullah, A.S.; Mohammed, S.A.; Basem, A.; Majdi, H.S.; Omara, Z.M. Enhancing water evaporation rate in hemispherical solar distillers through innovative modifications and Nano-PCM integration. *Sol. Energy* **2024**, *271*, 112453. [CrossRef]
185. Paul, U.K.; Mohtasim, M.S.; Kibria, M.G.; Das, B.K. Nano-material based composite phase change materials and nanofluid for solar thermal energy storage applications: Featuring numerical and experimental approaches. *J. Energy Storage* **2024**, *98*, 113032. [CrossRef]
186. Elamy, M.I.; Mohammed, S.A.; Basem, A.; Alawee, W.H.; Abdullah, A.S.; Majdi, H.S.; Omara, Z.M.; Essa, F.A. Optimizing cord pyramid solar distillers: A comprehensive study on square baffles, reflectors, and phase transition materials. *Case Stud. Therm. Eng.* **2024**, *63*, 105304. [CrossRef]
187. Attia, M.E.H.; Abdelgaied, M.; Harby, K. New graphite fins efficacy on conical solar distiller performance: Experimental study with thermal, exergy, economic, and sustainability analysis. *J. Water Process Eng.* **2024**, *64*, 105652. [CrossRef]
188. El-Gazar, E.F.; Yousef, M.S.; Elshaer, A.M.; Khattab, M.A.; Mouneer, T.A.; Hawwash, A.A. Enhancing solar still performance with hybrid nanofluid: A comprehensive assessment of energy, exergy, economics, and environmental impact using a novel fractional model. *Environ. Dev. Sustain.* **2024**, 1–28. [CrossRef]
189. van Daalen, K.R.; Jung, L.; Dada, S.; Othman, R.; Barrios-Ruiz, A.; Malolos, G.Z.; Wu, K.T.; Garza-Salas, A.; El-Gamal, S.; Ezzine, T.; et al. Bridging the gender, climate, and health gap: The road to COP29. *Lancet Planet Health* **2024**, *8*, e1088–e1105. [CrossRef]
190. Fearnside, P.M.; Filho, W.L. COP 30: Brazilian policies must change. *Science* **2025**, *387*, 1237. [CrossRef]
191. Tilsted, J.P.; Newell, P. Synthetic transitions: The political economy of fossil fuel as feedstock. *Rev. Int. Political Econ.* **2025**, *32*, 1214–1238. [CrossRef]

192. Adom, P.K. Global energy efficiency transition tendencies: Development phenomenon or not? *Energy Strategy Rev.* **2024**, *55*, 101524. [CrossRef]
193. Shannak, S.d.; Cochrane, L.; Bobarykina, D. Strategic analysis of metal dependency in the transition to low-carbon energy: A critical examination of nickel, cobalt, lithium, graphite, and copper scarcity using IEA future scenarios. *Energy Res. Soc. Sci.* **2024**, *118*, 103773. [CrossRef]
194. Suárez-Cuesta, D.; Latorre, M.C. Governance and policy implications of the Inflation Reduction Act: A European perspective. *Glob. Policy* **2024**, *15*, 37–43. [CrossRef]
195. Rizos, V.; Urban, P. Barriers and policy challenges in developing circularity approaches in the EU battery sector: An assessment. *Resour. Conserv. Recycl.* **2024**, *209*, 107800. [CrossRef]
196. Sabyrbekov, R.; Overland, I. Small and large friends of the EU's carbon border adjustment mechanism: Which non-EU countries are likely to support it? *Energy Strategy Rev.* **2024**, *51*, 101303. [CrossRef]
197. Götz, V.; Harnesk, D. Transnational contestation for local communities within the formation of the European Union's Critical Raw Materials Act—A critical appraisal. *Environ. Politics* **2025**, 1–27. [CrossRef]
198. Froidevaux, J.S.P.; Le Viol, I.; Barré, K.; Bas, Y.; Kerbiriou, C. A modeling framework for biodiversity assessment in renewable energy development: A case study on European bats and wind turbines. *Renew. Sustain. Energy Rev.* **2025**, *211*, 115323. [CrossRef]
199. Bang, G. The U.S. Inflation Reduction Act: Climate policy as economic crisis response. *Environ. Politics* **2025**, 1–22. [CrossRef]
200. Sigman, H. Climate Change Adaptation in Government Decisions: Empirical Evidence from Superfund Remedies. *Land Econ.* **2025**, *101*, 18. [CrossRef]
201. Chen, H.; Niu, D.; Gao, Y. Research on the impact of energy transition policies on green total factor productivity of Chinese high-energy-consuming enterprises. *Energy* **2025**, *319*, 135066. [CrossRef]
202. Xinhua. China Unveils Draft of Its First-Ever Environmental Code. Available online: http://english.scio.gov.cn/in-depth/2025-04/28/content_117848751.html (accessed on 1 May 2025).
203. Jindal, A.; Puri, S.; Shrimali, G. Designing a prospective carbon trading market in India: Key properties, enabling features and linkages. *Appl. Energy* **2025**, *386*, 125595. [CrossRef]
204. Ha-Duong, M. Power system planning in the energy transition era: The case of Vietnam's power development plan 8. *Clim. Policy* **2025**, *25*, 562–577. [CrossRef]
205. Paulino de Azevedo, J.H.; Pradelle, F.; Botelho, V.; Torres Serra, E.; Nohra Chaar Pradelle, R.; Leal Braga, S. An integrated geospatial model for evaluating offshore wind-to-hydrogen technical and economic production potential in Brazil. *Int. J. Hydrogen Energy* **2025**, *100*, 800–815. [CrossRef]
206. Uratani, J.M.; Evensen, D.; Sovacool, B.K.; Griffiths, S. Benchmarking and tailoring electric vehicle policies to stimulate adoption. *Energy Res. Soc. Sci.* **2025**, *125*, 104074. [CrossRef]

MDPI AG
Grosspeteranlage 5
4052 Basel
Switzerland
Tel.: +41 61 683 77 34

Energies Editorial Office
E-mail: energies@mdpi.com
www.mdpi.com/journal/energies

www.ingramcontent.com/pod-product-compliance
Lightning Source LLC
Chambersburg PA
CBHW040815250926
51850CB00007BA/1173
* 9 7 8 3 7 2 5 8 5 4 5 9 2 *